Techniques in Applied Microbiology

Techniques in Applied Microbiology

Edited by **Ricky Parks**

SYRAWOOD
PUBLISHING HOUSE

New York

Published by Syrawood Publishing House,
750 Third Avenue, 9th Floor,
New York, NY 10017, USA
www.syrawoodpublishinghouse.com

Techniques in Applied Microbiology
Edited by Ricky Parks

Printed in the United States of America.

Contents

Preface

Every book is initially just a concept; it takes months of research and hard work to give it the final shape in which the readers receive it. In its early stages, this book also went through rigorous reviewing. The notable contributions made by experts from across the globe were first molded into patterned chapters and then arranged in a sensibly sequential manner to bring out the best results.

Microbiology is a significant branch of science with a diverse range of interdisciplinary applications. This book discusses the applications of microbiology in agriculture, pharmaceutics, food and supplements. It also covers the uses of microbiology in biopolymers, bionanotechnology, green chemistry, etc. The extensive content of this book provides the readers with a thorough understanding of the subject. As this field is developing at a rapid pace, the contents of this book will help the students and academicians associated with the discipline to understand the modern concepts and applications of the subject.

It has been my immense pleasure to be a part of this project and to contribute my years of learning in such a meaningful form. I would like to take this opportunity to thank all the people who have been associated with the completion of this book at any step.

Editor

Improved bioethanol production in an engineered *Kluyveromyces lactis* strain shifted from respiratory to fermentative metabolism by deletion of *NDI1*

María Isabel González-Siso,* Alba Touriño, Ángel Vizoso, Ángel Pereira-Rodríguez,[†] Esther Rodríguez-Belmonte,
Manuel Becerra and María Esperanza Cerdán
Grupo de Investigación EXPRELA, Departamento de Bioloxía Celular e Molecular, Facultade de Ciencias, Universidade da Coruña, Campus de A Coruña, 15071-A Coruña, Spain.

Summary

In this paper, we report the metabolic engineering of the respiratory yeast *Kluyveromyces lactis* by construction and characterization of a null mutant (*Δklndi1*) in the single gene encoding a mitochondrial alternative internal dehydrogenase. Isolated mitochondria of the *Δklndi1* mutant show unaffected rate of oxidation of exogenous NADH, but no oxidation of matrix NADH; this confirms that *Kl*Ndi1p is the only internal NADH dehydrogenase in *K. lactis* mitochondria. Permeabilized cells of the *Δklndi1* mutant do not show oxidation of matrix NADH, which suggests that shuttle systems to transfer the NADH from mitochondrial matrix to cytosol, for being oxidized by external dehydrogenases, are not functional. The *Δklndi1* mutation decreases the chronological life span in absence of nutrients. The expression of *KlNDI1* is increased by glutathione reductase depletion. The *Δklndi1* mutation shifts the *K. lactis* metabolism from respiratory to fermentative: the *Δklndi1* strain shows reduced respiration rate and increased ethanol production from glucose, while it does not grow in non-fermentable carbon sources such as lactate. The biotechnological benefit of the *Δklndi1* mutant for bioethanol production from waste cheese whey lactose was proved.

*For correspondence. E-mail migs@udc.es

Funding Information This research was supported by grant BFU2009-08854 from MICINN (Spain) and by FEDER. General support to the laboratory during 2012–14 was funded by Xunta de Galicia (Consolidación D.O.G. 10-10-2012. CN: 2012/118) also supported by FEDER.

Introduction

In contrast to mammals, but similar to other eukaryotes like filamentous fungi or plants, in the inner mitochondrial membrane of yeasts, there are the so-called alternative dehydrogenases, or class-2 dehydrogenases, which are able to oxidize cytosolic and mitochondrial matrix NAD(P)H coenzymes and to transfer the electrons to ubiquinone. As opposed to complex I, these are non-proton-pumping, rotenone-insensitive and single polypeptide enzymes. Their precise physiological role in the different organisms has not been completely established, although it is known that they constitute a main entry point for electrons in the mitochondrial respiratory chain, being involved in the regulation of intracellular redox balance and energy production (Tarrío *et al.*, 2006b). The characteristics of the mitochondrial alternative NAD(P)H dehydrogenases studied hitherto in yeasts are related to the preponderance of respiratory or fermentative metabolism in different species (González-Siso and Cerdán, 2007).

Saccharomyces cerevisiae and *Kluyveromyces lactis* are two yeasts that metabolize the glucose by oxidative and fermentative pathways, but *S. cerevisiae* is predominantly fermentative (Crabtree positive) and *K. lactis* is predominantly respiratory (Crabtree negative) (González-Siso *et al.*, 2000). Both species lack complex I, which is present in strictly aerobic yeasts, and contain one internal alternative dehydrogenase and two external alternative dehydrogenases (Tarrío *et al.*, 2006a; González-Siso and Cerdán, 2007).

The external alternative dehydrogenases show differential characteristics between the fermentative and the respiratory yeast. In *S. cerevisiae*, a decrease in expression of the external NADH dehydrogenases is observed at high glucose concentrations and this feature has been related to a higher amount of reoxidation of cytosolic NADH by the alcohol dehydrogenases and therefore with the prevalence of aerobic fermentation (the Crabtree-positive phenotype) (Luttik *et al.*, 1998). In *K. lactis*, the ability of the external dehydrogenases to use NADPH, in addition to NADH, supports an increased use of the pentose phosphate pathway (PPP) to metabolize glucose in detriment of glycolysis (González-Siso *et al.*, 1996a,b; Tarrío *et al.*, 2005; 2006b).

Differently to mitochondria from plants and filamentous fungi, where complex I and internal alternative dehydrogenase coexist (MØller, 2002; Duarte and Videira, 2007), in *S. cerevisiae* and *K. lactis*, complex I is replaced by the alternative enzyme. Therefore, in these yeasts, the internal dehydrogenase constitutes the metabolic equivalent of complex I. In accordance with this function and in both yeasts, the expression of the alternative internal dehydrogenase is reduced at high glucose concentrations when fermentation is favoured (De Risi *et al.*, 1997; Tarrío *et al.*, 2005).

The ability to complement complex I defects has been proved for the *S. cerevisiae* alternative internal dehydrogenase, *Sc*Ndi1p (Yagi *et al.*, 2006; Marella *et al.*, 2008). Deficiency of complex I is one of the most frequent dysfunctions of the mitochondrial respiratory chain in human, often encompassing a large spectrum of clinical presentations, and is closely related to the aetiology of sporadic Parkinson's disease (Marella *et al.*, 2008) and ageing (Balaban *et al.*, 2005). The potential of the yeast *NDI1* gene in therapy of complex I defects has been demonstrated in rat animal models (Marella *et al.*, 2008; 2010). This potential might be limited by the difference between the fermentative metabolism shown by *S. cerevisiae* cells and the oxidative metabolism shown by most tissues affected by complex I deficiency like nervous and cardiac; therefore, the use of the *NDI1* genes from Crabtree-negative yeasts could become advantageous (González-Siso *et al.*, 2009).

In *S. cerevisiae*, *Sc*Ndi1p is also the closest homologue of mammalian AMID (apoptosis-inducing factor-homologous mitochondrion-associated inducer of death). Glucose repression of respiration (and also of antioxidant defences like the mitochondrial superoxide dismutase) has been proposed to be linked to the cell death process in yeast (Li *et al.*, 2006). Besides, mitochondrial release (Cui *et al.*, 2012) or overexpression of *Sc*Ndi1p (Li *et al.*, 2006) has been directly related to apoptosis in *S. cerevisiae*.

In contraposition to *S. cerevisiae*, *K. lactis* shows an oxidative metabolism, is scarcely sensitive to glucose repression of respiration (González-Siso *et al.*, 2000) and does not present an increase in longevity or a decrease in reactive oxygen species (ROS) formation when cultivated in lower glucose concentrations (Oliveira *et al.*, 2008). These differences might involve the specific characteristics of the *NDI1* genes and their products of both yeasts, and their knowledge is important for their use in gene therapy of complex I disorders. Moreover, the regulation of *NDI1* genes is an important point to control the respirofermentative metabolism of yeasts and consequently their biotechnological use as cell factories (González-Siso *et al.*, 2009) for recombinant protein production (González-Siso *et al.*, 2008) or other productions, like bioethanol.

The above exposed reasons led us to study *Kl*Ndi1p, the alternative internal dehydrogenase of *K. lactis* mitochondria. In this work, we have constructed a null mutant in *KlNDI1*, the gene encoding this enzyme and used it to perform a functional analysis of *Kl*Ndi1p. The specific characteristics found in the null strain made it valuable to produce bioethanol from cheese whey lactose, a waste product, thus also proved in this work.

Results and discussion

Construction of the KlNdi1 *null mutant*

The *KlNDI1* gene was isolated from a *K. lactis* genomic DNA library as previously described (Tarrío *et al.*, 2005) and subcloned in a commercial *S. cerevisiae* episomal vector. The resulting plasmid was used for the construction of the deletion cassette in which the coding region of *KlNDI1* was replaced by the KanMX4 module. The deletion cassette was introduced into the PM5-3C strain and the transformants selected for G418 (geneticin) resistance. Analytical polymerase chain reaction (PCR) confirmed the correct integration of the cassette in one of the two G418-resistant transformants obtained, which was named Δ*klndi1* (Fig. 1) and used in the following analyses.

Isolated mitochondria of the KlNdi1 *null mutant show unaffected rate of oxidation of exogenous NADH but no oxidation of matrix NADH*

Mitochondria from the Δ*klndi1* mutant and from the corresponding wild-type PM5-3C strain were isolated and used to measure oxygen consumption as described in *Experimental procedures*. Both externally added NADH and a combination of malate and pyruvate were used as substrates. The combination of malate and pyruvate produces NADH in the mitochondrial matrix via the tricarboxylic acid cycle. Externally added NADH cannot pass through the membrane and enter the mitochondrial matrix; it can be oxidized by external alternative dehydrogenases in intact isolated yeast mitochondria.

Oxygen consumption by isolated mitochondria from the Δ*klndi1* mutant compared with the wild-type strain (Fig. 2) shows that oxidation of matrix NADH is impaired in the mutant whereas oxidation of externally added NADH remains equal. This confirms that *Kl*Ndi1p, depleted in the mutant, is the single internal mitochondrial NADH : ubiquinone oxidoreductase in the mitochondria of *K. lactis*, and that activity of external alternative dehydrogenases is not affected by the Δ*klndi1* mutation. A similar phenotype has been reported for isolated mitochondria from the *S. cerevisiae* Δ*ndi1* mutant (Marres *et al.*, 1991; Bakker *et al.*, 2000).

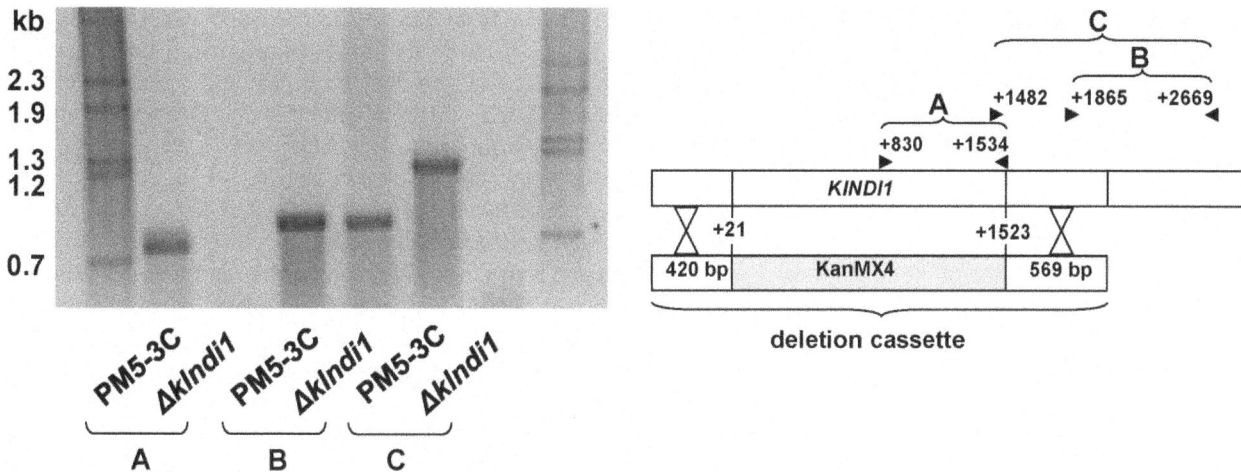

Fig. 1. Analytical PCRs with genomic DNA of the wild type PM5-3c and *ΔkIndi1* mutant strains as template to prove the correctness of the deletion: (A) primers amplify from +830 to +1534 of *KINDI1* sequence, (B) primers amplify from +1865 to +2669 of *KINDI1* sequence and (C) primers amplify from +1482 to +2669 of *KINDI1* sequence. Deleted region of the *KINDI1* gene extends from +21 to +1523.

In the KIndi1 *null mutant, there is no overexpression of the external alternative dehydrogenases*

Analysis of the expression of the three *K. lactis* mitochondrial alternative dehydrogenases in the *ΔkIndi1* mutant compared with the wild-type strain was done (Fig. 3). In the *ΔkIndi1* mutant, the band corresponding to the expression of the *KINDI1* gene disappears as expected and quantification of the bands corresponding to *KINDE1* and *KINDE2* expression, responsible of cytosolic NADH oxidation, shows no statistically significant differences in reference to those in the wild-type strain (*P* values of 0.99 and 0.11, respectively, in the *t*-test performed to compare the means). This is in accordance with the above exposed results measuring oxygen consumption using NADH form cytoplasmic or matrix origin

(Fig. 2) and corroborates that either expression or activity of external alternative dehydrogenases is not affected in the *ΔkIndi1* strain.

The growth of the KIndi1 *null mutant in glucose is unaffected and in lactate is impaired*

To test the effect of *KINDI1* disruption on *K. lactis* metabolism, we examined the growth of the *ΔkIndi1* mutant in comparison with that of the parental PM5-3C strain in liquid (shake flask cultures) and solid complete medium (CM) containing various carbon sources under the conditions described in *Experimental procedures*. The mutant was no longer able to utilize non-fermentable substrates such as lactate while showed the same growth rate than the

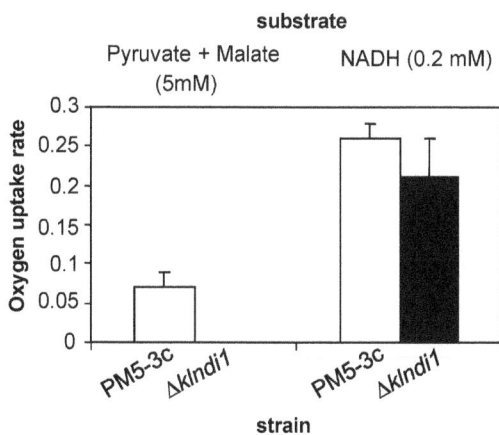

Fig. 2. Exogenous NADH and pyruvate + malate oxidation by isolated mitochondria from *K. lactis* wild type (PM5-3c) and *ΔkIndi1* mutant. The oxygen uptake rates (μmol O$_2$ min^{-1} mg protein^{-1}) were measured in the presence of 0.25 mM ADP. Results are the mean ± standard deviation (SD) of two independent experiments.

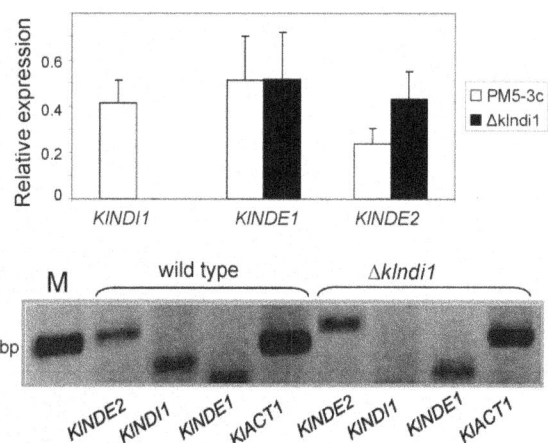

Fig. 3. Expression (down) of the three *K. lactis* mitochondrial alternative dehydrogenase genes in the *ΔkIndi1* mutant compared with the wild-type strain (PM5-3c). Relative expression values (divided by *KIACT1*) are the mean ± standard deviation (SD) of four (wild type) or two (*ΔkIndi1* mutant) independent experiments (up).

parental strain in fermentable substrates such us glucose (Fig. 4). A similar phenotype was reported for the *S. cerevisiae* Δ*ndi1* mutant (Marres *et al.*, 1991; Bakker *et al.*, 2001; Li *et al.*, 2006). Also Δ*ndi1* mutants of both *K. lactis* (this work) and *S. cerevisiae* (Marres *et al.*, 1991; Bakker *et al.*, 2001) exhibited similar growth rate as wild-type cells in galactose. However, a difference is that ethanol sustained a more limited growth of the Δ*klndi1* mutant than the wild type (Fig. 4), while growth of the *S. cerevisiae* Δ*ndi1* mutant in ethanol was unaffected with respect to the wild type (Marres *et al.*, 1991; Bakker *et al.*, 2001).

For growth on lactate and ethanol in the Δ*klndi1* mutant, with the NAD⁺-linked steps of the Krebs cycle inhibited, the glyoxylate cycle and glucogenogenesis pathways have to be induced. The amount of ATP generated in this metabolic condition is lower than when the Krebs cycle and respiratory chain are fully operative as occurs in the wild-type strain, which may explain the reduced growth in non-fermentable carbon sources of the Δ*klndi1* strain (Marres *et al.*, 1991).

The better growth of the Δ*klndi1* strain observed in ethanol than lactate may also be explained by energetic and redox balances. For lactate utilization, three genes have been characterized so far in *K. lactis*, namely *KlJEN1*, *KlDLD* and *KlCYB2*, which encode a permease and two lactate ferricytochrome *c* oxidoreductases respectively (Rodicio and Heinisch, 2013). Lactate dis-

similation by *S. cerevisiae* is also initiated by the oxidation of lactate to pyruvate via a cytochrome *c*-dependent mitochondrial lactate dehydrogenase and does not yield cytosolic NADH, while ethanol dissimilation by cytosolic alcohol dehydrogenases yields NADH that can be oxidized by the respiratory chain via both external mitochondrial dehydrogenases and the glycerol 3-phosphate shuttle, consisting of cytosolic NADH-linked glycerol 3-phosphate dehydrogenase and a membrane-bound FAD⁺-linked glycerol 3-phosphate : ubiquinone oxidoreductase (Bakker *et al.*, 2001). Both cytosolic NADH-oxidation systems are in concerted regulation (Rigoulet *et al.*, 2004). In *K. lactis*, two cytosolic isoforms of alcohol dehydrogenase, encoded by *KlADH1* and *KlADH2*, exist and these two mechanisms for cytosolic NADH oxidation and ATP generation in mitochondrial respiratory chain are also operative (González-Siso and Cerdán, 2007; Saliola *et al.*, 2010), as schematized in Fig. 5A. These metabolic characteristics may explain the growth of the Δ*klndi1* mutant in ethanol, limited but at a higher rate than in lactate.

Permeabilized cells of the Klndi1 *null mutant show no oxidation of mitochondrial matrix NADH*

In *S. cerevisiae*, there is experimental evidence that another important component of the redox-balancing

A

The generation time (h)

Carbon source	PM5-3C	Δ*klndi1*
Glucose	3.2 ± 0.6	2.9 ± 0.04
Galactose	3.2 ± 0.7	3.2 ± 0.1
Lactate + glucose 0.05%	6.1 ± 0.1	>12
Ethanol+ glucose 0.05%	4.3 ± 0.1	5.8 ± 0.2

B

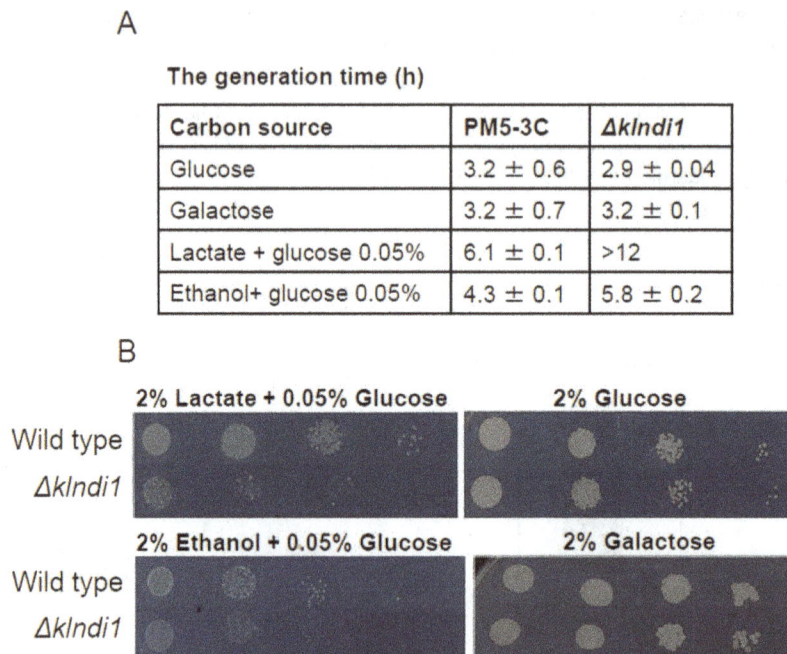

Fig. 4. (A) The generation time (hours) of the Δ*klndi1* mutant compared with the wild-type strain (PM5-3c) growing in liquid CM with the indicated carbon sources. Samples were taken each hour during the first 10 h of culture, inoculum at a OD₆₀₀ of 0.1. These data are represented as mean ± standard deviation (SD) from two independent experiments. (B) Growth of the Δ*klndi1* mutant compared with the PM5-3c strain (wild type for *NDI1*) in solid CM with the indicated carbon sources. Conditions as described in *Experimental procedures*. Photographs were taken after 1.5 days of growth on glucose, ethanol and lactate, and after 3 days of growth on galactose. *$P < 0.05$; **$P < 0.01$.

A

B

C

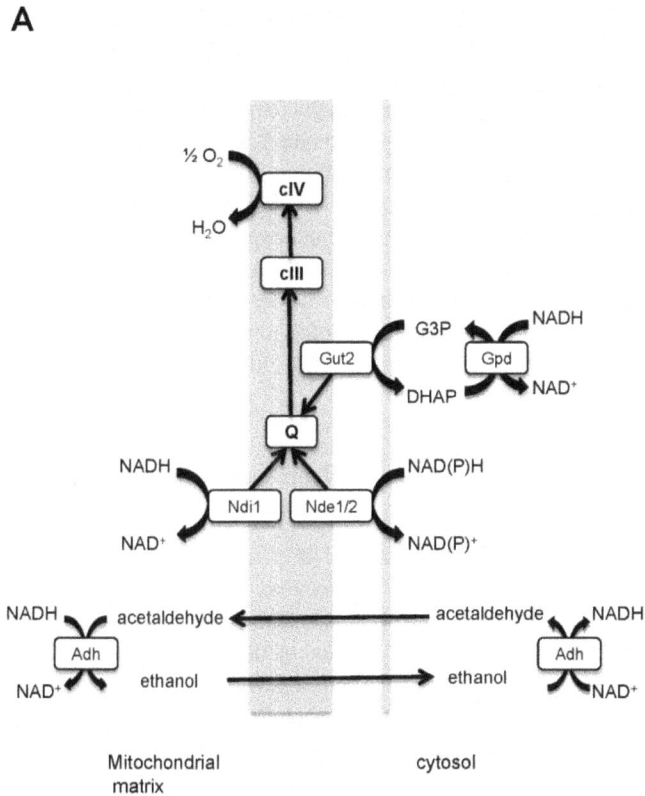

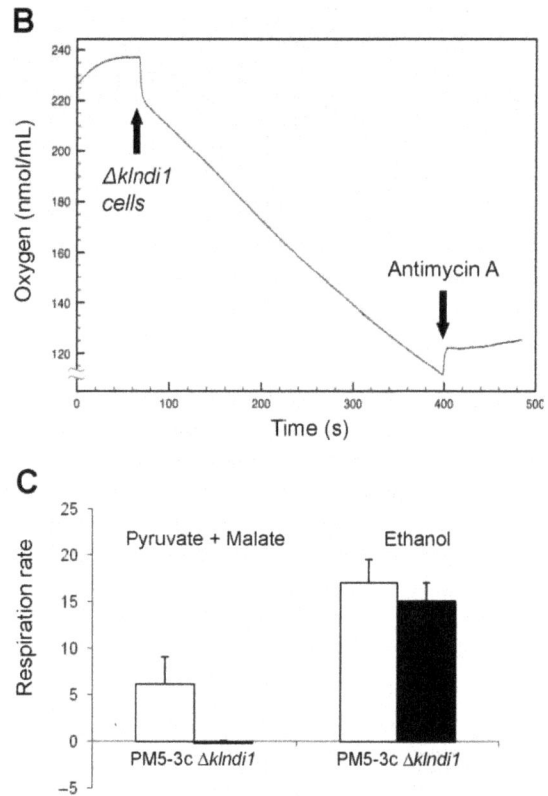

Fig. 5. (A) Scheme of the *K. lactis* electron transport chain and cytosol-mitochondria NADH shuttle systems. Adh, alcohol dehydrogenases; Ndi1, internal NADH dehydrogenase; Nde1/2 external NAD(P)H dehydrogenases; Q, ubiquinone; DHAP, dihydroxy acetone phosphate; G3P, glycerol 3-phosphate; Gpd, glycerol 3-phosphate dehydrogenase; Gut2, FAD$^+$-linked glycerol 3-phosphate dehydrogenase; cIII, complex III; cIV, complex IV. (B) Oxygen uptake by permeabilized cells of the *Δklndi1* mutant with 0.1 M ethanol as substrate stops upon antimycin A addition. (C) Respiration rate (nmoles O$_2$ min^{-1} ml^{-1}) by permeabilized cells of the *Δklndi1* mutant compared with the PM5-3c strain (wild type for *NDI1*) with the substrates pyruvate + malate and ethanol. Results are the mean ± standard deviation (SD) of three independent experiments. **$P < 0.01$.

system is the ethanol/acetaldehyde (Et–Ac) shuttle (Bakker *et al.*, 2001). These metabolites, which can freely diffuse through mitochondrial membranes, are oxidized or reduced by cytosolic or mitochondrial alcohol dehydrogenases transferring the NADH excess from one cellular compartment to the other. The Et–Ac shuttle allows respiratory growth of the *S. cerevisiae Δndi1* mutant at the same rate than the wild-type strain in both glucose and ethanol as carbon sources. This is because matrix NADH is exchanged by the reversible interconversion of acetaldehyde into ethanol to the cytosol, where it can be oxidized by external mitochondrial dehydrogenases and the glycerol 3-phosphate shuttle. Nevertheless, this system is limited, since the mutant does not grow on acetate (Bakker *et al.*, 2001). Saliola and colleagues (2010), based only on the fact that the addition of small amounts of ethanol, unable to sustain growth by itself, allowed a slight growth recovery in lactate medium of a *Δklndi1* mutant, have proposed that in *K. lactis*, the Et–Ac shuttle might be operative and could transfer the redox excess from the mitochondria to the cytoplasm, thus feeding the respiratory chain through

the external dehydrogenases *Kl*Nde1p and *Kl*Nde2p (Fig. 5A).

To verify if the Et–Ac shuttle is active in *K. lactis* and, in complete absence of *KlNDI1* expression, might exchange matrix NADH to be oxidized by cytosolic systems, we measured oxygen consumption by permeabilized cells of *K. lactis* wild type and *Δklndi1* mutant with the combination of pyruvate and malate as substrate. The use of this system instead of isolated mitochondria allows to test reoxidation of internal produced NADH in conditions in which the cytoplasmic component of the shuttle is active. Blockage of oxygen consumption by antimycin A in *Δklndi1*-permeabilized cells proved that measured respiration rate corresponds to the mitochondrial respiratory chain (Fig. 5B). However, permeabilized cells of the *Δklndi1* mutant do not consume oxygen in the presence of the combination of pyruvate and malate as substrate (Fig. 5C). As a control, we measured oxygen consumption with ethanol as substrate and there is no significant difference between permeabilized cells of the *Δklndi1* mutant in comparison to the wild-type strain (Fig. 5C). This may be because ethanol metabolism produces

NADH in the mitochondrial matrix but also in the cytosol since *K. lactis* has mitochondrial and cytosolic alcohol dehydrogenases and a functional glyoxylate cycle (Rodicio and Heinisch, 2013), and cytosolic NADH can be oxidized by the external alternative dehydrogenases of the respiratory chain. Therefore, the Et–Ac shuttle seems to be inactive in the *Δklndi1* mutant cells, at least in the conditions assayed.

The expression of the KINDI1 gene is regulated by KIGLR1 function

In previous papers, we proved the involvement of *Kl*Glr1p, the *K. lactis* NADPH-dependent glutathione reductase (GLR), not only in the defence against oxidative stress but also in the regulation of the glucose metabolism in *K. lactis*. This second role was inferred both from the improved growth on glucose caused by the *Δklglr1* mutation and by the increased GLR activity in the *Δklgcr1* strain (García-Leiro *et al.*, 2010a) being *Kl*Gcr1p a positive transcriptional regulator of glycolytic genes, as well as by the fact that enzymatic activities of GLR and glucose 6-phosphate dehydrogenase (G6PDH) from the PPP are positively correlated (Tarrío *et al.*, 2008). Thus, G6PDH is positively regulated by an active respiratory chain and GLR plays a role in the reoxidation of the NADPH from the PPP in these conditions (Tarrío *et al.*, 2008). A further proteomic analysis provided additional results; H_2O_2 addition causes downregulation of enzymes from the glycolytic pathway and Krebs cycle in wild-type *K. lactis*, whereas *Δklglr1* deletion prevents this effect and actually causes upregulation of the glycolytic, Krebs cycle and oxidative PPPs, which led us to propose that the redox imbalance produced by the *Δklglr1* deletion controls glucose metabolic fluxes in *K. lactis* (García-Leiro *et al.*, 2010b). New data obtained in this work also support this proposal (Fig. 6). Thus, RNA was extracted from cells grown in CM with 2% glucose (OD_{600} = 0.8) and *KINDI1*

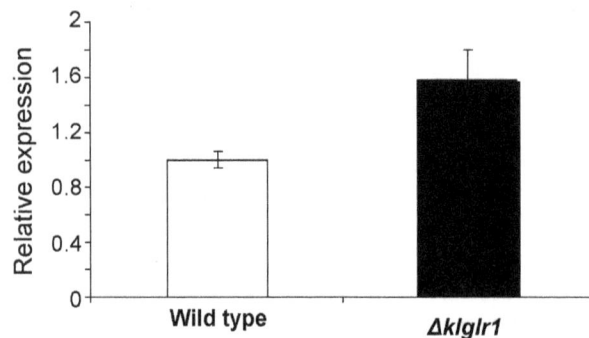

Fig. 6. Relative expression of *KINDI1* measured by qRT-PCR in the wild type NRRL-Y1140 and *Δklglr1* mutant strains. Results are the mean ± standard deviation (SD) of two independent cultures analysed in duplicate. **$P < 0.01$.

mRNA levels were measured by quantitative reverse transcription polymerase chain reaction (qRT-PCR) in a *Δklglr1* mutant (García-Leiro *et al.*, 2010a) and the isogenic parental strain. Significant differences were found: *KINDI1* mRNA levels being 1.58-fold higher in the *Δklglr1* mutant. The interpretation of these results is that GLR depletion, by reducing NADPH oxidation rate and therefore $NADP^+$ availability, necessary for G6PDH activity, could redirect glucose flow from the PPP which is preponderant in *K. lactis* for glucose metabolism (González-Siso *et al.*, 2000; 2009) to glycolysis and Krebs cycle, which will be associated to increased *KINDI1* expression.

The chronological life span of the KIndi1 null mutant is decreased

It has been reported that *Sc*Ndi1p is the closest homologue of mammalian AMID in yeast and is involved in chronological ageing (Li *et al.*, 2006). Overexpression of *Sc*Ndi1p causes apoptosis-like cell death. This effect can be repressed by increased respiration on glucose-limited media, being the apoptotic effect of *ScNDI1* overexpression associated with increased production of ROS in mitochondria (Li *et al.*, 2006). We have determined that also in *K. lactis*, the closest homologue to human AMID is *Kl*Ndi1p (González-Siso and Cerdán, 2007), and therefore studied its putative role in ageing as following described.

Chronological life span of the *Δklndi1* mutant and parental strains was measured using the protocols described in the *Experimental procedures*. For cells grown in a CM and further maintained in distilled water, significant differences in viability (quantified as half-life) were found between the *Δklndi1* mutant and wild-type strain (Fig. 7A), the viability being decreased in the mutant. This situation is different to the one reported in *S. cerevisiae* where disruption of *NDI1* decreases ROS production and elongates the chronological life span of yeast, accompanied by the loss of survival fitness (Li *et al.*, 2006). However, considering that these authors measured life span with culture medium renewal each 3 days, while our conditions are survival in absence of nutrients (distilled water), we also measured the life span of the *Δklndi1* mutant and parental strains using the same protocol than Li and colleagues (2006) without finding increased life span in the mutant (Fig. 7B). In the latter conditions but before medium renewal, there was neither significant difference in ROS levels (in exponential and stationary culture phase, measured as in Tarrío *et al.*, 2008), nor DNA and nuclear fragmentation between the *Δklndi1* mutant and wild-type strain (data not shown). The differences caused by *NDI1* deletion in the life span of the two species might be related to differences in

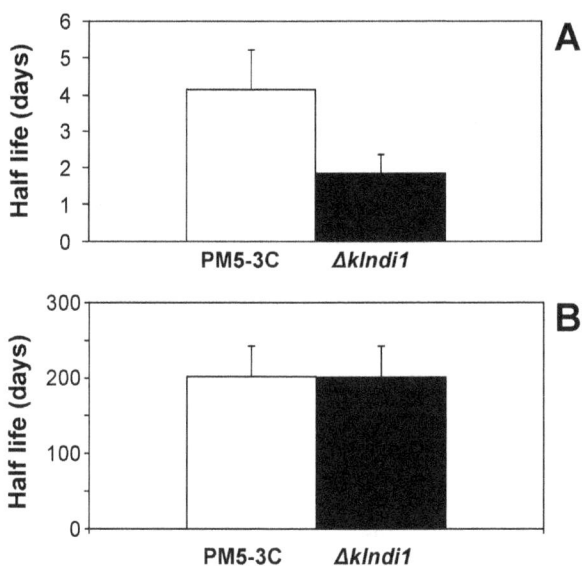

Fig. 7. Life span of the wild type PM5-3c and Δklndi1 mutant strains. Conditions as described in *Experimental procedures*. (A) Survival in distilled water. Results are the average of four independent cultures. (B) Survival with culture medium renewal. Results are the average of two independent cultures. **$P < 0.01$.

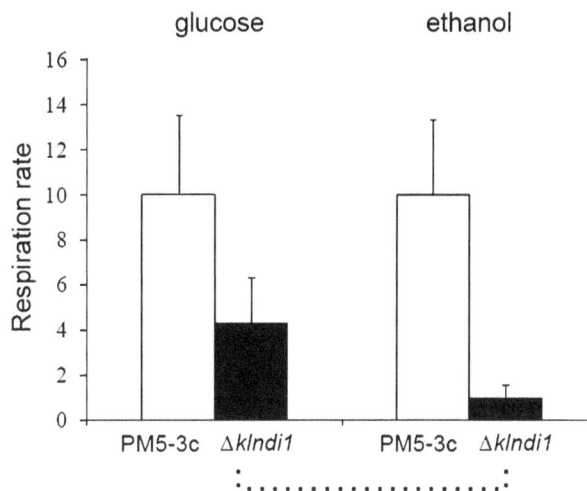

Fig. 8. Cell respiration rate (nmol O_2 ml^{-1} min^{-1} OD$_{600}$$^{-1}$) of *K. lactis* PM5-3c and Δklndi1 mutant cells cultured in glucose media and with glucose or ethanol as substrates. Results are the mean ± standard deviation (SD) of four independent cultures analysed in duplicate. *$P < 0.05$; **$P < 0.01$.

respiratory metabolism and ROS production (reviewed in González-Siso *et al.*, 2009), and their dependence on *NDI1* function, but further work is necessary to clarify this issue.

Interestingly, the Δnde1 mutation has been reported to shorten replicative life span in *S. cerevisiae* and the authors proposed that this effect results from multiple mechanisms including the low respiration rate and low energy production rather than just from a ROS-dependent pathway; however, regarding chronological life span, measured in rich medium, Δnde1 cells were able to survive better than wild-type cells in the stationary phase (Hacioglu *et al.*, 2012).

The Klndi1 null mutant shows overall reduced respiration rate and increased ethanol production from glucose

To increase supporting evidence of the effect of *KlNDI1* disruption on yeast metabolism, we measured the overall respiration rate, with glucose and ethanol as substrates, in cells of the Δklndi1 mutant and parental strains, sampled at two OD$_{600}$ values (2 and 6) from glucose (YPD) cultures. Figure 8 shows that overall respiration rate of the mutant compared with the parental strain is reduced both with glucose or ethanol as substrates. There is also a significant difference in cell respiration rate of the Δklndi1 mutant (but not in the case of the wild-type strain) being lower in ethanol compared with glucose as substrate.

We also measured ethanol production in glucose liquid cultures of the Δklndi1 mutant and wild-type strains, obtaining increased ethanol production in the mutant. Thus, after 2 days of culture in CM with 2% glucose, when glucose concentration in the medium was lower than 1 g l^{-1} and OD$_{600}$ about 10, ethanol in the culture medium was higher for the mutant strain than for the PM5-3C strain (1 ± 0.18 g l^{-1} versus 3.3 ± 0.17 g l^{-1}, $n = 2$). Similar results were obtained in a Δklndi1 mutant obtained in a different genetic background (Saliola *et al.*, 2010). Higher ethanol levels are due to the lower respiration rate of the mutant (Fig. 8) and, moreover, to ethanol accumulation due to its limited use as carbon source in the Δklndi1 mutant, also proved in this work (Fig. 4).

Both respiration rate and ethanol production measurements indicate a more fermentative glucose metabolism in the case of the Δklndi1 mutant compared with the parental strain that has biotechnological application in whey utilization as following reported.

Ethanol production from cheese whey lactose by the Klndi1 null mutant

Actual market trends point to a worldwide gradual increase in cheese production that generates whey as waste product (9 kg of whey per 1 kg of cheese). Whey disposal constitutes a pollution problem even though systems for its treatment have been developed. The most abundant component of cheese whey is the disaccharide lactose that remains in the permeate after whey ultrafiltration for protein recovery. The treatment of whey permeate by fermenting lactose to ethanol has received wide attention to date and there are industrial plants in operation (Guimarães *et al.*, 2010). However, the development of a high productivity lactose fermenting process continues to be of prime importance. *Saccharomyces cerevisiae* is unable to ferment directly lactose to ethanol and the microorganisms mainly used for this purpose are selected *Kluyveromyces* spp. To reduce distillation costs, it is recommended to use concentrated whey up to 120 g lactose l^{-1} (González-Siso, 1996; Arif *et al.*, 2008).

In this work, we took advantage of the predominantly fermentative metabolism shown by the *Δklndi1* mutant and assayed the use of this strain for the bioconversion of lactose to ethanol in concentrated cheese whey permeate. Figure 9 shows that the *Δklndi1* mutant, grown in the fermenter culture conditions described in *Experimental procedures*, fully degrades the lactose in 8–9 days and produces up to 50 g l^{-1} ethanol which gives a yield (80.1% of the theoretical maximum of 4 mol of ethanol from 1 mol of lactose) close to the maximum reported for industrial strains (Della-Bianca *et al.*, 2013; Diniz *et al.*, 2013), and higher than the NRRL-Y1140 laboratory strain that is wild type for the *KINDI1* gene (the yield ranged between 62.6% and 50% of the theoretical maximum in the same conditions, according to the highest value and the value adjusted to the curve respectively). Although the yield obtained with the NRRL-Y1140 strain seems high for a respiratory yeast, the distribution between respiratory and fermentative metabolism in wild-type *K. lactis* is dependent on oxygen availability (González-Siso *et al.*, 1996a; Guimarães *et al.*, 2010) and low levels of oxygen in these cultures favour fermentation. Similar ethanol yields in whey than those reached by the NRRL-Y1140 strain in this work were obtained by other authors in fermentative

conditions with other *Kluyveromyces* strains (Szczodrak *et al.*, 1997; Guimarães *et al.*, 2010).

Once we proved the ethanol production performance of the *Δklndi1* mutant, we decided to transform this strain with the multicopy plasmid pSPGK1-LAC4 (Becerra *et al.*, 2001) to improve its capability to grow in lactose and therefore ethanol production. The pSPGK1-LAC4 plasmid expresses the *LAC4* gene that codes for *K. lactis* beta-galactosidase. The results show that the *Δklndi1* mutant, in the sealed tube conditions described in *Experimental procedures*, degrades the 79% of the lactose in 4 days (a 56% more than the untransformed strain) producing ethanol with a yield (quantified as percentage of the theoretical maximum of 4 mol of ethanol from 1 mol of lactose) 16.4% higher for the transformed versus untransformed strain. Overexpression of beta-galactosidase increases both lactose consumption rate and yield of ethanol from lactose.

These results support the biotechnological use of *Δklndi1* and its derivative pSPGK1-LAC4 transformants for efficient fermentation of cheese whey to bioethanol.

Conclusions

In this paper, we have constructed and characterized an engineered *K. lactis* strain by deletion of the *KINDI1* gene. We proved that *KINDI1* encodes the single internal NADH : ubiquinone oxidoreductase in the mitochondria of *K. lactis* (*KI*Ndi1p), and that activity of external alternative dehydrogenases is not affected by the *Δklndi1* mutation. Similarly to data previously reported for the null in *S. cerevisiae*, the *Δklndi1* mutant grows at the same rate

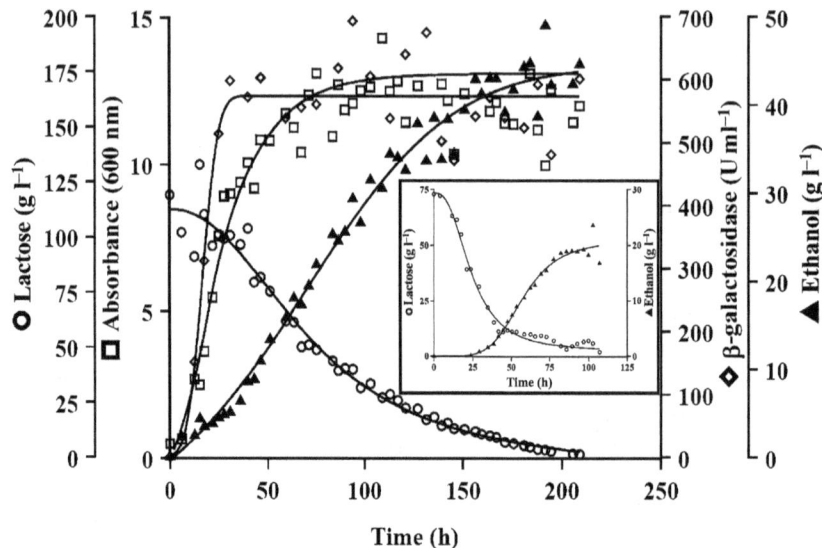

Fig. 9. Parameters of the *Δklndi1* respiratory mutant growing in a 2 l bioreactor with threefold concentrated cheese whey permeate as substrate. Insert figure depicts the parameters of NRRL-Y1140 growing in a 2 l bioreactor with twofold concentrated cheese whey permeate as substrate. Conditions as described in *Experimental procedures*.

than the wild-type strain in glucose but is unable to grow in lactate. Differently to the phenotypes reported for the $\Delta ndi1$ in *S. cerevisiae*, the $\Delta klndi1$ strain grows in ethanol with a lower rate and shows reduced chronological life span in absence of nutrients compared with the wild-type strain. Besides, shuttle systems transferring NADH from the mitochondrial matrix to the cytosol for oxidation by external dehydrogenases do not operate in permeabilized *K. lactis* cells. As a result of multiple functional differences, the $\Delta klndi1$ mutation shifts glucose metabolism to fermentative, thus increasing the performance of the $\Delta klndi1$ strain for bioethanol production from lactose present in cheese whey. This is a biotechnological advantage versus *S. cerevisiae*, which is unable to metabolize lactose, and versus *K. lactis* expressing *NDI1*, which is highly respiratory. The $\Delta klndi1$ strain combines three important characteristics: the ability to use lactose, high levels of ethanol production and low use of the ethanol produced as carbon source.

Experimental procedures

Strains, media and culture conditions

The *K. lactis* strain PM5-3C (*MATa uraA Rag*[+]) (Wésolowski-Louvel *et al.*, 1992) was used.

Growth and handling of yeasts were carried out according to standard procedures (Kaiser *et al.*, 1994). The yeast cells were cultivated, unless otherwise stated, in Erlenmeyer flasks at 30°C and 150–250 r.p.m. in the synthetic CM (Zitomer and Hall, 1976), the dropout medium CM-ura (without uracil) or YP medium (2% bactopeptone, 1% yeast extract) containing one of the following carbon sources: 0.1–2% glucose, 2% lactate, 2% ethanol and 2% galactose. Lactate and ethanol media were supplemented with 0.05% glucose. The flasks were filled with 40% volume of culture medium. Solid growth media also contained 1.5% agar.

For growth analysis on solid CM, cells from overnight Erlenmeyer cultures in the same medium were diluted to $OD_{600} = 0.1$ and then several serial 1:10 dilutions were made. Drops (0.005 ml) of the dilutions were added to the solid media, incubated at 30°C and photographed after 1.5 and 3 days.

Construction of null mutant

Deletion of nucleotides +21 to +1523 in the *KINDI1* gene (Tarrío *et al.*, 2005) was carried out by the one-step method (Rothstein, 1991). The template of the deletion cassette was constructed as described in Zaragoza (2003) in the plasmid YEplac195, and included the kanamycin resistance gene (the kanMX4 module) flanked by 420 bp homologous to the *KINDI1* 5′ region and 569 bp homologous to the *KINDI1* 3′ region. The deletion cassette was PCR amplified and the product transformed to the *K. lactis* wild-type strain PM5-3C. For selection of transformants, geneticin was added at a final concentration of 0.3 mg ml^{-1} to YPD medium. Correct *KINDI1* gene deletion was verified by analytical PCR. Genomic DNA

as template and several combinations of inward primers, binding outside the deletion cassette, and outward primers, binding within the deleted region, were used. Also, the absence of *KINDI1* mRNA in the null mutant was verified by reverse transcriptase-polymerase chain reaction (RT-PCR).

Isolation of RNA, production of cDNA and RT-PCR

The kits NucleoSpin (Macherey-Nagel) and SuperScript II (Promega) were used for RNA extraction from the *K. lactis* cells and for cDNA synthesis and RT-PCR, respectively, following the instructions of the suppliers. Twenty cycles of PCR were performed with primers, which amplified the cDNA from positions +1020 to +1380 of the *KINDI1* cds, +1088 to +1603 of the *KINDE2* cds and +31 to +315 of the *KINDE1* cds. *KINDE1* and *KINDE2* are the genes encoding the two *K. lactis* external alternative dehydrogenases (Tarrío *et al.*, 2005; 2006b). The gene of the *K. lactis* actin (KLLA0D05357g) was used as control (positions +1290 to +1728 of the cds). Quantification of the products, separated on agarose gels, stained with ethidium bromide and photographed with a digital system, was performed in tagged image file format with densitometry analysis software (ImageMaster Total Laboratory, version 2.00, GE Healthcare).

Analysis of expression by qRT-PCR

Total RNA isolated using the AURUM kit (Bio-Rad) was converted into cDNA and labelled with the KAPA SYBR FAST universal one-step qRT-PCR kit (Kappa Biosystems, Woburn, MA, USA).

PCR primers for *KINDI1* were designed with the tool 'Primer Blast' at NCBI to generate 80–120 base pairs amplicon. Primer length of 18–24 bp, Tm of 59 or 60°C and G/C content not higher than 50% were selected. The sequence of primers chosen is GAATACTTGGCTAA GGTCTTCGAC (forward) and CTTGAATGGCTTGAAA CCGTTTTC (reverse). The specificity of the primers for the *KINDI1* cds was verified.

The ECO Real-Time PCR System was used for the experiments (Illumina, San Diego, CA, USA) and calculations were made by the 2-$\Delta\Delta$Ct method (Livak and Schmittgen, 2001). Two independent cultures and RNA extractions were assayed for each strain or condition. The mRNA levels of the *KINDI1* gene were corrected by the geometric mean of the mRNA levels of the gene *ACT1*, a control gene also used for RT-PCR experiments as above described, which was previously verified to be constitutive in the assayed conditions. A *t*-test was applied to evaluate the differences between ΔCt values (Ct values normalized with reference genes) of control and treated samples with a *P* value of 0.05.

Cell respiration rate

The respiration rate of *K. lactis* PM5-3c and *KINDI1* null mutant cells was measured at 25°C using a Clark-type oxygen electrode (Hansatech) following the procedure previously described (García-Leiro *et al.*, 2010a). The cells grown in liquid YPD medium were collected by centrifugation (previously OD_{600} was measured) and suspended in one-tenth volume of 50 mM potassium phosphate (pH 6.5) buffer with 0.1 M glucose or ethanol. 50–100 µl of the cell suspension was introduced into the chamber filled with 1 ml of the same buffer and the amount of oxygen consumed was recorded. Data are expressed as nmol O_2 ml^{-1} min^{-1} OD_{600}^{-1}.

Mitochondrial respiration rate in permeabilized cells

Five millilitres of *K. lactis* cells of the wild-type PM5-3c and *Klndi1* null mutant strains grown in liquid medium (10 ml of culture in a 50 ml Erlenmeyer flask incubated at 28°C and 250 r.p.m.) up to $OD_{600} = 1$ were collected by centrifugation and suspended in 5 ml of phosphate buffer 25 mM pH = 7.2 with $MgCl_2$ 5 mM and sorbitol 0.65 M. For permeabilization 0.1 ml of 50× protease inhibitor Complete Mini, EDTA Free (Roche) and 0.025 ml of digitonin 10 mM were added, tubes were gentle agitated during 1 min, cells collected again by centrifugation and suspended in 0.5 ml of the same buffer.

The respiration rate (oxygen uptake) of permeabilized cells was measured at 28°C using a Clark-type oxygen electrode (Hansatech). In the electrode cuvette, 0.8 ml of buffer was added plus 0.05 ml of adenosine diphosphate (ADP) 50 mM and 0.05 ml of the substrate (1 M pyruvate-malate or 2 M ethanol). After baseline stabilization, 0.1 ml of the permeabilized cell suspension was added and the amount of oxygen consumed was recorded. The basal respiration of cells in the same conditions but without substrate was subtracted. Data are expressed as nmol O_2 ml^{-1} min^{-1}. All respiration was found to be sensitive to antimycin A.

Glucose, lactose and ethanol

Analyses were done as previously described (González-Siso and Suárez Doval, 1994) with a refractive index detector coupled to an isocratic high-performance liquid chromatography system operating with a Sugar-Pack column (Waters, Millipore) at 90°C, Milli-Q water was the mobile phase at 0.5 ml min^{-1}.

DNA and nuclear fragmentation

To evaluate genomic DNA and nuclear integrity, cultures of *K. lactis* PM5-3c and *Klndi1* null mutant in liquid YPD media (2% glucose) were inoculated at $OD_{600} = 0.1$ from fresh overnight pre-cultures and every day until 4 days an aliquot was taken. Genomic DNA was extracted and visualized after electrophoresis on agarose gels and ethidium bromide staining. Nuclear staining was performed, after fixing the yeast cells with 2.5% glutaraldehyde on ice for 10 min, with 10 µg ml^{-1} propidium iodide and 2.5 µg ml^{-1} 4,6-diamidino-2-phenylindole. The stained cells were viewed with a fluorescent microscope and photographed with a digital camera.

Life span

Chronological life span of *K. lactis* PM5-3c and *Δklndi1* mutant was determined by measurement of colony-forming ability [colony-forming units (cfu)]. Cultures in liquid YPD and

CM media with 2% glucose (two cultures in YPD and two cultures in CM) were inoculated at $OD_{600} = 0.1$ from fresh overnight pre-cultures in the same medium; after 2 days of growth, the cells were washed and transferred to distilled water at $OD_{600} = 1$ and further incubated in the same conditions (at 30°C and 150–250 r.p.m.). The first sample was taken at the moment of transfer and considered 100% viability. Aliquots of the cultures containing approximately 100 cells were plated to YPD medium. Every 2 days, new samples were taken to score viability until less than 10% of the culture was viable. The half-life of each strain was calculated by exponential fit of the data with Excel software.

Survival of *K. lactis* PM5-3c and *Δklndi1* mutant was also determined with culture medium renewal (Li *et al.*, 2006). Percentage of viable versus total cell number was measured, by methylene blue staining and counting in Neubauer chamber. Only unviable cells are stained by this dye. Cultures were performed as described above in YPD but cells were transferred to fresh medium every 3–4 days during 6 weeks. Just before medium renewal, samples were taken and cellular viability was measured.

Isolation of mitochondria

Kluyveromyces lactis mitochondria were isolated as previously described (Tarrío *et al.*, 2005; 2006b). Cells were aerobically grown in CM with 1% glucose as carbon source.

Oxygen uptake studies with mitochondrial preparations

Substrate-dependent oxygen consumption rates of freshly isolated mitochondria were determined at 30°C using a Clark-type oxygen electrode (Hansatech) as previously described (Tarrío *et al.*, 2005; 2006b). Respiratory substrates were NADH (0.2 mM) and pyruvate + L-malate (5 mM), which generate intramitochondrial NADH. Measurements were made in the absence or presence of 0.25 mM ADP, and all respiration was found to be sensitive to antimycin A.

Bioethanol production from cheese whey

Cheese whey ultrafiltrated permeate, threefold concentrated, was obtained from a local cheese factory (Queizúar S.L.). Cultures of wild type and *Klndi1* null mutant strains were performed in a 2 l vessel bioreactor (Biostat MD). Preinocula were grown in YPD at 30°C. Temperature, pH, agitation and aeration were maintained at 30°C, 5.6, 500 r.p.m. and 1–1.3 l min^{-1} respectively for 40 h when aeration was stopped and agitation reduced to 150 r.p.m. Samples were taken at determined time intervals for analysis of growth (OD_{600}), beta-galactosidase activity, lactose and ethanol.

For cultures in sealed tubes, preinocula were grown in 500 ml flasks at 30°C in 200 ml YPD. When cells reached an $OD_{600} = 3$ at 600 nm, they were centrifuged and added as inoculum to 100 ml sealed tubes containing 40 ml of three-times concentrated cheese whey permeate. They were grown at 30°C and 250 r.p.m. Aliquots were taken at different times for analysis of growth (OD_{600}), beta-galactosidase activity, lactose and ethanol.

Beta-galactosidase activity

Beta-galactosidase activity was measured using the method of Guarente (1983) as previously described (Becerra et al., 2001). One enzyme unit (EU) was defined as the quantity of enzyme that catalyses the liberation of 1 nanomol of orthonitrophenol from orthonitrophenyl-β-D-galactopyranoside per min under assay conditions. EU are expressed per ml of culture medium.

Protein determination

Protein concentration was measured by the method of Bradford (1976) using bovine serum albumin as a standard.

Other procedures

Standard procedures for manipulation of nucleic acids were essentially those of Sambrook and colleagues (1989). *Escherichia coli* DH-10B was used for plasmid amplification. Yeast transformation was performed using a lithium acetate procedure (Gietz and Woods, 1994).

Statistical analysis

Data are expressed as mean ± standard deviation. The statistical significance of differences found between means was evaluated by the unpaired Student's *t*-test, two-tailed *P* values calculated with Excel software.

Acknowledgements

We thank Dr. M. Wésolowski-Louvel for providing the strain PM5-3c and QUEIZUAR S.L. for the cheese whey. Aida Sánchez rendered technical assistance. We also thank Mariana Fernández, Ana García and Fátima Rodríguez for collaboration in preliminary studies with this mutant.

References

Arif, K., Islam, M.A., Maroof, A., Shazia, K., Umber, K., Deepa, B., and Owais, M. (2008) Applications of whey in biotechnology. In *Advances in Cheese Whey Utilization.* Cerdán, M.E., González-Siso, M.I., and Becerra, M. (eds). Kerala, India: Transworld Research Network, pp. 1–34.

Bakker, B.M., Bro, C., Kötter, P., Luttik, M.A.H., Van Dijken, J.P., and Pronk, J.T. (2000) The mitochondrial alcohol dehydrogenase Adh3p is involved in a redox shuttle in *Saccharomyces cerevisiae. J Bacteriol* **182**: 4730–4737.

Bakker, B.M., Overkamp, K.M., Van Maris, A.J.A., Kötter, P., Luttik, M.A.H., Van Dijken, J.P., and Pronk, J.T. (2001) Stoichiometry and compartmentation of NADH metabolism in *Saccharomyces cerevisiae. FEMS Microb Rev* **25**: 15–37.

Balaban, R.S., Remoto, S., and Finkel, T. (2005) Mitochondria, oxidants and aging. *Cell* **120**: 483–495.

Becerra, M., Díaz Prado, S., González-Siso, M.I., and Cerdán, M.E. (2001) New secretory strategies for *Kluyveromyces lactis* beta-galactosidase. *Protein Eng* **14**: 379–386.

Bradford, M.M. (1976) A rapid and sensitive method for the quantitation of microgram quantities of protein utilizing the principle of protein-dye binding. *Anal Biochem* **72**: 248–254.

Cui, Y., Zhao, S., Wu, Z., Dai, P., and Zhou, B. (2012) Mitochondrial release of the NADH dehydrogenase Ndi1 induces apoptosis in yeast. *Mol Biol Cell* **23**: 4373–4382.

De Risi, J.L., Iyer, V.R., and Brown, P.O. (1997) Exploring the metabolic and genetic control of gene expression on a genomic scale. *Science* **278**: 680–686.

Della-Bianca, B.E., Basso, T.O., Stambuk, B.U., Basso, L.C., and Gombert, A.K. (2013) What do we know about the yeast strains from the Brazilian fuel ethanol industry? *Appl Microbiol Biotechnol* **97**: 979–991.

Diniz, R.H.S., Rodrigues, M.Q.R.B., Fietto, L.G., Passos, F.M.L., and Silveira, W.B. (2013) Optimizing and validating the production of ethanol from cheese whey permeate by *Kluyveromyces marxianus* UFV-3. *Biocatal Agric Biotechnol* **3**: 111–117.

Duarte, M., and Videira, A. (2007) Mitochondrial NAD(P)H dehydrogenases in filamentous fungi. In *Complex I and Alternative Dehydrogenases.* González-Siso, M.I., and Cerdán, M.E. (eds). Kerala, India: Transworld Research Network, pp. 55–68.

García-Leiro, A., Cerdán, M.E., and González-Siso, M.I. (2010a) A functional analysis of *Kluyveromyces lactis* glutathione reductase. *Yeast* **27**: 431–441.

García-Leiro, A., Cerdán, M.E., and González-Siso, M.I. (2010b) Proteomic analysis of the oxidative stress response in *Kluyveromyces lactis* and effect of glutathione reductase depletion. *J Proteome Res* **9**: 2358–2376.

Gietz, R.D., and Woods, R.A. (1994) High efficiency transformation in yeast. In *Molecular Genetics of Yeast: Practical Approaches.* Johnston, J.A. (ed.). Oxford, UK: Oxford University Press, pp. 121–134.

González-Siso, M.I. (1996) The biotechnological utilization of cheese whey: a review. *Biores Technol* **57**: 1–11.

González-Siso, M.I., and Cerdán, M.E. (2007) Mitochondrial alternative NAD(P)H-dehydrogenases and respiro-fermentative metabolism in yeasts. In *Complex I and Alternative Dehydrogenases.* González-Siso, M.I., and Cerdán, M.E. (eds). Kerala, India: Transworld Research Network, pp. 69–84.

González-Siso, M.I., and Suárez Doval, S. (1994) *Kluyveromyces lactis* immobilization on corn grits for milk whey lactose hydrolysis. *Enzyme Microb Technol* **16**: 303–310.

González-Siso, M.I., Ramil, E., Cerdan, M.E., and Freire Picos, M.A. (1996a) Respirofermentative metabolism in *Kluyveromyces lactis*: ethanol production and the Crabtree effect. *Enzyme Microb Technol* **18**: 585–591.

González-Siso, M.I., Freire-Picos, M.A., and Cerdán, M.E. (1996b) Reoxidation of the NADPH produced by the pentose phosphate pathway is necessary for the utilization of glucose by *Kluyveromyces lactis rag2* mutants. *FEBS Lett* **387**: 7–10.

González-Siso, M.I., Freire-Picos, M.A., Ramil, E., González-Domínguez, M., Rodríguez-Torres, A., and Cerdán, M.E. (2000) Respirofermentative metabolism in *Kluyveromyces lactis*: insights and perspectives. *Enzyme Microb Technol* **26:** 699–705.

González-Siso, M.I., Pereira, A., Fernández Leiro, R., Cerdán, M.E., and Becerra, M. (2008) Production of recombinant proteins from cheese whey. In *Advances in Cheese Whey Utilization*. Cerdán, M.E., González-Siso, M.I., and Becerra, M. (eds). Kerala, India: Transworld Research Network, pp. 195–209.

González-Siso, M.I., García-Leiro, A., Tarrío, N., and Cerdán, M.E. (2009) Sugar metabolism, redox balance and oxidative stress response in the respiratory yeast *Kluyveromyces lactis*. *Microb Cell Fact* **8:** 46.

Guarente, L. (1983) Yeast promoters and lacZ fusions designed to study expression of cloned genes in yeast. *Methods Enzymol* **101:** 181–191.

Guimarães, P.M.R., Teixeira, J.A., and Domingues, L. (2010) Fermentation of lactose to bio-ethanol by yeasts as part of integrated solutions for the valorisation of cheese whey. *Biotechnol Adv* **28:** 375–384.

Hacioglu, E., Demir, A.B., and Koc, A. (2012) Identification of respiratory chain gene mutations that shorten replicative life span in yeast. *Exp Gerontol* **47:** 149–153.

Kaiser, C., Michaelis, S., and Mitchell, A. (1994) *Methods in Yeast Genetics, A Cold Spring Harbor Laboratory Course Manual*. New York, NY, USA: Cold Spring Harbor Laboratory Press.

Li, W., Sun, L., Liang, Q., Wang, J., Mo, W., and Zhou, B. (2006) Yeast AMID homologue Ndi1p displays respiration-restricted apoptotic activity and is involved in chronological aging. *Mol Biol Cell* **17:** 1802–1811.

Livak, K.J., and Schmittgen, T.D. (2001) Analysis of relative gene expression data using real-time quantitative PCR and the $2(-\Delta\Delta C(T))$ method. *Methods* **25:** 402–408.

Luttik, M.A.H., Overkamp, K.M., Kötter, P., de Vries, S., van Dijken, J.P., and Pronk, J.T. (1998) The *Saccharomyces cerevisiae NDE1* and *NDE2* genes encode separate mitochondrial NADH dehydrogenases catalyzing the oxidation of cytosolic NADH. *J Biol Chem* **38:** 24529–24534.

Marella, M., Seo, B.B., Nakamaru-Ogiso, E., Greenamyre, J.T., Matsuno-Yagi, A., and Yagi, T. (2008) Protection by the NDI1 Gene against neurodegeneration in a rotenone rat model of Parkinson's disease. *PLoS ONE* **3:** e1433.

Marella, M., Seo, B.B., Thomas, B.B., Matsuno-Yagi, A., and Yagi, T. (2010) Successful amelioration of mitochondrial optic neuropathy using the yeast NDI1 gene in a rat animal model. *PLoS ONE* **5:** e11472.

Marres, C.A.M., de Vries, S., and Grivell, L.A. (1991) Isolation and inactivation of the nuclear gene encoding the rotenone-insensitive internal NADH:ubiquinone oxidoreductase of mitochondria from *Saccharomyces cerevisiae*. *Eur J Biochem* **195:** 857–862.

MØller, I.M. (2002) A new dawn for plant mitochondrial NAD(P)H dehydrogenases. *Trends Plant Sci* **7:** 235–237.

Oliveira, G.A., Tahara, E.B., Gombert, A.K., Barros, M.H., and Kowaltowski, A.J. (2008) Increased aerobic metabolism is essential for the beneficial effects of caloric restriction on yeast life span. *J Bioenerg Biomembr* **40:** 381–388.

Rigoulet, M., Aguilaniu, H., Avéret, N., Bunoust, O., Camougrand, N., Grandier-Vazeille, X., *et al.* (2004) Organization and regulation of the cytosolic NADH metabolism in the yeast *Saccharomyces cerevisiae*. *Mol Cell Biochem* **256:** 73–81.

Rodicio, R., and Heinisch, J.J. (2013) Yeast on the milky way: genetics, physiology and biotechnology of *Kluyveromyces lactis*. *Yeast* **30:** 165–177.

Rothstein, R. (1991) Targeting, disruption, replacement, and allele rescue: integrative DNA transformation in yeasts. *Methods Enzymol* **194:** 281–301.

Saliola, M., D'Amici, S., Sponziello, M., Mancini, P., Tassone, P., and Falcone, C. (2010) The transdehydrogenase genes *KlNDE1* and *KlNDI1* regulate the expression of *KlGUT2* in the yeast *Kluyveromyces lactis*. *FEMS Yeast Res* **10:** 518–526.

Sambrook, J., Maniatis, T., and Fritsch, E.F. (1989) *Molecular Cloning: A Laboratory Manual*. New York, NY, USA: Cold Spring Harbor Laboratory Press.

Szczodrak, J., Szewczuk, D., Rogalski, J., and Fiedurek, J. (1997) Selection of yeast strain and fermentation conditions for high-yield ethanol production from lactose and concentrated whey. *Acta Biotechnol* **17:** 51–61.

Tarrío, N., Díaz-Prado, S., Cerdán, M.E., and González-Siso, M.I. (2005) The nuclear genes encoding the internal (*KlNDI1*) and external (*KlNDE1*) alternative NAD(P)H:ubiquinone oxidoreductases of mitochondria from *Kluyveromyces lactis*. *BBA – Bioenergetics* **1707:** 199–210.

Tarrío, N., Becerra, M., Cerdán, M.E., and González-Siso, M.I. (2006a) Reoxidation of cytosolic NADPH in *Kluyveromyces lactis*. *FEMS Yeast Res* **6:** 371–380.

Tarrío, N., Cerdán, M.E., and González-Siso, M.I. (2006b) Characterization of the second external alternative dehydrogenase from mitochondria of the respiratory yeast *Kluyveromyces lactis*. *BBA – Bioenergetics* **1757:** 1476–1484.

Tarrío, N., García-Leiro, A., Cerdán, M.E., and González-Siso, M.I. (2008) The role of glutathione reductase in the interplay between oxidative stress response and turnover of cytosolic NADPH in *Kluyveromyces lactis*. *FEMS Yeast Res* **8:** 597–606.

Wésolowski-Louvel, M., Prior, C., Bornecque, D., and Fukuhara, H. (1992) Rag⁻ mutations involved in glucose metabolism in yeast: isolation and genetic characterization. *Yeast* **8:** 711–719.

Yagi, T., Seo, B.B., Nakamaru-Ogiso, E., Marella, M., Barber-Singh, J., Yamashita, T., and Matsuno-Yagi, A. (2006) Possibility of transkingdom gene therapy for Complex I diseases. *BBA – Bioenergetics* **1757:** 708–714.

Zaragoza, O. (2003) Generation of disruption cassettes *in vivo* using a PCR product and *Saccharomyces cerevisiae*. *J Microbiol Methods* **52:** 141–145.

Zitomer, R.S., and Hall, B.D. (1976) Yeast cytochrome c messenger RNA. *In vitro* translation and specific immunoprecipitation of the *CYC1* gene product. *J Biol Chem* **251:** 6320–6326.

2

Four-stage dissolved oxygen strategy based on multi-scale analysis for improving spinosad yield by *Saccharopolyspora spinosa* ATCC49460

Yun Bai,[1,2] Peng-Peng Zhou,[1,2] Pei Fan,[1,2]
Yuan-Min Zhu,[1,2] Yao Tong,[1,2] Hong-bo Wang[1,2] and
Long-Jiang Yu[1,2,*]

[1]*Institute of Resource Biology and Biotechnology,
Department of Biotechnology, College of Life Science
and Technology and* [2]*Key Laboratory of Molecular
Biophysics Ministry of Education, Huazhong University
of Science and Technology, Wuhan 430074, China.*

Summary

Dissolved oxygen (DO) is an important influencing factor in the process of aerobic microbial fermentation. Spinosad is an aerobic microbial-derived secondary metabolite. In our study, spinosad was used as an example to establish a DO strategy by multi-scale analysis, which included a reactor, cell and gene scales. We changed DO conditions that are related to the characteristics of cell metabolism (glucose consumption rate, biomass accumulation and spinosad production). Consequently, cell growth was promoted by maintaining DO at 40% in the first 24 h and subsequently increasing DO to 50% in 24 h to 96 h. In an in-depth analysis of the key enzyme genes (*gtt*, *spn* A, *spn* K and *spn* O), expression of spinosad and specific Adenosine Triphosphate (ATP), the spinosad yield was increased by regulating DO to 30% within 96 h to 192 h and then changing it to 25% in 192 h to 240 h. Under the four-phase DO strategy, spinosad yield increased by 652.1%, 326.1%, 546.8%, and 781.4% compared with the yield obtained under constant DO control at 50%, 40%, 30%, and 20% respectively. The proposed method provides a novel way to develop a precise DO strategy for fermentation.

*For correspondence. E-mail yulongjiang@mail.hust.edu.cn

Funding Information No funding information provided.

Introduction

Dissolved oxygen (DO) plays a significant role in aerobic fermentation (Xu and Zhong, 2011). Using a relatively high DO in the fermentation process would lead to high energy consumption; on the contrary, relatively low DO would negatively affect cell growth (Chen *et al.*, 2012). Researchers (Xu *et al.*, 2009; Wang *et al.*, 2010; Song *et al.*, 2013) have studied two or more stages of a DO strategy in aerobic fermentation processes. Such stages improved the corresponding product yield. However, other influencing factors, such as intracellular nucleotides and gene expression, also affect fermentation products (Xavier *et al.*, 2007). Therefore, a multi-scale analysis should be involved in the study of a DO strategy in the fermentation process.

Spinosad, an aerobic microbial-derived secondary metabolite of soil actinomycete *Saccharopolyspora spinosa* (*S. spinosa*) (Mertz and Yao, 1990), is a mixture of spinosyns A and spinosyns D, which are two major components in *S. spinosa* fermentation products (Huang *et al.*, 2009). As a highly effective targeted pesticide, spinosad was awarded the Presidential Green Chemistry Challenge Award in 1999 for its low environmental impact, low risk to non-target species and low mammalian toxicity (Sparks *et al.*, 2001). Oxygen is an indispensable raw material that must be supplied in large amounts in industrial spinosad production. Luo and colleagues (2012) have integrated the expression of *Vitreoscilla* haemoglobin gene in *S. spinosa*, thereby improving oxygen uptake. Such integration was effective for the genetic improvement of *S. spinosa* fermentation. However, information on the integrated DO strategy for efficient spinosad production by *S. spinosa*-submerged fermentation is lacking. Fortunately, details of the spinosad biosynthesis pathway and the genome information of *S. spinosa* have been elucidated recently (Madduri *et al.*, 2001; Hong *et al.*, 2006; Huang *et al.*, 2008; Yang *et al.*, 2014). Many other microbial DO strategies were established successfully (Garcia-Ochoa *et al.*, 2010; Cao *et al.*, 2013; Huang *et al.*, 2013). Those strategies were useful for the study of DO strategy on spinosad production.

In this paper, the reactor levels of DO conditions related to the characteristic of cell metabolism (spinosad yield, glucose consumption rate, biomass accumulation, and ATP production) and expressions of four key enzyme genes (*gtt*, *spn* A, *spn* K and *spn* O) for spinosad production were studied. We investigated DO in relation to the three scales mentioned above to establish a multi-stage DO control, reduce energy consumption, save costs and set up an efficient fermentation production process.

Results

Spinosad production at different DO levels

To investigate the effects of DO on spinosad production, different oxygen conditions were analysed at constant DO (20%, 30%, 40% and 50%). Any other fermentation conditions were not changed in these processes.

As shown in Fig. 1, the glucose consumption rate was the highest before 120 h and then decreased along with the glucose concentration at 50% DO, and almost no glucose was used after 168 h. When at 30% and 20% DO, cell growth and glucose consumption rate decreased obviously, at the same time the dry cell weight (DCW) were lower than that at 50% DO. The maximum DCW and

maximum $Y_{P/X}$ (mg l^{-1} spinosad yield / g l^{-1} DCW), were achieved at 50% and 30% DO during 24–96 h respectively. The result indicates that the high DCW and $Y_{P/X}$ could not be achieved simultaneously by controlling a constant DO during the whole culture process. This process could be at least divided into two stages, which are cell growth and product synthesis. Accordingly, cell growth at 50% DO and spinosad production at 30% DO are likely to be optimal. Furthermore, the DCW under 50% DO was not significantly different with that under 40% DO before 24 h, but during 24–96 h the former (DCW 19.43 ± 0.70) became higher than the latter (DCW 17.40 ± 0.85) ($P < 0.05$). In consideration of power saving, maintaining 40% DO at the first 24 h and then changing it to 50% from 24 h to 96 h could be a better choice. In order to realize a more precisely DO-controlling spinosad production process, a further investigation should be conducted.

Based on above experimental results, we can ensure 0–24 h DO 40%, 24–96 h DO 50%, but DO is undetermined yet at 96–240 h. According to the Table 1 results, the different three-phase DO strategy were studied. Table 1 shows the 35%, 30%, 25%, 20% DO, respectively,

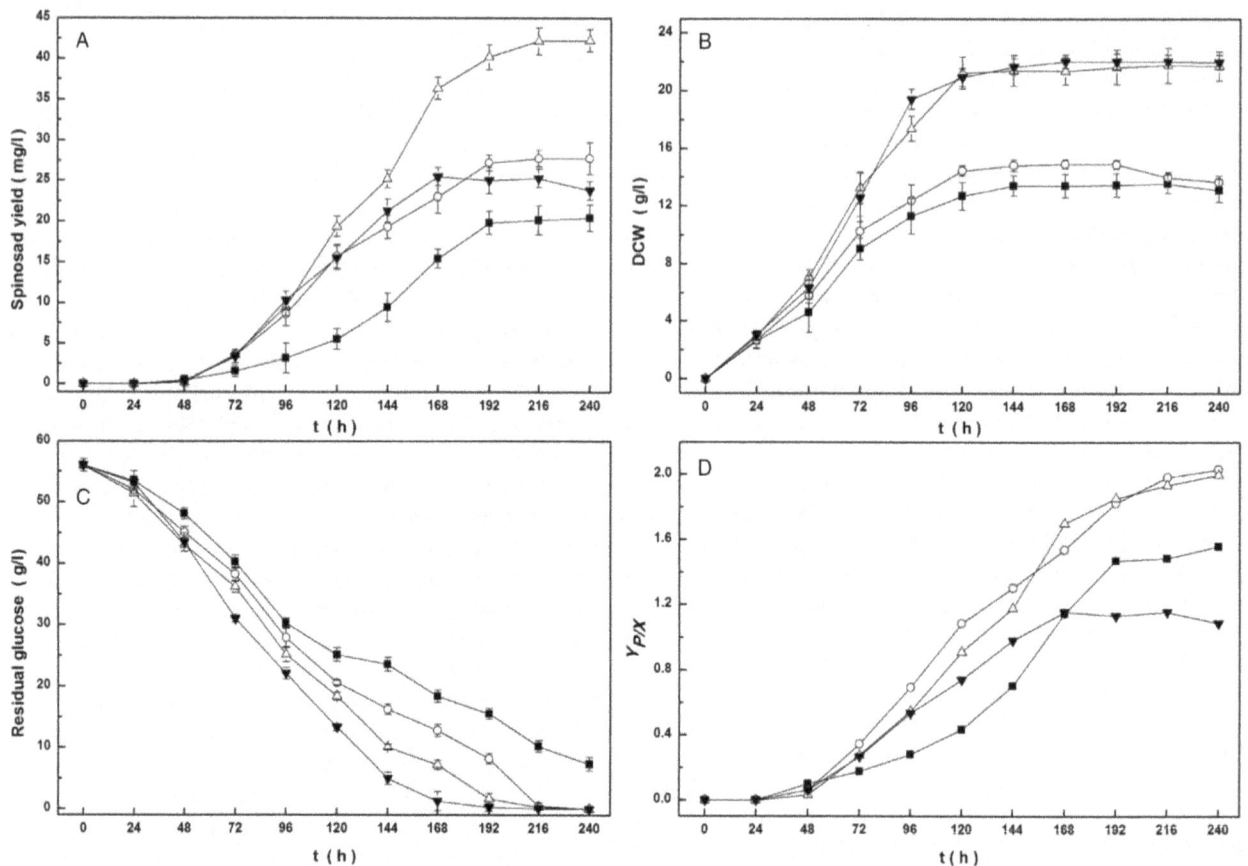

Fig. 1. Spinosad production at different DO level of (■) 20%, (○) 30%, (△) 40% and (▼) 50%. Changes over time in (A) spinosad yield, (B) DCW, (C) residual glucose and (D) $Y_{P/X}$. Each value is the mean of three parallel replicates.

Table 1. Comparison of spinosad production at different three-phase DO strategy.

Different DO strategy	DCW(g l^{-1})	Spinosad yield (mg l^{-1})	Residul glucose (g l^{-1})	$Y_{P/X}$
40%[a]–50%[b]–35%[c]	21.82 ± 0.10	80 ± 1.01	0 ± 0.10	3.67
40%[a]–50%[b]–30%[c]	22.00 ± 0.11	138 ± 1.23	0 ± 0.12	6.27
40%[a]–50%[b]–25%[c]	22.02 ± 0.10	135 ± 1.32	0 ± 0.11	6.13
40%[a]–50%[b]–20%[c]	21.20 ± 0.15	91 ± 1.77	0 ± 0.13	4.26

a. Means fermentation during 0–24 h.
b. Means fermentation during 24–96 h.
c. Means fermentation during 96–240 h.

during 96–240 h; the DCW and spinosad yield were no difference between 25% and 30% DO at 96–240 h; the spinosad yield at 35%; and 20% DO were lower than 30% and 25% DO. The results show that it is beneficial for spinosad production when DO is controlled at 25% or 30% at 96–240 h. To detect the optimal DO level during 96–240 h of the fermentation process, we need in-depth research.

Key enzyme gene expression under different three-phase DO strategy

Spn A, spn K spn O and gtt involved in spinosad biosynthesis pathway were selected as target genes in this research because these four genes encode rate limited enzymes in charge of spinosad biosynthesis (Xue et al., 2013a). The expression of selected genes involved in spinosad biosynthesis was monitored using quantitative reverse transcriptase polymerase chain reaction (QRT-PCR) to investigate how DO influences spinosad production. The expression ratios of all these target genes were compared by 2$^{-\Delta\Delta CT}$ method.

As shown in Fig. 2, the transcript levels of spn A (Fig. 2A), spn K (Fig. 2B), spn O (Fig. 2C) and gtt (Fig. 2D) under 30% DO significantly increased compared with 25% DO in 96 h–168 h. The transcript levels of spn A (Fig. 2A), spn K (Fig. 2B), spn O (Fig. 2C) and gtt (Fig. 2D) under 25% DO significantly increased compared with 30% DO in 168–240 h. In Fig. 2, we illustrate that it is more conducive to the accumulation of spinosad 30% DO in 96–168 h and with 25% DO in 192–240 h.

The SATP levels with different three-phase DO strategy

Table 1 shows that the DCW and spinosad yield at 30% DO were not significantly different with that at 25% DO. But the expression ratios of key enzyme genes were significantly different (Fig. 2). Generally, traditional techniques for quantifying biomass are not able to provide information about the quality, activity or viability of the microorganisms. Nevertheless, SATP (mg ATP g^{-1} DCW) can be as a biomass quality indicator (Gikas and Livingston, 1998).

The SATP between different three-phase DO strategies (20%, 25%, 30% and 35% DO during 96–240 h) were shown in Fig. 3. Compared with other DO conditions, SATP level was the maximum under 30% DO during

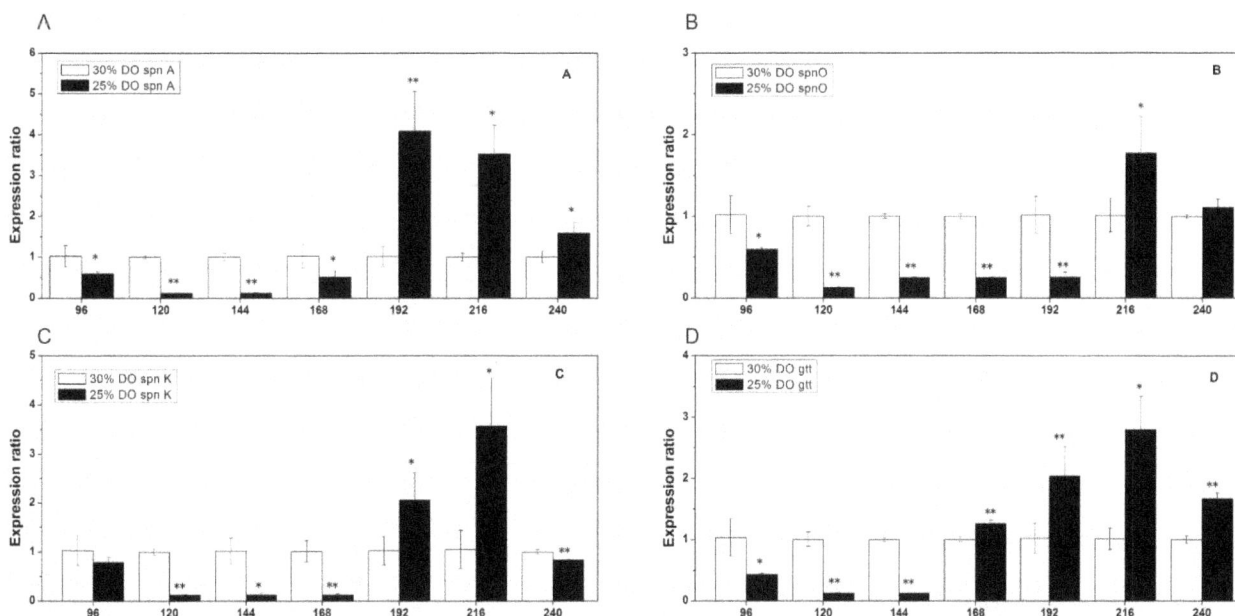

Fig. 2. Relative expression ratios (experimental group$_{(40\%–50\%–25\%\ DO)}$/control group$_{(40\%–50\%–30\%\ DO)}$) of four rate-limited spinosad biosynthesis genes – (A) spn A, (B) spn K, (C) spn O and (D) gtt – at 96, 120, 144, 168, 192, 216 and 240 h. The *$P < 0.05$ and **$P < 0.01$ were obtained from the t-test for statistical analysis.

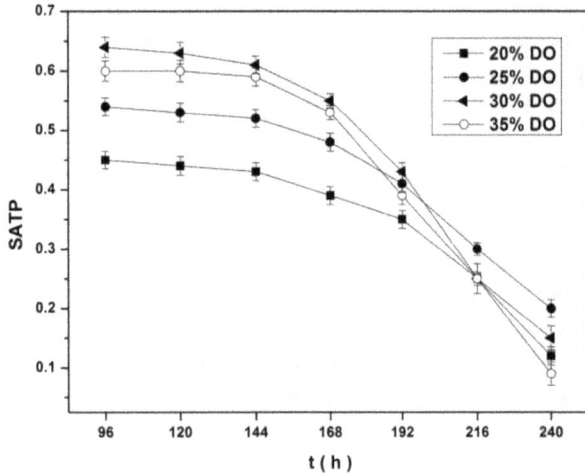

Fig. 3. The comparison of SATP among 20%, 25%, 30% and 35% DO strategy after 96 h, which DO are all under 40% during 0–24 h and 50% during 24–96 h.

96–192 h, after when it dropped more rapidly than the others. In the meantime, under 25% Do, the SATP level was slowly reduced. The result shows that at the beginning of the stationary phase, 30% DO is better than 25% DO, and in the later period of the stationary phase, it is proper to regulate DO to a lower level for maintaining cell vitality.

Four-phase DO strategy for spinosad production

According to above results, a four-phase DO strategy was investigated by controlling DO at 40% during 0–24 h, 50%

during 24–96 h, 30% during 96–192 h, 25% until the end of the fermentation. Spinosad fermentation in 10 l of bioreactors using four-phase DO strategy was shown in Fig. 4. Each phase corresponds to the particular physiological characteristics of spinosad biosynthesis. At Phase I (0–24 h) with DO 40%, the nutrients were used for cell growth. At Phase II (24–96 h) with DO 50%, the nutrients were used for cell growth and a small quantity of spinosad was produced. At Phase III (96–192 h) with DO 30%, cell growth went into stationary state, and spinosad production rate increased rapidly. At the Phase IV (192–240 h), spinosad production was still in high rate by adjusting DO to 25%. The result of the four-phase DO control shows that spinosad yield was 179.8 ± 8.0 mg l^{-1}, which is increased by 652.1%, 326.1%, 546.8% and 781.4%, compared with constant DO control at 50%, 40%, 30% and 20% respectively.

Discussion

DO affects cell growth and product biosynthesis in numerous microorganisms (Li *et al.*, 2012; Chang *et al.*, 2014). An excessive oxygen supply would decrease productivity because of the direct aerobic respiration of substrates, thereby dramatically elevating the costs of production in industrial fermentation (Käß *et al.*, 2014). However, a relatively low DO would negatively affect cell growth. A successful industrial fermentation process involves higher productivity and a cost-effective production strategy. Cell growth of *S. spinosa* ATCC49460 depends on the TCA

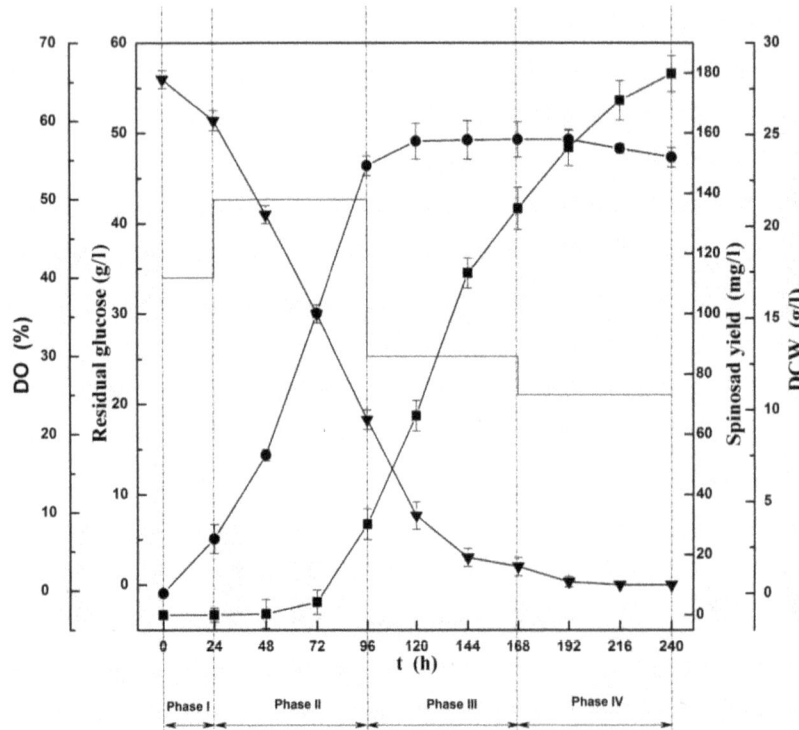

Fig. 4. Changes over time in spinosad fermentation with four-stage DO strategy in a 10-l bioreactor. (■) Spinosad yield, (●) DCW, (▼) residual glucose and (——) DO. The error bars indicate the standard deviations of three independent samples.

cycle (Yang *et al.*, 2014), in which DO concentration needs to remain at a high level. Spinosad production begins to increase at the start of the stationary phase of *S. spinosa* ATCC49460 growth. This phase does not require a high DO concentration, but the synthesis of spinosyns A ($C_{41}H_{65}NO_{10}$) and D ($C_{42}H_{67}NO_{10}$) requires oxygen. Accordingly, the DO needs to remain at a proper level. DO affects cell growth and spinosad biosynthesis, and a higher concentration of DO is required for the former than for the latter. A two-stage oxygen supply strategy can ensure spinosad production by *S. spinosa* ATCC 49460. Researchers (Jin *et al.*, 2008; Xu *et al.*, 2009; Rosa *et al.*, 2010) used a similar method to establish a two-phase DO strategy.

A DO strategy can be established by determining the DO tension that is related to biomass and productive yield. However, other influencing factors, such as intracellular nucleotides and gene expression, also affect the fermentation products (Zhong *et al.*, 2004). In our study, spinosyns were used to establish a DO strategy by multi-scale analysis, which includes reactor, cell and gene scales. Similar results were obtained by Liang and colleagues (2004) who presented an online data association analysis performed on multiple scales and a fermentation regulation method that was based on the analysis and control of the metabolic flux.

The spinosad biosynthetic genes are 80 kb (Waldron *et al.*, 2000), including five large genes (*spn* A, B, C, D and E) encoding type I polyketide synthase, four genes (*spn* F, J, L and M) involved in intramolecular C–C bond formation, four genes (*spn* G, I, K and H) involved in rhamnose attachment and methylation and six genes (*spn* P, O, N, Q, R and S) involved in forosamine biosynthesis. The genes for synthesis of rhamnose (*gtt*, *gdh*, *epi* and *kre*) are not closely linked to the spinosyn gene cluster (Waldron *et al.*, 2001; Hong *et al.*, 2006; 2008). Four key enzyme genes (*gtt*, *spn* A, *spn* K and *spn* O) for spinosad production were studied, and the results demonstrated how DO changes affect spinosad production. Similar results were obtained by Xue and colleagues (2013b), who compared the expression of four key enzyme gene (*gtt*, *spn* A, *spn* K and *spn* O) at 120, 144, 168, 192, 216 and 240 h with and without additional exogenous fatty acids. Results elucidated why the change of exogenous fatty acids could affect spinosad production.

Bacterial viability is difficult to define accurately by DCW. Thus, biomass activity was determined to measure the ability of biomass to metabolize a particular substrate, thereby describing the potential activity of biomass. SATP has been frequently used as a biomass quality indicator because ATP is present in all living organisms and is involved in most biochemical pathways. The result in Fig. 3 illustrates the slowing down of cell metabolism toward the end of the stability phase and the decrease in oxygen demand. If DO remained at a high level, reduced cell vitality and the production of spinosad would be observed. Similar results were reported by Zhang and colleagues (2012), who found that curdlan production had a positive relationship with intracellular levels. Thus, a simple and reproducible two-stage DO control process was developed.

Optimizing the industrial production of spinosad by *S. spinosa* is difficult because large numbers of physiological regulatory mechanisms occur during the process. A holistic concept cannot be obtained by just a single reactor regulatory factor. In our study, spinosyns were used to establish a DO strategy by multi-scale analysis, which included reactor, cell and gene scales. The characteristics of cell metabolism under different DO conditions were investigated. Oxygen demand was higher before the stationary phase, during which cell growth and glucose consumption rate were rapid. Starting from the late logarithmic phase of *S. spinosa* ATCC49460, synthesis of a large amount of spinosad began, whereas the demand for DO and the glucose consumption rate started to decrease. This finding indicated that the process could be divided into two stages, namely, cell growth and spinosad synthesis. Subsequently, in an in-depth analysis of the key enzymes gene expression of spinosad synthesis and SATP, the expressions of four key enzyme genes (*gtt*, *spn* A, *spn* K and *spn* O) resulted in high DO, which could lead to negative spinosad biosynthesis. The SATP data showed that cell metabolism slowed toward the end of the stability phase and oxygen demand decreased. A more precise oxygen supply strategy can be ensured for spinosad synthesis. Finally, a four-phase DO strategy was established.

To our knowledge, this is the first report on the effects of various DO concentrations combined with multi-scale analysis on spinosad fermentation. Our results would help elucidate spinosad production and metabolic response of *S. spinosa* ATCC49460 under various DO conditions. Furthermore, applying the proposed four-stage DO strategy on fermentation models would help develop and improve the production processes of other aerobic microbial-derived secondary metabolites.

Experimental procedures

Microorganism and medium

Microorganism: *S. spinosa* ATCC49460. The plate medium was composed of the following (g l⁻¹): glucose, 10.0; N-Z Amine Type A, 2.0; yeast extract, 1.0; and agar, 15.0, and incubated at 30°C up to 8 days. The spores from the plate culture were inoculated into a 250 ml Erlenmeyer flask containing 25 ml of seed medium. The seed medium was composed of the following (g l⁻¹): glucose, 10.0; N-Z Amine Type

Table 2. List of gene and gene primers used in this study.

Gene name	Gene function	Primers	Primers Sequence 5′ → 3′
gtt	Rhamnose synthesis and cell synthesis	*gtt*-F	ACTTCCGGTGTACGACAAGC
		gtt-R	AAGCCCTGCCCGTAAAAGAT
spn A	Polyketide synthesis	*spn* A-F	TTGGTACGCCACACTGTCTC
		spn A-R	GAGCGCATCCAGATAGGCAT
spn K	Rhamnose methylation	*spn* K-F	GTCAGGACGAAGTCAACGGT
		spn K-R	ATGTCCACAACGCACGAGAT
spn S	Forosamine synthesis	*spn* S-F	GGAAACCACCAGGGTTCGAT
		spn S-R	ATTCGCAGAAACCTCCTCGG
16S rRNA	Conserved sequence	16S rRNA-F	CCTACGAGCTCTTTACGCCC
		16S rRNA-R	AGAAGCACCGGCTAACTACG

A, 2.0; yeast extract, 1.0; K$_2$HPO$_4$, 0.2; and MgSO$_4$, 2.0; the pH was adjusted to 7.0 before autoclaving. After incubation at 30°C on a rotary shaker at 220 r.p.m. for 60 h, a 2 ml portion of the seed culture was used to inoculate 25 ml of production medium into a 250 ml Erlenmeyer flask fermentation medium. The fermentation medium was composed of the following (g l^{-1}): glucose, 60.0; cottonseed protein, 20.0; yeast extract, 8.0; CaCO$_3$, 1.0; and rapeseed oil, 10.0. In all cases, the medium was sterilized using an autoclave for 20 min at 121°C. The flasks were incubated for 60 h on an orbital shaker (Infors Zhi Cheng, Shanghai, China) at 200 r.p.m. and 30°C. Approximately, 10% (v/v) of the seed culture was inoculated into the 10 l bioreactor. The culture was incubated for 240 h at 30°C.

Fermentation in a 10 l bioreactor

The oxygen supply was also analysed during batch fermentation in a 10 l bioreactor (Biotech-10JS, Shanghai, China) with 7 l working volume. The DO probe (Mettler-Toledo GmbH, Switzerland) and pH probe (Mettler-Toledo GmbH, Switzerland) measured the DO, pH and temperature, agitation speed were measured online. The pH was adjusted using 2.0 mol l^{-1} NaOH or 2.0 mol l^{-1} HCl. The temperature was controlled automatically. Different constant DO 20%, 30%, 40% and 50% in 10 l bioreactors were studied. The three-stage controls were performed as follows: during 0–24 h, the DO was set to 40%; during 24–96 h, the DO was set to 50%; and during 96–240 h, the DO was set to 30%. The four-stage oxygen controls were performed as follows: at Phase I (0–24 h), the DO was set to 40%; at the subsequent culture Phase II (24–96 h), the DO was set to 50%; at Phase III (96–192 h), the DO was set to 30%; at Phase IV (192–240 h), the DO was set to 25%. The DO level (percentage of air saturation) was controlled at constant values (20%, 30%, 40% and 50%) or varying values by cascading different agitation speeds (from 150 up to 500 r.p.m.), and the ventilation was 1 vvm throughout. The other culture conditions were the same as in the above experiments. Three batches were repeated for each experiment.

Determinations

The spinosad was determined as described by Yáñez and colleagues (2014).

To determine the DCW, 10 ml of the fermentation broth was centrifuged at 8000 × *g* for 10 min, and the supernatant was discarded. After washing twice with distilled water, the supernatant was discarded and the cell precipitate was dried to a constant weight at 60°C for 48 h. All cell pellets were weighed after drying. The glucose concentration was measured using the 3,5-dinitrosalicylic acid spectrometric method (Miller, 1959).

Extraction and determination of the intracellular ATP levels: the concentrations of intracellular ATP were determined by an in vitro procedure based on inactivation of metabolism, which method according to the method described by Dhople and Hanks (1973). All experimental values were presented as the means of three replicates ± standard deviation.

RNA extraction, cDNA synthesis and qPCR

The method of RNA extraction, cDNA synthesis and qPCR according to the method (Song *et al.*, 2014) described with some modifications. Total RNA were isolated from the samples at 96, 120, 144, 168,192, 216 and 240 h. For each time point, three biological replicates were obtained and analysed. Specific primers were designed for genes involved in spinosad biosynthesis (Table 2 lists the primers used in this study). The internal control used was 16S rRNA. 16S rRNA was a suitable reference gene in this study, which has been determined in another study. The data obtained were analysed by applying the $2^{-\Delta\Delta CT}$ method, $-\Delta\Delta C_T = (C_{T(trager\ gene)}-C_{T(reference\ gene)})_{test} - (C_{T(trager\ gene)}-C_{T(reference\ gene)})_{calibrator}$, where C_T is the threshold cycle number of the target gene at any collection time, $C_{T(reference\ gene)}$ is the expression quantity of 16S rRNA at the same collection time as Ct, $(C_{T(trager\ gene)}-C_{T(reference\ gene)})_{test}$ is the expression quantity of the target gene on day 25% DO and $(C_{T(trager\ gene)}-C_{T(reference\ gene)})_{calibrator}$ is the expression quantity of the target gene on day 30% DO. Statistical analyses were performed using Microsoft Excel (2007). All experiments were repeated three times.

Conflict of interest

None declared.

References

Cao, W., Qi, B., Zhao, J., Qiao, C., Su, Y., and Wan, Y. (2013) Control strategy of pH, dissolved oxygen concentration and stirring speed for enhancing β-poly (malic acid) production by. *J Chem Technol Biotechnol* **88:** 808–817.

Chang, G., Wu, J., Jiang, C., Tian, G., Wu, Q., Chang, M., and Wang, X. (2014) The relationship of oxygen uptake rate and kLa with rheological properties in high cell density cultivation of docosahexaenoic acid by *Schizochytrium* sp. S31. *Bioresour Technol* **152:** 234–240.

Chen, X., Liu, L., Li, J., Du, G., and Chen, J. (2012) Improved glucosamine and N-acetylglucosamine production by an engineered *Escherichia coli* via step-wise regulation of dissolved oxygen level. *Bioresour Technol* **110:** 534–538.

Dhople, A.M., and Hanks, J.H. (1973) Quantitative extraction of adenosine triphosphate from cultivable and host-grown microbes: calculation of adenosine triphosphate pools. *Appl Environ Microbiol* **27:** 399–403.

Garcia-Ochoa, F., Gomez, E., Santos, V.E., and Merchuk, J.C. (2010) Oxygen uptake rate in microbial processes: an overview. *Biochem Eng J* **49:** 289–307.

Gikas, P., and Livingston, A.G. (1998) Use of specific ATP concentration and specific oxygen uptake rate to determine parameters of a structured model of biomass growth. *Enzyme Microb Technol* **22:** 500–510.

Hong, L., Zhao, Z., and Liu, H. (2006) Characterization of SpnQ from the spinosyn biosynthetic pathway of *Saccharopolyspora spinosa*: mechanistic and evolutionary implications for C-3 deoxygenation in deoxysugar biosynthesis. *J Am Chem Soc* **128:** 14262–14263.

Hong, L., Zhao, Z., Melançon, C.E., Zhang, H., and Liu, H. (2008) In vitro characterization of the enzymes involved in TDP-D-forosamine biosynthesis in the spinosyn pathway of *Saccharopolyspora spinosa*. *J Am Chem Soc* **130:** 4954–4967.

Huang, K., Zahn, J., and Han, L. (2008) SpnH from *Saccharopolyspora spinosa* encodes a rhamnosyl 4′-O-mothyltransferase for biosynthesis of the insecticidal macrolide, spinosyn A. *J Ind Microbiol Biotechnol* **35:** 1669–1676.

Huang, K., Xia, L., Zhang, Y., Ding, X., and Zahn, J.A. (2009) Recent advances in the biochemistry of spinosyns. *Appl Microbiol Biotechnol* **82:** 13–23.

Huang, M., Wan, J., Hu, K., Ma, Y., and Wang, Y. (2013) Enhancing dissolved oxygen control using an on-line hybrid fuzzy-neural soft-sensing model-based control system in an anaerobic/anoxic/oxic process. *J Ind Microbiol Biotechnol* **40:** 1393–1401.

Jin, M., Huang, H., Xiao, A., Zhang, K., Liu, X., Li, S., and Peng, C. (2008) A novel two-step fermentation process for improved arachidonic acid production by *Mortierella alpina*. *Biotechnol Lett* **30:** 1087–1091.

Käß, F., Hariskos, I., Michel, A., Brandt, H., Spann, R., Junne, S., et al. (2014) Assessment of robustness against dissolved oxygen/substrate oscillations for *C. glutamicum* DM1933 in two-compartment bioreactor. *Bioprocess Biosyst Eng* **37:** 1151–1162.

Li, K., Zhou, J., Cheng, X., and Wei, S. (2012) Study on the dissolved oxygen control strategy in large-scale vitamin B. *J Chem Technol Biotechnol* **87:** 1648–1653.

Liang, Z.S., Ju, C., Ping, Z.Y., and Zhi, H.M. (2004) Studies on the multi-scale problems of guanosine fermentation process. In. 23–29

Luo, Y., Kou, X., Ding, X., Hu, S., Tang, Y., Li, W., et al. (2012) Promotion of spinosad biosynthesis by chromosomal integration of the *Vitreoscilla* hemoglobin gene in *Saccharopolyspora spinosa*. *Sci China Life Sci* **55:** 172–180.

Madduri, K., Waldron, C., Matsushima, P., Broughton, M.C., Crawford, K., Merlo, D.J., and Baltz, R.H. (2001) Genes for the biosynthesis of spinosyns: applications for yield improvement in *Saccharopolyspora spinosa*. *J Ind Microbiol Biotechnol* **27:** 399–402.

Mertz, F., and Yao, R. (1990) *Saccharopolyspora spinosa* sp. nov. isolated from soil collected in a sugar mill rum still. *Int J Syst Bacteriol* **40:** 34–39.

Miller, G. (1959) Use of dinitrosalicylic acid reagent for determination of reducing sugar. In. 426–428.

Rosa, S.M., Soria, M.A., Vélez, C.G., and Galvagno, M.A. (2010) Improvement of a two-stage fermentation process for docosahexaenoic acid production by *Aurantiochytrium limacinum* SR21 applying statistical experimental designs and data analysis. *Bioresour Technol* **101:** 2367–2374.

Song, G.H., Zhao, C.F., Zhang, M., Fu, C.H., Zhang, H., and Yu, L.J. (2014) Correlation analysis of the taxane core functional group modification, enzyme expression, and metabolite accumulation profiles under methyl jasmonate treatment. *Biotechnol Progr* **30:** 269–280.

Song, P., Chen, C., Tian, Q., Lin, M., Huang, H., and Li, S. (2013) Two-stage oxygen supply strategy for enhanced lipase production by *Bacillus subtilis* based on metabolic flux analysis. *Biochem Eng J* **71:** 1–10.

Sparks, T.C., Crouse, G.D., and Durst, G. (2001) Natural products as insecticides: the biology, biochemistry and quantitative structure–activity relationships of spinosyns and spinosoids. *Pest Manag Sci* **57:** 896–905.

Waldron, C., Madduri, K., Crawford, K., Merlo, D.J., Treadway, P., Broughton, M.C., and Baltz, R.H. (2000) A cluster of genes for the biosynthesis of spinosyns, novel macrolide insect control agents produced by *Saccharopolyspora spinosa*. *Antonie Van Leeuwenhoek* **78:** 385–390.

Waldron, C., Matsushima, P., Rosteck, P.J., Broughton, M.C., Turner, J., Madduri, K., et al. (2001) Cloning and analysis of the spinosad biosynthetic gene cluster of *Saccharopolyspora spinosa*. *Chem Biol* **8:** 487–499.

Wang, Z., Wang, H., Li, Y., Chu, J., Huang, M., Zhuang, Y., and Zhang, S. (2010) Improved vitamin B12 production by step-wise reduction of oxygen uptake rate under dissolved oxygen limiting level during fermentation process. *Bioresour Technol* **101:** 2845–2852.

Xavier, J.B., de Kreuk, M.K., Picioreanu, C., and van Loosdrecht, M.C.M. (2007) Multi-scale individual-based model of microbial and bioconversion dynamics in aerobic granular sludge. *Environ Sci Technol* **41:** 6410–6417.

Xu, H., Dou, W., Xu, H., Zhang, X., Rao, Z., Shi, Z., and Xu, Z. (2009) A two-stage oxygen supply strategy for enhanced l-arginine production by *Corynebacterium crenatum* based on metabolic fluxes analysis. *Biochem Eng J* **43:** 41–51.

Xu, Y., and Zhong, J. (2011) Significance of oxygen supply in production of a novel antibiotic by *Pseudomonas* sp. SJT25. *Bioresour Technol* **102:** 9167–9174.

Xue, C., Duan, Y., Zhao, F., and Lu, W. (2013a) Stepwise increase of spinosad production in *Saccharopolyspora spinosa* by metabolic engineering. *Biochem Eng J* **72:** 90–95.

Xue, C., Zhang, X., Yu, Z., Zhao, F., Wang, M., and Lu, W. (2013b) Up-regulated spinosad pathway coupling with the increased concentration of acetyl-CoA and malonyl-CoA contributed to the increase of spinosad in the presence of exogenous fatty acid. *Biochem Eng J* **81:** 47–53.

Yáñez, K.P., Martín, M.T., Bernal, J.L., Nozal, M.J., and Bernal, J. (2014) Determination of spinosad at trace levels in bee pollen and beeswax with solid-liquid extraction and LC-ESI-MS. *J Sep Sci* **37:** 204–210.

Yang, Q., Ding, X., Liu, X., Liu, S., Sun, Y., Yu, Z., *et al.* (2014) Differential proteomic profiling reveals regulatory proteins and novel links between primary metabolism and spinosad production in *Saccharopolyspora spinosa*. *Microb Cell Fact* **13:** 27.

Zhang, H., Zhan, X., Zheng, Z., Wu, J., English, N., Yu, X., and Lin, C. (2012) Improved curdlan fermentation process based on optimization of dissolved oxygen combined with pH control and metabolic characterization of *Agrobacterium* sp. ATCC 31749. *Appl Microbiol Biotechnol* **93:** 367–379.

Zhong, J., Zhang, S., Chu, J., and Zhuang, Y. (2004) A multi-scale study of industrial fermentation processes and their optimization. *Adv Biochem Eng Biotechnol* **87:** 97–150.

3

A preliminary and qualitative study of resource ratio theory to nitrifying lab-scale bioreactors

Micol Bellucci,[1,2*†] Irina D. Ofiţeru,[3,4] Luciano Beneduce,[2] David W. Graham,[1] Ian M. Head[1] and Thomas P. Curtis[1]

[1]School of Civil Engineering and Geosciences, Newcastle University, Newcastle upon Tyne, NE1 7RU, UK.

[2]Dipartimento di Scienze Agrarie, Alimentari ed Ambientali, Università di Foggia, via Napoli 25, Foggia, 71122 Italy.

[3]School of Chemical Engineering and Advanced Materials, Merz Court, Newcastle University, Newcastle upon Tyne, NE1 7RU, UK.

[4]Chemical Engineering Department, University Politehnica of Bucharest, Polizu 1-7, Bucharest 011061, Romania.

Summary

The incorporation of microbial diversity in design would ideally require predictive theory that would relate operational parameters to the numbers and distribution of taxa. Resource ratio-theory (RRT) might be one such theory. Based on Monod kinetics, it explains diversity in function of resource-ratio and richness. However, to be usable in biological engineered system, the growth parameters of all the bacteria under consideration and the resource supply and diffusion parameters for all the relevant nutrients should be determined. This is challenging, but plausible, at least for low diversity groups with simple resource requirements like the ammonia oxidizing bacteria (AOB). One of the major successes of RRT was its ability to explain the 'paradox of enrichment' which states that diversity first increases and then decreases with resource richness. Here, we demonstrate that this pattern can be seen in lab-scale-activated sludge reactors and parallel simulations that incorporate the principles of RRT in a floc-based system. High and low ammonia and oxygen were supplied to continuous flow bioreactors with resource conditions correlating with the composition and diversity of resident AOB communities based on AOB 16S rDNA clone libraries. Neither the experimental work nor the simulations are definitive proof for the application of RRT in this context. However, it is sufficient evidence that such approach might work and justify a more rigorous investigation.

Introduction

There is growing consensus on the key role of species richness of ecosystem function (Hooper et al., 2005; Cardinale et al., 2006). Empirical and theoretical studies in ecology suggest that elevated species richness improves ecosystem functional stability, especially system resilience to perturbations (Pimm, 1984; Tilman and Downing, 1994; Tilman et al., 2001). In principle, more diverse systems have a greater pool of physiological and genetic traits, which provide them the capacity to change and sustain function under varying environmental conditions.

This observation is of particular relevance to the engineering of open biological systems. Activated sludge and biofilm systems with high diversity have been shown to sustain functionality aftershock loading of mercury for example (von Canstein et al., 2002; Saikaly and Oerther, 2011). Similarly, studies on nitrifying systems suggested that a more diverse ammonia-oxidizing microbial community is more resistant to operational variability and can enhance the reliability of the process (Daims et al., 2001; Egli et al., 2003; Rowan et al., 2003). However, the exact mechanism underlying the beneficial effect of diversity is still uncertain. Wittebolle and colleagues (2009) argued that the initial evenness of the community rather than species richness per se is the key factor for the functional stability of the system. Nevertheless, it seems likely that increasing microbial diversity will do no harm and may improve stability.

The rational engineering of microbial diversity would be possible if one could incorporate theoretical ecology into process design (Curtis et al., 2003), which is both intrinsically fascinating and deeply practical. However, the body of available theory is modest and largely untested.

One plausible approach is Tilman's resource-ratio theory (RRT; Tilman, 1982). Tilman used the Monod

†For correspondence. E-mail micol.bellucci@gmail.com

Funding Information The authors thank ECOSERV for financial support under the EU Marie Curie Excellence Programme (MEXT-CT-2006-023469). I.D.O. was supported also by the European Reintegration Grant FLOMAS (FP7/2007-2013 no. 256440).

kinetics, familiar to engineers, to explain the variation of diversity with the quantity and ratio of resources supplied. According to the RRT, individual populations consume resources and increase in size until one resource becomes limiting. In heterogeneous environments, the theory predicts that (i) high species diversity and evenness occur when low to intermediate amount of resources are supplied, while resource enrichment leads to decreased species richness and evenness, (ii) high variations of the two resources lead to increased species richness and (iii) the resource supply ratio determines which species are dominant. In addition, the RRT provides one of the most elegant theoretical explanations for the 'paradox of enrichment' noted by Rosenzweig (1971). In essence this paradox is that increasing the amount of resource causes and initial increase and then a substantial decrease of the number of taxa within a system, leading to a destabilization of the ecosystem. Resource ratio-theory has been used to explain the biological community structure in terrestrial and aquatic plants, plankton, dental plaque and microbial populations in hydrocarbon-contaminated systems (Tilman, 1982; Smith, 1983; 1993; Smith *et al.*, 1998). A critique of these studies has suggested that the results of most of these tests are inconclusive (Miller *et al.*, 2005).

In principle, engineered biological processes are ideal systems for testing both the principles of the RRT and the paradox of enrichment because pollutants (growth limiting resources) supplied to the microbial community can be easily controlled. One important case where RRT might be used is nitrification in activated sludge treatment plants. Nitrification is a critical step in many wastewater treatment plants (WWTPs) including nitrogen removal process. During a conventional nitrification process, the ammonia oxidizing bacteria (AOB) oxidize the ammonia to nitrite, which is then converted to nitrate by nitrite oxidizing bacteria (NOB). However, nitrification often fails unexpectedly. If the predictions of RRT are correct, low oxygen and ammonia inputs should select for a more diverse AOB community. This, in turn, should increase the reliability of the biological ammonia oxidation process. Hence, reducing oxygen levels and prospective energy consumption for aeration might have the simultaneous benefit of reducing the risk of nitrification failure. A truly rigorous test of RRT in a nitrifying wastewater treatment plant is a daunting prospect. It would require us to know the maximum specific growth rates and the half saturation coefficients for ammonia and oxygen of dozens of taxa, as well as the parameters required to describe the diffusion and consumption of ammonia and oxygen across the population of flocs.

Our goal was more limited, to qualitatively evaluate the ability of RRT to describe variation of the AOB diversity in lab-scale nitrifying bioreactors. The experimental studies were complemented by a parallel simulation that incorporates the principles of RRT in a floc-based system where the resource ratio is controlled by diffusion into the biomass. High and low ammonia and oxygen supplies (2 × 2 design) were provided to different units with resource conditions and correlated with the composition and diversity of resident AOB communities based on AOB 16S ribosomal ribonucleic acid (rRNA) clone libraries. The 'paradox of enrichment' was qualitatively demonstrated both in lab reactors and in the parallel computational simulations.

Results

Reactor's performance

Four reactors were set up (Table 1) and designate (R1) had low nitrogen and high oxygen ($L_N H_O$), R2, low nitrogen and low oxygen ($L_N L_O$), R3 high nitrogen and high oxygen ($H_N H_O$) and R4 high nitrogen and low oxygen ($H_N L_O$).

Ammonia oxidation was achieved in all systems, although time scales and rates varied (Fig. 1). Complete ammonium removal occurred within 4 and 37 days from inoculation in R1-$L_N H_O$ and R2-$L_N L_O$, respectively. In R3-$H_N H_O$, the ammonia consumption was 22.3% ± 4.4% of the total ammonium concentration of the influent, while in R4-$H_N L_O$ an average of 29.1% ± 4.4% of ammonium was removed after 28 days of operation. Free ammonia and free nitrous acid concentration were almost negligible in R1-$L_N H_O$, while they varied over time in the other bioreactors (Fig. S2 and SI).

Dissolved oxygen (DO) concentrations varied in the systems according to the mode of aeration and ammonia levels in the influent [analysis of variance (ANOVA), $n = 25$, P-value < 0.05]. The DO concentration averaged 3.58 ± 1.08 mg/l in R1-$L_N H_O$ and R3-$H_N H_O$, and 0.22 ± 0.22 mg/l in R2-$L_N L_O$ and R4-$H_N L_O$. The different resources supplied affected also the pH (ANOVA, $n = 25$, P-value < 0.05) that was 8 ± 0.4 in R1-$L_N H_O$, 8.05 ± 1.54 in R2-$L_N L_O$ and 7.5 ± 0.72 in R4-$H_N L_O$; a drastic decrease of pH to 6.08 ± 0.86 was observed in R3-$H_N H_O$. The chemical oxygen demand (COD) removal rates were very high during the entire experiment. In R1-$L_N H_O$ and R3-$H_N H_O$, the COD removal stabilized at an average level of 93.7% ± 2% after 4 days of operation. In the low DO reactors (R2-$L_N L_O$ and R4-$H_N L_O$), the COD removal was higher than 89.5% ± 4.9% until the ammonia oxidation activity started. Then, it decreased to 79.9% ± 6.3% and 71.6% ± 5.1% in R2-$L_N L_O$ and R4-$H_N L_O$, respectively. The volatile suspended solid (VSS) concentration stabilized 10 days after inoculation, averaging 257 ± 47 mg/l in all configurations.

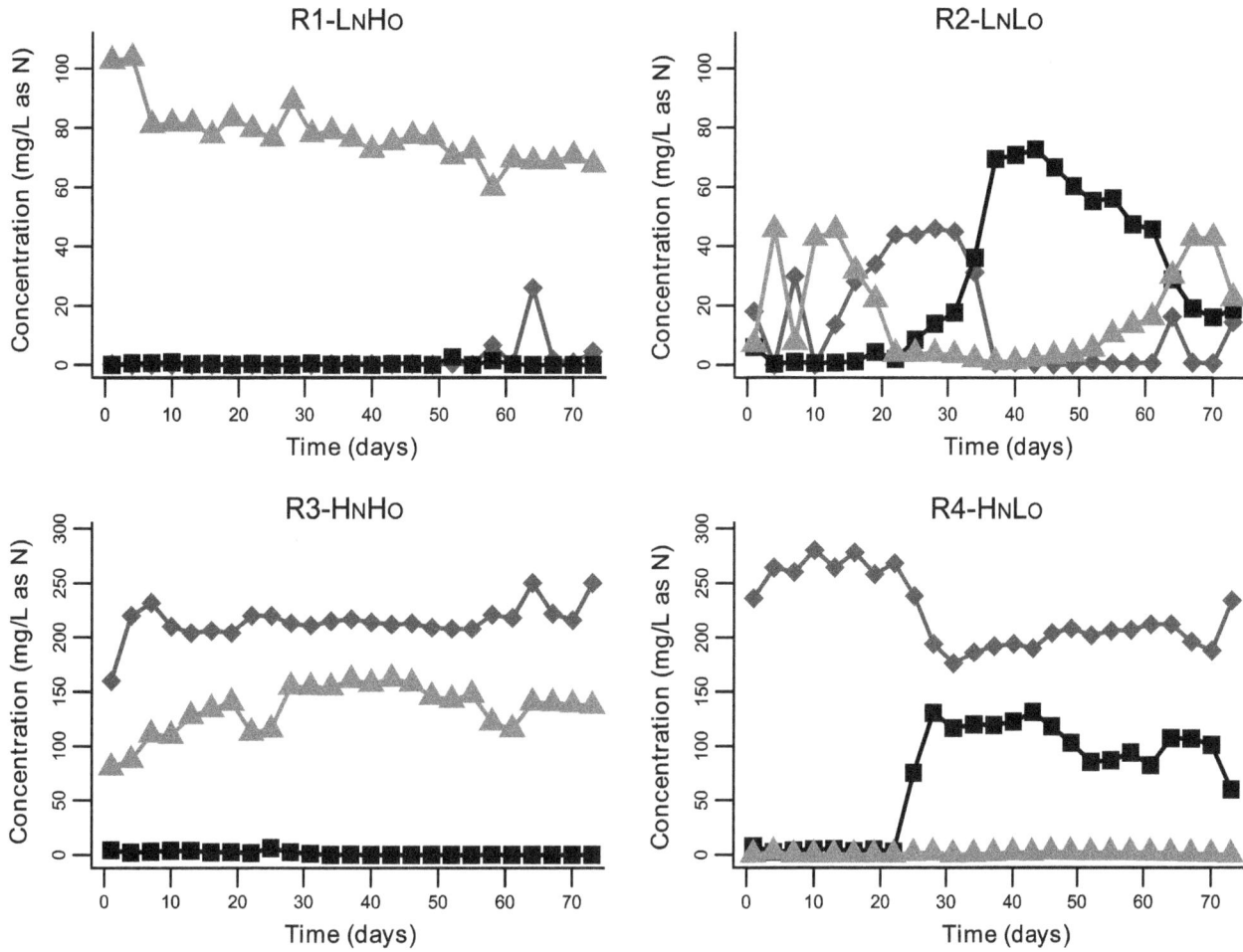

Fig. 1. Concentration of ammonia (♦), nitrite (■) and nitrate (▲) detected in the reactors over time.

Abundance and dynamics of the AOB

The abundance of AOB fluctuated over time (Fig. 2). However, there was no significant difference between configurations (ANOVA, $n = 8$, P-value = 0.053). Overall, the number of AOB ranged between 2.45×10^5 and 3.96×10^6 cells per ml. We should acknowledge that the AOB quantification could be overestimated

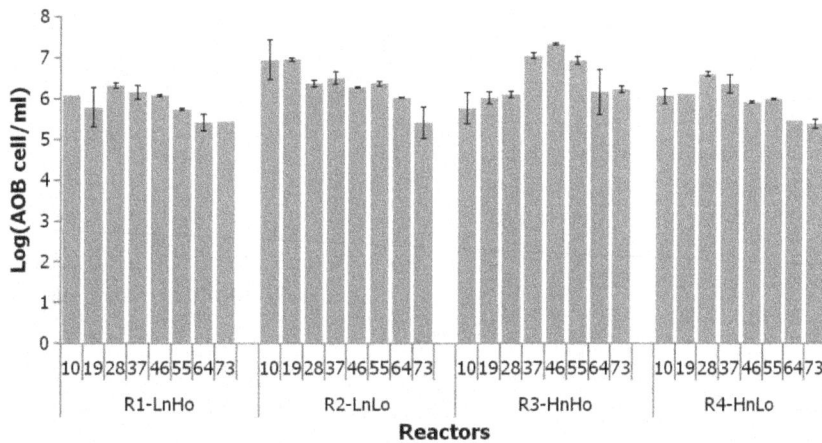

Fig. 2. Ammonia oxidizing bacteria abundance in the reactors over time assessed by qPCR. Bars show the mean of duplicate sample, while error bars represent standard deviation.

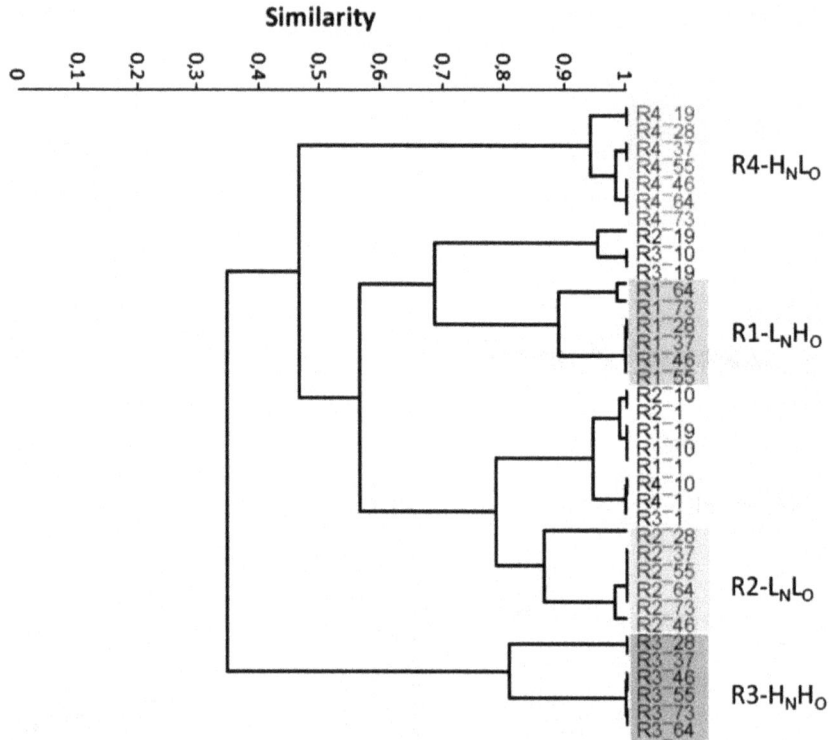

Fig. 3. Cluster analysis of the AOB DGGE profiles based on Raup and Crick coefficient. The band patterns of the samples taken 28 days after inoculation from the same unit had grouped together as shown by the shaded boxes.

as the CTO primers could anneal non-AOB DNA template.

Temporal changes in the denaturant gradient gel electrophoresis (DGGE) profiles of the AOB 16S rRNA gene fragments were used to observe AOB dynamics in all configurations.

The DGGE patterns of samples from each reactor were analysed using Raup and Crick diversity indices (S_{RC}) before and after 28 days of operation. The results indicate that before day 28, the samples are clustering according to time, while after day 28, they are clustering according to the reactor. This suggests that the AOB community structure in each reactor changed according to the supply of oxygen and ammonia to the systems (Fig. 3).

Temporal changes in the AOB community were evident from moving window analysis (Fig. 4). Significant shifts were observed in the two reactors supplied with low oxygen between days 19 and 28 ($S_{RC} = 0.025$ in R4-H_NL_O and $S_{RC} = 0.05$ in R2-L_NL_O). In R3-H_NH_O, the lowest S_{RC} value observed (0.225) was detected between the first and the 10th day of the experiment. The AOB community in R2-L_NL_O, R3-H_NH_O and R4-H_NL_O reached an apparent dynamic equilibrium during the last period of the experiment (between days 55 and 73) as the DGGE profiles were significantly similar ($S_{RC} > 0.95$). In contrast, the AOB community in R1-L_NH_O

was quite stable until day 64, when a change occurred; however, the difference between the DGGE profile of

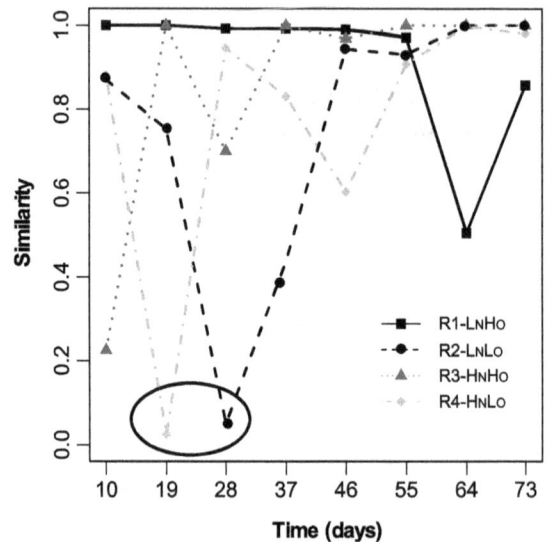

Fig. 4. Ammonia oxidizing bacteria community drifts detected by moving window analyses. The similarity between DGGE profiles of samples taken from the same reactor at two consecutive sampling days was calculated based on the Raup and Crick index. The circle highlights that the AOB community changed significantly 19 and 28 days after inoculation in R4-H_NL_O and R2-L_NL_O, respectively.

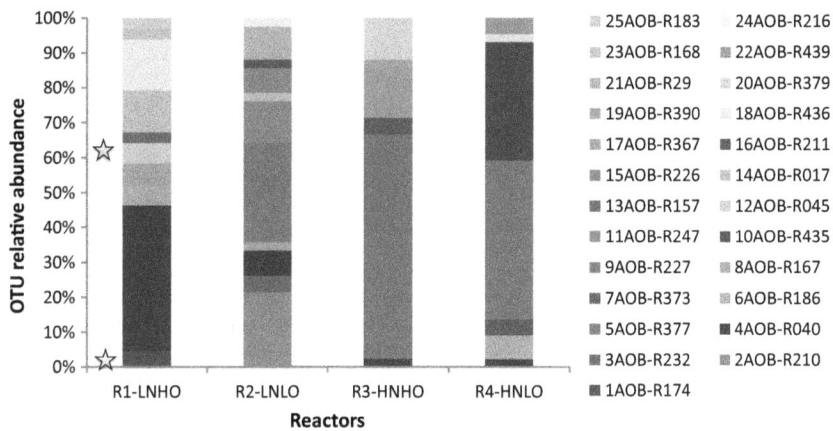

Fig. 5. Distribution of AOB 16S rRNA gene clones retrieved from the four reactors after 73 days. The 209 clones were sorted into 25 groups represented by a single OTU (bars in different colours). 12AOB-R045 and 1AOB-R174 (stars) are non-AOB OTUs.

days 55 and 64 was modest and no greater than would be expected by random association of bands in the DGGE patterns.

Composition of the AOB community assessed by 16S rRNA gene clone libraries

More detailed AOB community analyses were conducted on samples collected from the four reactors at the end of the experiment (day 73) by constructing AOB 16S rRNA gene clone libraries. The 169 partial 16S rRNA gene sequences were grouped in 25 operational taxonomic units (OTUs), which were distributed in the reactors as reported in Fig. 5. Representative sequences of each OTU were analysed to identify the most closely related sequences in the GenBank database (Table 2), and phylogenetic analysis was conducted (Fig. 6). Of the 25 OTUs, 22 showed between 96% and 100% identity to AOB 16S rRNA sequences present in the database, one had 95% of identity with *Nitrosomonas europaea*, and two were associated with non-AOB. Most of the sequences (80%) were closely related to sequences from members of the genus *Nitrosomonas*, while the rest were associated with *Nitrosospira* spp.. In R1-L_NH_O, members of *Nitrosomonas oligotropha* were dominant, though they coexisted with

Nitrosomonas ureae and *Nitrosomonas* sp. Nm51. The sequences recovered from R2-L_NL_O were related mostly to *Nitrosococcus mobilis*, *Nitrosomonas eutropha*, and *N. europaea*, but a few sequences were related to *N. oligotropha* and *Nitrosomonas marina*. In R4-H_NL_O, an *N. eutropha*-like bacterium was the dominant AOB, but organisms related to *N. oligotropha*, *N. ureae*, *Nitrosospira* and *N. europaea* were also detected. In contrast with the other reactors, most of the sequences retrieved from R3-H_NH_O were related to *Nitrosospira* spp., and only one clone was associated with the *N. oligotropha* lineage.

Species richness

The number of OTUs can be used as an estimation of species richness. The 16S rRNA gene clone library data showed that the AOB communities selected in each reactor were made up of different numbers of species (S) as reported in Table 3. However, the observed diversity resulted slightly lower than the estimated one by non parametric diversity indices, Chao1 and ACE, as the rarefaction curves revealed that the AOB communities were under sampled in the AOB clone libraries, though the rates of OTU accumulation were beginning to decelerate (Fig. S3).

Table 1. Main operational parameters of the bioreactors.

	R1-L_NH_O	R2-L_NL_O	R3-H_NH_O	R4-H_NL_O
COD $_{infl}$ (mg/l)	528.2 ± 32.5	528.2 ± 32.5	528.2 ± 32.5	528.2 ± 32.5
NH$_4^+$-N $_{infl}$ (mg/l)	24.2 ± 5.5	24.2 ± 5.5	280 ± 17	280 ± 17
Oxygen supplied (%)	21	2	21	2
Air flow rate (l/min)	0.2	0.2	0.2	0.2
DO (mg/l)	3.79 ± 0.49	0.26 ± 0.2	3.37 ± 1.43	0.18 ± 0.24
Volume (l)	3	3	3	3
SRT (days)	5	5	5	5
Temperature (°C)	20 ± 1.6	20 ± 1.6	20 ± 1.6	20 ± 1.6

Table 2. Closest matches (BLASTN) of the sequences of the representative OTUs retrieved from the AOB clone libraries. The query coverage (QC), max identity (MI) and reference are also reported.

OTU	QC (%)	MI (%)	Accession	Description
1AOB-R174	100	99	EU924130	Uncultured *Dechloromonas* sp., clone AOB_B12 16S ribosomal RNA gene
2AOB-R210	100	99	AF527019	Uncultured *Nitrosomonas* sp., 36Fb 16S ribosomal RNA gene
3AOB-R232	100	99	AL954747	*N. europaea* ATCC 19718, complete genome
4AOB-R040	100	99	EU285320	Uncultured *Nitrosomonas* sp., clone AOB17 16S ribosomal RNA gene
5AOB-R377	99	99	EU285326	Uncultured *Nitrosospira* sp., clone AOB23 16S ribosomal RNA gene
6AOB-R186	100	98	FM201082	Uncultured *Nitrosomonas* sp., partial 16S rRNA gene, clone MBR-8_LF_BF68
7AOB-R373	98	96	GQ451713	*N. europaea* strain ATCC 25978 16S ribosomal RNA gene
8AOB-R167	100	97	AF527014	Uncultured *Nitrosomonas* sp, 21Fb 16S ribosomal RNA gene
9AOB-R227	100	99	CP000450	*Nitrosomonas eutropha* C91, complete genome
10AOB-R435	98	100	CP000450	*Nitrosomonas eutropha* C91, complete genome
11AOB-R247	98	99	AJ298728	*Nitrosococcus mobilis* partial 16S rRNA gene, isolate Nc2.
12AOB-R045	100	99	EU924131	Uncultured *Dechloromonas* sp., clone AOB_B3 16S ribosomal RNA gene
13AOB-R157	98	98	AY543074	Uncultured *Nitrosomonas* sp., clone 15BAFln1 16S ribosomal RNA gene
14AOB-R017	100	99	EU285320	Uncultured *Nitrosomonas* sp., clone AOB17 16S ribosomal RNA gene
15AOB-R226	100	98	AF386750	*Nitrosomonas* sp. R5c88 16S ribosomal RNA gene, partial sequence.
16AOB-R211	99	97	CP000450	*Nitrosomonas eutropha* C91, complete genome
17AOB-R367	100	99	EU285326	Uncultured *Nitrosospira* sp., clone AOB23 16S ribosomal RNA gene
18AOB-R436	98	99	DQ002458	Uncultured *Nitrosomonas* sp., clone EZS-3 16S ribosomal RNA gene
19AOB-R390	100	99	AY138531	*Nitrosospira* sp., DNB_E1 16S ribosomal RNA gene
20AOB-R379	100	98	X84661	*Nitrosospira* sp., 16S rRNA gene, isolate T7.
21AOB-R29	98	99	AL954747	*N. europaea* ATCC 19718, complete genome
22AOB-R439	100	95	AL954747	*N. europaea* ATCC 19718, complete genome
23AOB-R168	100	97	EF016119	*N. oligotropha* isolate AS1 16S ribosomal RNA gene
24AOB-R216	100	97	AF386753	*Nitrosomonas* sp. R7c140 16S ribosomal RNA gene, partial sequence.
25AOB-R183	100	98	AF527021	Uncultured *Nitrosomonas* sp., 11Fb 16S ribosomal RNA gene

AOB community composition and RRT

The AOB species richness values (as defined by OTUs) observed in the reactors were compared with the simulations of the RRT. In the first set of simulations, the activated sludge floc was considered as a heterogeneous environment, where the ammonia and oxygen levels vary with the depth of the floc. The variation in ammonia and oxygen through the floc was modelled by taking into consideration the diffusion–consumption of the two resources. The mathematical computation of the diffusion–consumption of the oxygen revealed a non-dimensional variation (σ) of oxygen equal to 0.014, whereas the variation of the ammonia concentration in the floc was negligible ($\sigma = 0$). Tilman (1982) defined heterogeneity in the environment as the 0.99 probability contour of a bivariate normal distribution of resource concentrations, which is obtained at $2.58 \cdot \sigma$ and by assuming that the two resources are independent ($r = 0$) and have the same variance ($\sigma = \sigma_1 = \sigma_2$). The simulated species richness-resource abundance curve is humped (Fig. 7A), suggesting that habitats with very low resource levels cannot support large numbers of species, but in general low resources lead to maximal diversity, while a high level of resources leads to a decrease in the number of species. The standard deviation of the species richness is high in the same gradient of resources, suggesting that it is intrinsically more difficult to predict AOB diversity at low resource levels (Fig. 7B). The experimental data (closed dots in Fig. 7A) were very close to the simulated range.

The second set of simulations was run taking into account the minimum microhabitat containing all the species. Overall, the trend of the resulting species richness-resource gradient curve and relative standard deviation concurred with the initial simulation, though the predicted AOB species richness was generally lower (Fig. 7B).

Discussion

In this study, we successfully manipulated the AOB community structure of lab-scale nitrifying systems according to the two growing limiting resources supplied. Different amounts of oxygen and ammonia selected for AOB communities different in richness and composition as shown by DGGE and clone libraries analyses of AOB 16S rRNA genes. This suggests that the principles of RRT have the potential to help predicting and possibly managing some characteristics of microbial communities in WWTPs, in particular, and open microbial systems, in general. The qualitative agreement with the simulation was also gratifying in what was a simple first attempt with parameters taken from the literature.

We would however, offer substantial *caveats*. First, our reactors were not replicated. Nevertheless, in a previous

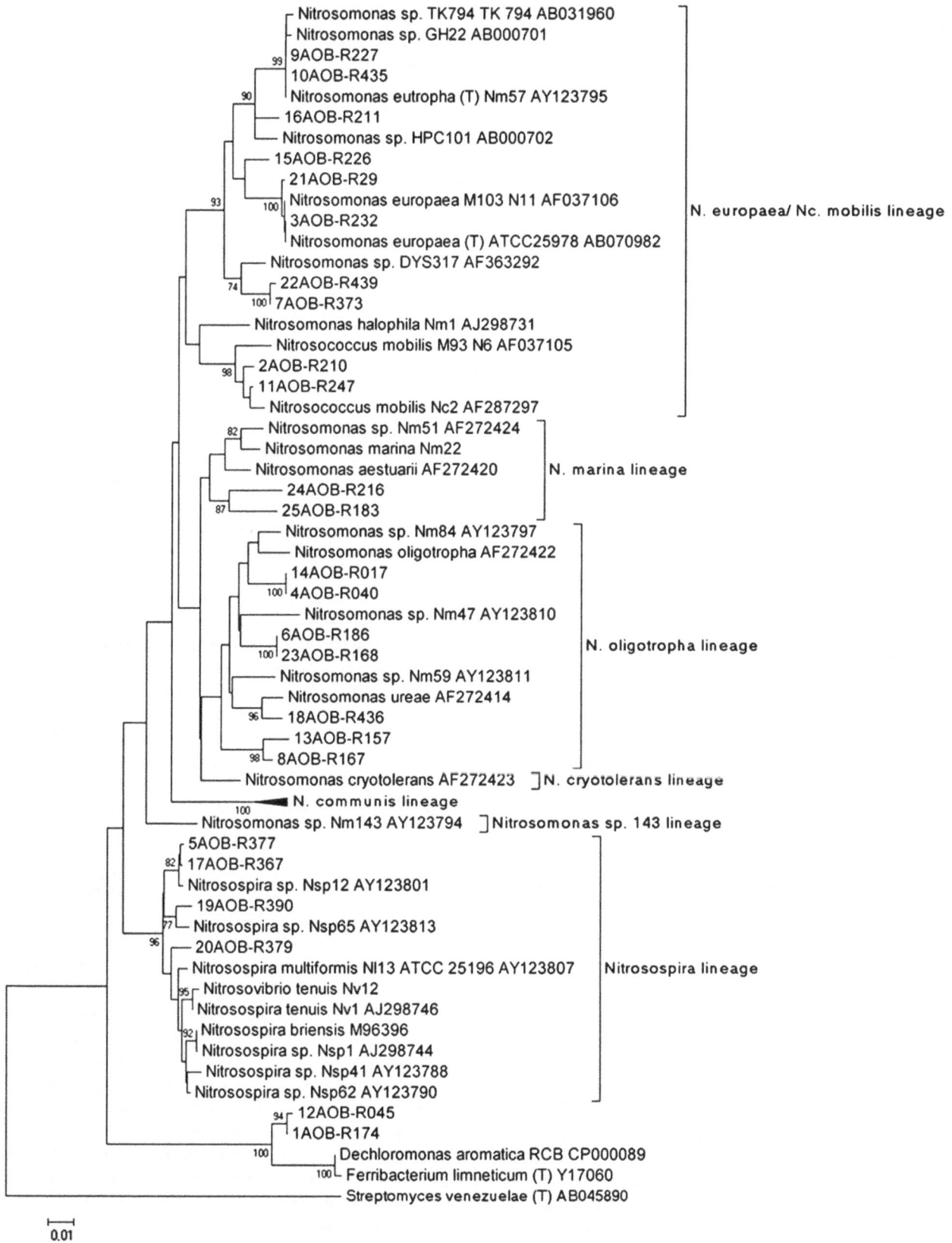

Fig. 6. Phylogenetic tree of the AOB 16S rRNA gene sequences of the representative OTUs and their phylogenetic relatives constructed with MEGA5. The tree is based on minimum evolution method.

Table 3. Observed (AOB 16S rRNA gene clone libraries and DGGE) and estimated AOB species richness.

	Reactors			
	R1-$L_N H_O$	R2-$L_N L_O$	R3-$H_N H_O$	R4-$H_N L_O$
Number of OTUs[a]	8	11	6	7
Chao1 (standard error)	9 (3.4)	14 (11.7)	6.5 (3.7)	7.3 (1.9)
ACE (standard error)[b]	10.7 (1.5)	15.14 (2)	7.6 (1.2)	8.4 (1.4)
Number of bands (standard deviation)[c]	9 (1.5)	7 (0.8)	11.75 (0.5)	5.5 (0.6)

a. Number of OTUs observed in the AOB 16S rRNA gene clone libraries without considering the non-AOB OTUs.

b. Abundance-based coverage estimation (ACE).

c. Mean values of the number of bands detected in the DGGE profiles of the samples collected the last four sampling days (46, 55, 64, 73) in each reactor.

study where the same bioreactors were operated in duplicate but with different operating conditions from this study, we demonstrated that AOB communities are highly replicable with a coefficient of variation between 0.04 and 0.1 (Bellucci et al., 2011). Though replication would have been ideal, and it is proposed for a future study, a statistically robust replication of the pattern would not prove the underlying mechanism envisaged in RRT. That would require us to demonstrate that the kinetic parameters of the microbes were consistent with diversities observed.

This in turn would require a robust calibration of our preliminary simulation. This is possible but challenging. However, even with a perfect calibration of the model, accurate prediction of diversity would be difficult as the computational simulations in our study suggest that the diversity is intrinsically more variable in habitats supplied with low amounts of resources than in enriched environments. This, together with analogous findings in a recent study, modelling the outcome of competition between nine phytoplankton species for nitrogen, phosphorus and light (Brauer et al., 2012), might explain why several experimental works failed to fully corroborate the predictions of RRT (Miller et al., 2005). This caveat should be carefully considered when applying RRT in the design of WWTPs.

The modest discrepancy between experimental and predicted data is not surprising. The model used in this study requires large number of unknown parameters, and, therefore, provides only a crude estimation of the diversity. The number of species detected by the model is affected by the size of the microhabitat given by the variance of the resources through the activated floc. It would have been possible, but not meaningful, to adapt the literature values used to fit our experimental findings.

Nevertheless, the low concentration of resources in R2-$L_N L_O$ selected for a community with the highest number of species, while the AOB community in the reactor with high oxygen and N input (R3-$H_N H_O$) was the least diverse, in agreement with the predictions of RRT. The low diversity retrieved from R3-$H_N H_O$, which was supplied with the highest amount of resources suggests that the 'paradox of enrichment' (Rosenzweig, 1971) can be found in engineered systems, at least in autotrophs.

In so far as we can tell, the presence of the taxa in the systems is consistent with the physiologies associated with these phylotypes. For example, our work is consistent with an earlier study in chemostats (Park and Noguera, 2007) as the low oxygen and ammonia conditions in R2-$L_N L_O$ favoured the parallel growth of members of the N. oligotropha lineage and N. europaea. Apparently, the contrasting affinity for oxygen and ammonia of these AOB represents a strategy for their coexistence in the system (Park and Noguera, 2007). The ability of N. europaea to use nitrite as electron acceptor (Chain et al., 2003) could also be a selective advantage in system with high nitrite levels, such as R2-$L_N L_O$. On the other hand, the high concentration of ammonia apparently promoted the establishment of an AOB community dominated by Nitrosospira and N. eutropha in R3-$H_N H_O$ and R4-$H_N L_O$, respectively. From an ecological prospective, our findings run counter to the consensus that Nitrosospira spp. are typical K-strategists with high affinity for ammonia and low growth rate compared with N. europaea spp., which are considered to be r-strategists (Andrews and Harris, 1986). However, this definition resulted from evidence of Nitrosomonas spp. rather than Nitrosospira in ammonium-rich activated sludge and biofilm reactors in previous studies where the pH was maintained near neutral values (Mobarry et al., 1996; Schramm et al., 1996; 1999; 2000; Terada et al., 2013). In our experiment the pH was not adjusted, and it decreased to ~ 6 in R3-$H_N H_O$. Probably, the low pH, together with the high ammonia concentration of the influent, selected for Nitrosospira sp. as its capacity for urease production between pH 5 and 6 (Koops et al., 2003) provides a competitive advantage in such conditions. One the other hand, N. eutropha-like AOB are often found in the SHARON (Stable High rate Ammonia Removal Over Nitrite) processes in which high ammonia loads are imposed to promote nitrite accumulation (Logemann et al., 1998; van Dongen et al., 2000). Nitrosomonas eutropha can tolerate high concentrations of ammonium (600 mM) and compete with other AOB in systems where competition for oxygen with heterotrophic organisms is acute; thanks to the presence of a nirK gene in the genome (Stein et al., 2007). It could be speculated

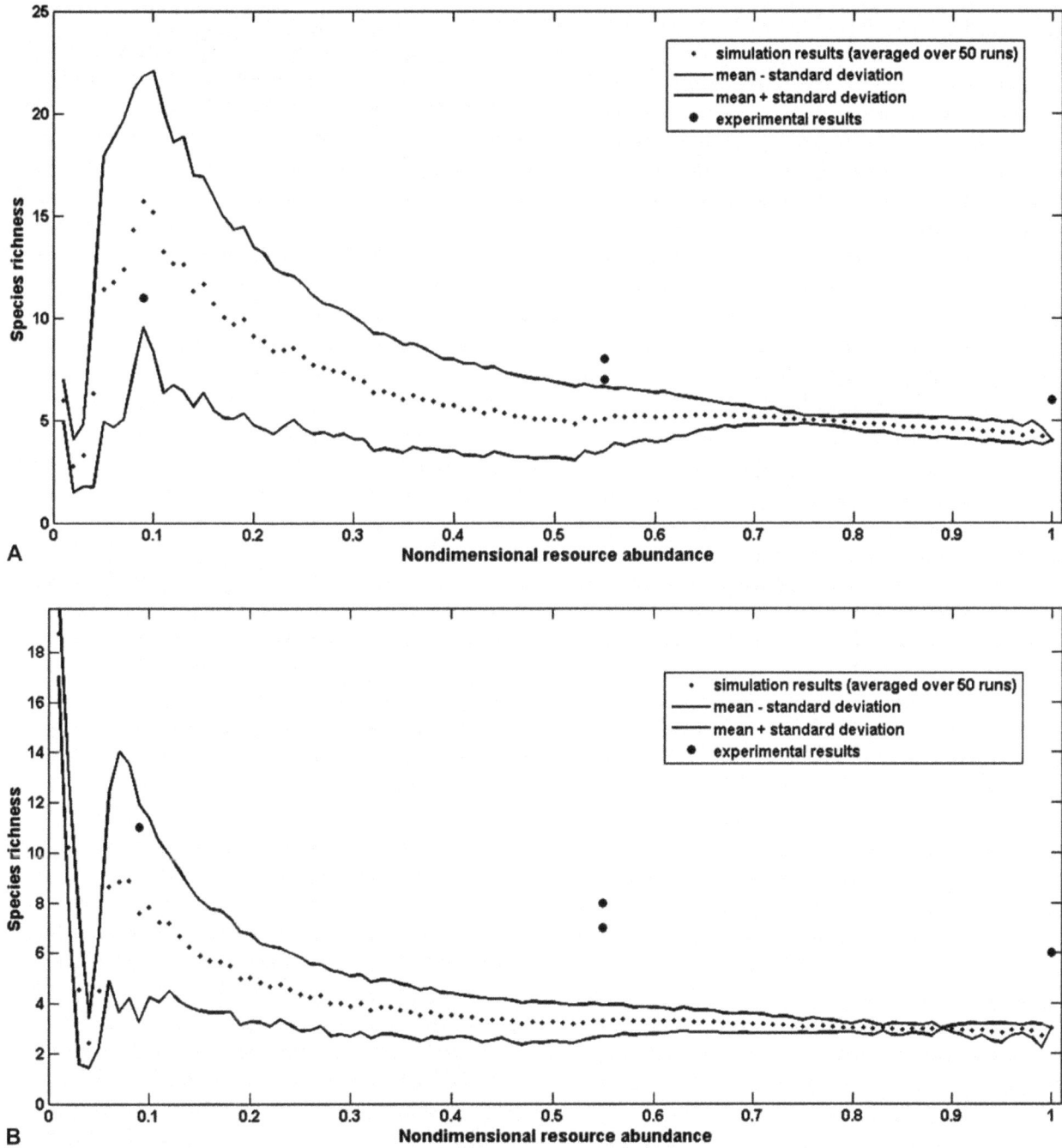

Fig. 7. Comparison between empirical and theoretical results when the variation of oxygen consumption (A) and the minimum microhabitat containing all the species (B) are taken into account.

that the ammonia oxidizing archaea (AOA), which possess the *amoA* gene encoding for the ammonia monooxygenase (Schleper *et al.*, 2005), might have contributed to the performance of our bioreactors, as they have been already found in natural and engineered systems with low DO concentrations (Park *et al.*, 2006; Gómez-Silván *et al.*, 2009; Zhang *et al.*, 2009; Pitcher *et al.*, 2011). However, AOA were not detected in our systems by polymerase chain reaction (PCR; results not shown); presumably, their growth and activity could not be sustained in our bioreactors that were operated with relatively high ammonia loading and short retention time (Hatzenpichler, 2012). This hypothesis can be confirmed only when the mechanisms selecting for AOA, and their real role, in activated sludge are better understood.

Overall, deterministic factors appeared to have a major role in shaping the AOB community of the reactors. Shifts in the community occurred as a function of the ammonia and oxygen supplied to the systems as most of the AOB community DGGE profiles taken from the same reactor formed distinct groups in cluster analysis (Fig. 3). This is in agreement with previous studies, suggesting that an AOB community with particular ecological requirements emerged from a common pool (the seed) and became dominant over time (Juretschko *et al.*, 1998; Figuerola and Erijman, 2007; Ayarza *et al.*, 2010; Bellucci *et al.*, 2011; 2013; Beneduce *et al.*, 2014). Oxygen exerted a stronger effect than ammonia on the selection of the AOB community as only the DGGE patterns of the samples collected after 28 days in R2-$L_N L_O$ and R4-$H_N L_O$ were significantly dissimilar to the original common pool of bacteria. Therefore, as Park and Noguera (2004; 2007) have previously reported, the oxygen supplied in R2-$L_N L_O$ and R4-$H_N L_O$ dictated the AOB community structure. The identification of the operational conditions required to establish the AOB species that can function efficiently in low oxygen environments is fundamental for the development of less costly WWTPs.

In conclusion, if we could find an authoritative and reliable method for calibrating the model, the RRT approach would contribute to the management of WWTPs and to the refinement of ecological theory. However, even with perfect calibration, at some resource ratios the predictions may be more variable than previously recognized.

Experimental procedures

Experimental set-up

Bench scale reactors. Four continuous flow lab-scale reactors were operated in parallel for 73 days. The reactors consisted of glass cylinders with a funnel bottom in which a porous glass grid was placed in the centre. The systems were inoculated (day 1) with 3 liters of return activated sludge from a Municipal Wastewater Plant in Spennymoor (County Durham, UK, 54° 42′ 0″ N, 1° 35′ 24″ W), and the liquor was constantly mixed by a stirrer (~ 100 rpm). The reactors were set up using a 2×2 factorial design; the two factors considered were the percentage of oxygen supply and the inorganic ammonia concentration in the influent (Table 1). Two of the reactors were supplied with air (R1-$L_N H_O$ and R3-$H_N H_O$), while the other two systems were supplied with a gas mixture containing 2% of oxygen and 98% of nitrogen (R2-$L_N L_O$ and R4-$H_N L_O$). The gas flow rate was fixed at 0.2 l/min. Synthetic wastewater kept at 4°C was continuously pumped into the reactors at a fixed solid and hydraulic retention time (SRT and HRT) of 5 days. Synthetic wastewater consisted of either 66 mg/l $(NH_4)_2SO_4$ in R1-$L_N H_O$ and R2-$L_N L_O$ or 1322 mg/L of $(NH_4)_2SO_4$ R3-$H_N H_O$ and R4-$H_N L_O$, 320 mg/l peptone, 190 mg/l meat extract, 30 mg/l yeast extract, 30 mg/l urea, 28 mg/l K_2HPO_4, 2 mg/l $CaCl_2 \cdot 2H_2O$ and 2 mg/l of $MgSO_4 \cdot 7H_2O$. The media were autoclaved (120°C for 20 min)

prior to use and, after autoclaving, 1 ml/l of trace element solution (0.75 g/l $FeCl_3 \cdot 6H_2O$, 0.075 g/l H_3BO_3, 0.015 g/l $CuSO_4 \cdot 5H_2O$, 0.09 g/l KI, 0.06 g/l $MnCl_2 \cdot 4H_2O$, 0.03 g/l $NaMoO_4 \cdot 2H_2O$, 0.06 g/l $ZnSO_4 \cdot 7H_2O$, 0.075 g/l $CoCl_2 \cdot 6H_2O$, 0.5 g/l Ethylenediaminetetraacetic acid (EDTA) and 1 ml/l concentrated hydrochloric acid) and 0.7 mg/l $NaHCO_3$ (Knapp and Graham, 2007) were added. Such solutions yielded different Total Kjeldahl Nitrogen (TKN), NH_4^+-N and COD concentrations, which are reported in Table 1. Ideally, replicate reactors could have been used, but we showed that performance and community structure was highly replicable in a previous study using identical reactors (Bellucci *et al.*, 2011). Samples (200 ml) from each reactor and the feed synthetic wastewaters were collected every 3 days for chemical, physical and microbial community analyses.

Physical and chemical analyses. The dissolved oxygen concentration, temperature and pH were constantly monitored with specific probes (Broadley Technologies, UK). Nitrification performance and COD removal were assessed by analysing NH_4^+-N, NO_2^--N, NO_3^--N and COD levels in the bulk solution over time. Ammonium (NH_4^+-N) and COD concentrations were determined using the Ammonium Cell Test Kit and COD Cell Test (MERCK KGaA, Germany), respectively, according to the manufacturer's instructions. Measured were NO_2^- and NO_3^- by ion chromatography (DW-100 Ion Chromatography, Dionex, Sunnyvale, CA., USA); the system had an IonPac AS14A Analytical column, flow rate equal to 1 ml/min, the eluent was a 8.0 mM Na_2CO_3/1.0 mM $NaHCO_3$ solution, and the injection loop was 25 µl. Free ammonia and free nitrous acid were calculated in function of the experimental values of total ammonia and nitrite concentrations, pH and temperature using the equation described by Anthonisen and colleagues (1976) and reported in the Supporting Information. Total suspended solid (TSS) and VSS, as well as the TKN in the medium supplied to the systems, were evaluated according to standard methods (APHA, 1998).

DNA extraction and quantitative PCR. DNA was extracted from the mixed liquor (250 µL from the four reactors every 9 days starting from day 1 to day 73) and stored at –20°C until further analyses. After mechanical cell lysis using Lysing Matrix E (MP Biomedicals, Solon, USA) in a Ribolyser (Hybaid, UK), DNA extraction was performed using the FastDNA SPIN kit For Soil (MP Biomedicals, Solon, USA) in accordance with manufacturer's instructions.

The abundance of AOB in the mixed liquor was evaluated by primers CTO 189fA/B (5′-GGAGRAAAGCAGGGGA TCG-3′), CTO189fC (5′-GGAGGAAAGTAGGGGATCG-3′) (Kowalchuk *et al.*, 1997) and RT1r (5′-CGTCCTCTCAGA CCARCTACTG-3′) (Hermansson and Lindgren, 2001). The quantitative PCR (qPCR) reactions were performed in a BioRad iCycler equipped with an iCycler iQ fluorescence detector and associated software (BioRad Version 2.3). The 15 µl PCR reaction mixture contained 2 µl of DNA template, 1 µl of primer mixture (7.5 pmol each) and 12 µl of PCR Precision Mastermix PCR reagent (Primer Design, UK). For fluorescence detection, SYBR green I (10,000 x, Sigma, UK) was first diluted 1/100 in sterile and autoclaved molecular water, and such solution was added to the PCR reaction mixture as 1% of the total volume of the reaction mix (vol/vol).

The thermal cycling was carried out as previously reported by Hermansson and Lindgren (2001). The 16S rRNA gene copy numbers were converted to equivalent cell numbers, assuming that one rRNA operon exists per AOB cell (Klappenbach et al., 2001).

Nested PCR and DGGE. A nested PCR approach, followed by DGGE was performed to monitor AOB population dynamics. In the first PCR, a 465 bp fragment of the 16S rRNA gene of AOB was amplified using CTO189F (5′-GAGRAAAGYAG GGGATCG-3′) and CTO654R (5′-CTAGCYTTGTAGTTTCAA ACGC-3′) (Kowalchuk et al., 1997). The PCR products from this reaction were used as a template for a second PCR reaction with primer 3 (5′-CGCCCGCCGCGCGCGGCGGGC GGGGCGGGGGCACGGGGGGGCCTACGGGAGGCAGCAG-3′) and primer 2 (5′-ATTACCGCGGCTGCTGG-3′) (Muyzer et al., 1993). All PCRs were carried out in 25 µl reactions containing 23.5 µl of PCR buffer (MegaMix-BLUE, Microzone, UK), 0.5 µl of each primer (10 pmol) and 0.5 µl of DNA template. The PCR conditions for the CTO primer set and primer 3 and primer 2 set were previously described (Muyzer et al., 1993; Kowalchuk et al., 1997), and all the PCR reactions were performed using a Px2 Thermal Cycler (Thermo Hybaid, Hybaid, UK).

Nested PCR amplified fragments were separated using a D-Code DGGE system (Bio-Rad Laboratories, UK) as described previously (Bellucci and Curtis, 2011; Bellucci et al., 2011, Bellucci et al., 2013, Beneduce et al., 2014). The gels were processed using Bionumerics 4.0 (Applied Maths BVBA, Saint-Martens-Latem, Belgium). Gels were normalized using a reference marker that was loaded into the polyacrilamide gels every eight samples and co-migrated with them. This approach allows comparing community profiles within the same gel and among differing gels. Pairwise similarities between DGGE profiles were calculated with Raup and Crick coefficients for cluster (single linkage clustering) and moving window analyses. The Raup and Crick similarity index (S_{RC}) is defined as the probability that the expected similarity (randomized data) would be greater than or equal to the observed similarity (Raup and Crick, 1979). Similarity values between 0.05 and 0.95 are indicative of random occurrence of the same organism (DGGE band) in two samples, whereas S_{RC} above 0.95 and below 0.05 indicate significant similarity and significant dissimilarity, respectively. The algorithm assumes the taxa are equally likely to be selected. Cluster analysis was conducted using PAST (Hammer et al., 2001). In the moving windows analysis the similarity between DGGE profiles of a given day with the previous sampling day was calculated for each configuration.

The number of bands detected in the DGGE profiles could be considered as a rough estimation of AOB diversity, though it should be kept in mind that two amplicons differing in sequences may migrate together, and the same species might contain several copies of the same gene that differ slightly in sequence producing several bands in the DGGE. Nevertheless, the number of bands in the profiles of the samples collected the last four sampling days (46, 55, 64, 73) were counted and compared to confirm further that the AOB community richness was stable at the end of the reactor operation.

Clone library preparation. To evaluate and compare the AOB community structure selected in the four reactor configurations after 73 days of operation, gene clone libraries were constructed from AOB partial 16S rRNA genes amplified from samples collected from each reactor at the end of the experiment. Polymerase chain reaction products amplified with the CTO primer set were cloned with a TOPO TA cloning kit (Invitrogen, UK). Following PCR amplification, the correct size of the fragment was checked by electrophoresis in a 1.5% agarose gel run in 1 × TAE buffer stained with ethidium bromide. The bands were excised and purified with a Qiaquick PCR Gel Extraction Kit (QIAGEN, UK). The PCR products and the vector provided by the kit were ligated and used to transform OneShot competent cells according to the manufacturer's instructions (Invitrogen, UK). Clones (37–42) were randomly picked and transferred to a 30 µl PCR mixture containing the primers T3 (5′-ATTAACCCTCACTAAAG GGA-3′) and T7 (5′-TAATACGACTCACTATAGGG-3′). Polymerase chain reaction products of the correct size were prepared for sequencing using ExoSAP-IT (GE Healthcare, UK), and the sequences of the partial 16S rRNA gene fragment (ca 465 pb) were determined by Genevision, UK.

The sequences obtained were aligned using CLUSTALX v1.83 (Thompson et al., 1997), and vector and primer sequences were removed. The triggered sequences with > 97% identity were grouped by FASTGROUPII (Yu et al., 2006) into OTUs and the representative sequences of each group were analysed using BLASTN. The representative OTUs were also checked for putative chimeras with the online software BELLEROPHON (Huber et al., 2004) using a window size of 200 pb. Molecular Evolutionary Genetics Analysis (MEGA) version 5 (Tamura et al., 2011) was used to conduct phylogenetic analysis. The evolutionary history was inferred using the Minimum Evolution (ME) method (Rzhetsky and Nei, 1992) using the OTUs found in this study and selected AOB and *Ferribacterium* sequences (62 in total) for which 16S rRNA sequences longer than 1200 bp are available in the database Ribosomal Database Project (Maidak et al., 1999). The bootstrap consensus tree was inferred from 1000 replicates (Felsenstein, 1985). Branches corresponding to partitions reproduced in less than 50% bootstrap replicates were collapsed. The evolutionary distances were computed using the maximum composite likelihood method (Tamura et al., 2004) and are in the units of the number of base substitutions per site. The ME tree was searched using the close-neighbour-interchange algorithm (Nei and Kumar, 2000) at a search level of 1. The Neighbour-joining algorithm (Saitou and Nei, 1987) was used to generate the initial tree. On the basis of the OTUs detected in the AOB clone libraries, the AOB diversity was assessed as observed richness and non-parametric estimates of AOB richness, Chao1 and ACE, with the software R (R Development Core Team, 2011) using the package VEGAN and the function estimator (Oksanen et al., 2013). The AOB observed and estimated richness was compared with the rarefied one by building individual rarefaction curve with PAST (Hammer et al., 2001).

Nucleotide sequence accession numbers. The sequences determined in this study have been deposited in the GenBank database under accession number KC346976-KC347000.

Model description

Resource-ratio theory states that the quantity of the growth limiting resources available in a heterogeneous environment determines the species richness of a biological community (Tilman, 1982). The underpinning theory is provided in the Supporting Information, including a numerical model developed to simulate the theory. Specifically, an AOB community of 23 species was mathematically simulated based on the competition for oxygen and ammonia as primary resources using the RRT model. The model assumes microhabitats are best represented as explicit circles (with defined radius) around resource-intersection points associated with mean values of the two resources (see Fig. S1 and SI). The radius of each microhabitat was defined (i) by inferring the diffusion and consumption of a given resource and (ii) by considering the minimum resource variance suitable to comprise all the potential species. Model assumptions are provided in the Supporting Information (Table S1). For each given radius, 50 replicates were generated. The averages of the species richness and the relative standard deviations were then plotted against the resource.

Acknowledgement

We would like to thank Professor Vasile Lavric for fruitful discussion and advice.

Conflict of Interest

None declare.

References

Andrews, J., and Harris, R. (1986) r- and K-selection and microbial ecology. In *Advances in Microbial Ecology.* Marshall, K.C. (ed.). New York, USA: Springer, pp. 99–147.

Anthonisen, A.C., Loehr, R.C., Prakasam, T.B.S., and Srinath, E.G. (1976) Inhibition of nitrification by ammonia and nitrous acid. *J Water Pollut Control Fed* **48**: 835–852.

Ayarza, J., Guerrero, L., and Erijman, L. (2010) Nonrandom assembly of bacterial populations in activated sludge flocs. *Microb Ecol* **59**: 436–444.

Bellucci, M., and Curtis, T.P. (2011) Ammonia-Oxidizing Bacteria in Wastewater. In *Methods in Enzymology*, Vol. **496**. Klotz, M.G. and Stein L.Y. (eds). Burlington: Academic Press, pp. 269–286.

Bellucci, M., Ofiteru, I.D., Graham, D.W., Head, I.M., and Curtis, T.P. (2011) Low-dissolved-oxygen nitrifying systems exploit ammonia-oxidizing bacteria with unusually high yields. *Appl Environ Microbiol* **77**: 7787–7796.

Bellucci, M., Ofiţeru, I.D., Head, I.M., Curtis, T.P., and Graham, D.W. (2013) Nitrification in hybrid bioreactors treating simulated domestic wastewater. *J Appl Microbiol* **115**: 621–630.

Beneduce, L., Spano, G., Lamacchia, F., Bellucci, M., Consiglio, F., and Head, I. (2014) Correlation of seasonal nitrification failure and ammonia-oxidizing community dynamics in a wastewater treatment plant treating water from a saline thermal spa. *Ann Microbiol* **64**: 1671–1682.

Brauer, V.S., Stomp, M., and Huisman, J. (2012) The nutrient-load hypothesis: patterns of resource limitation and community structure driven by competition for nutrients and light. *Am Nat* **179**: 721–740.

von Canstein, H., Kelly, S., Li, Y., and Wagner-Döbler, I. (2002) Species diversity improves the efficiency of mercury-reducing biofilms under changing environmental conditions. *Appl Environ Microbiol* **68**: 2829–2837.

Cardinale, B.J., Srivastava, D.S., Emmett Duffy, J., Wright, J.P., Downing, A.L., Sankaran, M., *et al.* (2006) Effects of biodiversity on the functioning of trophic groups and ecosystems. *Nature* **443**: 989–992.

Chain, P., Lamerdin, J., Larimer, F., Regala, W., Lao, V., Land, M., *et al.* (2003) Complete genome sequence of the ammonia-oxidizing bacterium and obligate chemolithoautotroph *Nitrosomonas europaea. J Bacteriol* **185**: 2759–2773.

Curtis, T.P., Head, I.M., and Graham, D.W. (2003) Theoretical ecology for engineering biology. *Environ Sci Technol* **37**: 64A–70A.

Daims, H., Ramsing, N.B., Schleifer, K.-H., and Wagner, M. (2001) Cultivation-independent, semiautomatic determination of absolute bacterial cell Numbers in environmental samples by fluorescence *in situ* hybridization. *Appl Environ Microbiol* **67**: 5810–5818.

van Dongen, U., Jetten, M.S.M., and van Loosdrecht, M.C.M. (2000) *International Conference on Wastewater and EU Nutrient Guidelines*. Amsterdam, Netherlands: I W a Publishing.

Egli, K., Langer, C., Siegrist, H.R., Zehnder, A.J.B., Wagner, M., and van der Meer, J.R. (2003) Community analysis of ammonia and nitrite oxidizers during start-up of nitritation reactors. *Appl Environ Microbiol* **69**: 3213–3222.

Felsenstein, J. (1985) Confidence limits on phylogenies: an approach using the bootstrap. *Evolution* **39**: 783–791.

Figuerola, E.L.M., and Erijman, L. (2007) Bacterial taxa abundance pattern in an industrial wastewater treatment system determined by the full rRNA cycle approach. *Environ Microbiol* **9**: 1780–1789.

Gómez-Silván, C., Molina-Muñoz, M., Poyatos, J.M., Ramos, A., Hontoria, E., Rodelas, B., *et al.* (2009) Structure of archaeal communities in membrane-bioreactor and submerged-biofilter wastewater treatment plants. *Bioresour Technol* **101**: 2096–2105.

Hammer, Ø., Harper, D.A.T., and Ryan, P.D. (2001) Past: Paleontological Statistics Software Package for Education and Data Analysis. *Palaeontologia Electronica* **4**: art. 4, 9 pp., 178 kb.

Hatzenpichler, R. (2012) Diversity, physiology, and niche differentiation of ammonia-oxidizing archaea. *Appl Environ Microbiol* **78**: 7501–7510.

Hermansson, A., and Lindgren, P.-E. (2001) Quantification of ammonia-oxidizing bacteria in arable soil by real-time PCR. *Appl Environ Microbiol* **67**: 972–976.

Hooper, D.U., Chapin, F.S., Ewel, J.J., Hector, A., Inchausti, P., Lavorel, S., *et al.* (2005) Effects of biodiversity on ecosystem functioning: a consensus of current knowledge. *Ecol Monogr* **75:** 3–35.

Huber, T., Faulkner, G., and Hugenholtz, P. (2004) Bellerophon: a program to detect chimeric sequences in multiple sequence alignments. *Bioinformatics* **20:** 2317–2319.

Juretschko, S., Timmermann, G., Schmid, M., Schleifer, K.H., Pommerening-Roser, A., Koops, H.P., *et al.* (1998) Combined molecular and conventional analyses of nitrifying bacterium diversity in activated sludge: nitrosococcus mobilis and Nitrospira-like bacteria as dominant populations. *Appl Environ Microbiol* **64:** 3042–3051.

Klappenbach, J.A., Saxman, P.R., Cole, J.R., and Schmidt, T.M. (2001) rrndb: the ribosomal RNA operon copy number database. *Nucleic Acids Res* **29:** 181–184.

Knapp, C.W., and Graham, D.W. (2007) Nitrite-oxidizing bacteria guild ecology associated with nitrification failure in a continuous-flow reactor. *FEMS Microbiol Ecol* **62:** 195–201.

Koops, H.P., Purkhold, U., Pommerening-Roser, A., Timmermann, G., and Wagner, M. (2003) The lithoautotrophic ammonia-oxidizing bacteria. In *The Prokaryotes: An Evolving Electronic Resources for the Microbiological Community*, Third edn. Dworkin, M., *et al.* (eds). New York, NY, USA: Springer-Verlang, pp. 778–811.

Kowalchuk, G.A., Stephen, J.R., De Boer, W., Prosser, J.I., Embley, T.M., and Woldendorp, J.W. (1997) Analysis of ammonia-oxidizing bacteria of the beta subdivision of the class Proteobacteria in coastal sand dunes by denaturing gradient gel electrophoresis and sequencing of PCR-amplified 16S ribosomal DNA fragments. *Appl Environ Microbiol* **63:** 1489–1497.

Logemann, S., Schantl, J., Bijvank, S., van Loosdrecht, M., Kuenen, J.G., and Jetten, M. (1998) Molecular microbial diversity in a nitrifying reactor system without sludge retention. *FEMS Microbiol Ecol* **27:** 239–249.

Maidak, B.L., Cole, J.R., Parker, C.T., Garrity, G.M., Larsen, N., Li, B., *et al.* (1999) A new version of the RDP (Ribosomal Database Project). *Nucleic Acids Res* **27:** 171–173.

Miller, T.E., Burns, J.H., Munguia, P., Walters, E.L., Kneitel, J.M., Richards, P.M., *et al.* (2005) A critical review of twenty years' use of the resource-ratio theory. *Am Nat* **165:** 439–448.

Mobarry, B.K., Wagner, M., Urbain, V., Rittmann, B.E., and Stahl, D.A. (1996) Phylogenetic probes for analyzing abundance and spatial organization of nitrifying bacteria. *Appl Environ Microbiol* **62:** 2156–2162.

Muyzer, G., de Waal, E.C., and Uitterlinden, A.G. (1993) Profiling of complex microbial populations by denaturing gradient gel electrophoresis analysis of polymerase chain reaction-amplified genes coding for 16S rRNA. *Appl Environ Microbiol* **59:** 695–700.

Nei, M., and Kumar, S. (2000) *Molecular Evolution and Phylogenetics*. New York, NY, USA: Oxford University Press.

Oksanen, J., Blanchet, F.G., Kindt, R., Legendre, P., Minchin, P.R., O'Hara, R.B., *et al.* (2013) Package 'vegan'. [WWW document]. URL http://www.cran.r-project.org/web/packages/vegan/vegan.pdf.

Park, H.D., and Noguera, D.R. (2004) Evaluating the effect of dissolved oxygen on ammonia-oxidizing bacterial communities in activated sludge. *Water Res* **38:** 3275–3286.

Park, H.D., and Noguera, D.R. (2007) Characterization of two ammonia-oxidizing bacteria isolated from reactors operated with low dissolved oxygen concentrations. *J Appl Microbiol* **102:** 1401–1417.

Park, H.D., Wells, G.F., Bae, H., Criddle, C.S., and Francis, C.A. (2006) Occurrence of ammonia-oxidizing archaea in wastewater treatment plant bioreactors. *Appl Environ Microbiol* **72:** 5643–5647.

Pimm, S.L. (1984) The complexity and stability of ecosystems. *Nature* **307:** 321–326.

Pitcher, A., Villanueva, L., Hopmans, E.C., Schouten, S., Reichart, G.-J., and Sinninghe Damste, J.S. (2011) Niche segregation of ammonia-oxidizing archaea and anammox bacteria in the Arabian Sea oxygen minimum zone. *ISME J* **5:** 1896–1904.

R Development Core Team (2011) R: A Language and Environment for Statistical Computing. Vienna, Austria: the R Foundation for Statistical Computing. ISBN: 3-900051-07-0. Available online at http://www.R-project.org/.

Raup, D.M., and Crick, R.E. (1979) Measurement of faunal similarity in paleontology. *J Paleontol* **53:** 1213–1227.

Rosenzweig, M.L. (1971) Paradox of Enrichment: Destabilization of Exploitation Ecosystems in Ecological Time. *Science* **171:** 385–387.

Rowan, A.K., Snape, J.R., Fearnside, D., Barer, M.R., Curtis, T.P., and Head, I.M. (2003) Composition and diversity of ammonia-oxidising bacterial communities in wastewater treatment reactors of different design treating identical wastewater. *FEMS Microbiol Ecol* **43:** 195–206.

Rzhetsky, A., and Nei, M. (1992) A simple method for estimating and testing minimum-evolution trees. *Mol Biol Evol* **9:** 945–967.

Salkaly, P., and Oerther, D. (2011) Diversity of dominant bacterial taxa in activated sludge promotes functional resistance following toxic shock loading. *Microb Ecol* **61:** 557–567.

Saitou, N., and Nei, M. (1987) The neighbor-joining method: a new method for reconstructing phylogenetic trees. *Mol Biol Evol* **4:** 406–425.

Schleper, C., Jurgens, G., and Jonuscheit, M. (2005) Genomic studies of uncultivated archaea. *Nat Rev Micro* **3:** 479–488.

Schramm, A., Larsen, L.H., Revsbech, N.P., Ramsing, N.B., Amann, R., and Schleifer, K.H. (1996) Structure and function of a nitrifying biofilm as determined by in situ hybridization and the use of microelectrodes. *Appl Environ Microbiol* **62:** 4641–4647.

Schramm, A., de Beer, D., van den Heuvel, J.C., Ottengraf, S., and Amann, R. (1999) Microscale distribution of populations and activities of Nitrosospira and Nitrospira spp. along a macroscale gradient in a nitrifying bioreactor: quantification by in situ hybridization and the use of microsensors. *Appl Environ Microbiol* **65:** 3690–3696.

Schramm, A., De Beer, D., Gieseke, A., and Amann, R. (2000) Microenvironments and distribution of nitrifying bacteria in a membrane-bound biofilm. *Environ Microbiol* **2:** 680–686.

Smith, V.H. (1983) Low nitrogen to phosphorus ratios favor dominance by blue-green-algae in lake phytoplankton. *Science* **221:** 669–671.

Smith, V.H. (1993) Applicability of resource-ratio theory to microbial ecology. *Limnol Oceanogr* **38:** 239–249.

Smith, V.H., Graham, D.W., and Cleland, D.D. (1998) Application of resource-ratio theory to hydrocarbon biodegradation. *Environ Sci Technol* **32:** 3386–3395.

Stein, L.Y., Arp, D.J., Berube, P.M., Chain, P.S.G., Hauser, L., Jetten, M.S.M., *et al.* (2007) Whole-genome analysis of the ammonia-oxidizing bacterium, *Nitrosomonas eutropha* C91: implications for niche adaptation. *Environ Microbiol* **9:** 2993–3007.

Tamura, K., Nei, M., and Kumar, S. (2004) Prospects for inferring very large phylogenies by using the neighbor-joining method. *Proc Natl Acad Sci USA* **101:** 11030–11035.

Tamura, K., Peterson, D., Peterson, N., Stecher, G., Nei, M., and Kumar, S. (2011) MEGA5: molecular evolutionary genetics analysis using maximum likelihood, evolutionary distance, and maximum parsimony methods. *Mol Biol Evol* **28:** 2731–2739.

Terada, A., Sugawara, S., Yamamoto, T., Zhou, S., Koba, K., and Hosomi, M. (2013) Physiological characteristics of predominant ammonia-oxidizing bacteria enriched from bioreactors with different influent supply regimes. *Biochem Eng J* **79:** 153–161.

Thompson, J.D., Gibson, T.J., Plewniak, F., Jeanmougin, F., and Higgins, D.G. (1997) The CLUSTAL_X windows interface: flexible strategies for multiple sequence alignment aided by quality analysis tools. *Nucl Acids Res* **25:** 4876–4882.

Tilman, D. (1982) *Resource Competition and Community Structure. Monographs in Population Biology.* Princeton, USA: Princeton University Press.

Tilman, D., and Downing, J.A. (1994) Biodiversity and stability in grasslands. *Nature* **367:** 363–365.

Tilman, D., Reich, P.B., Knops, J., Wedin, D., Mielke, T., and Lehman, C. (2001) Diversity and productivity in a long-term grassland experiment. *Science* **294:** 843–845.

Wittebolle, L., Marzorati, M., Clement, L., Balloi, A., Daffonchio, D., Heylen, K., *et al.* (2009) Initial community evenness favours functionality under selective stress. *Nature* **458:** 623–626.

Yu, Y., Breitbart, M., McNairnie, P., and Rohwer, F. (2006) FastGroupII: a web-based bioinformatics platform for analyses of large 16S rDNA libraries. *BMC bioinformatics* **7:** 57.

Zhang, T., Jin, T., Yan, Q., Shao, M., Wells, G., Criddle, C., *et al.* (2009) Occurrence of ammonia-oxidizing Archaea in activated sludges of a laboratory scale reactor and two wastewater treatment plants. *J Appl Microbiol* **107:** 970–977.

Supporting information

Additional Supporting Information may be found in the online version of this article at the publisher's web-site:

Fig. S1. Competition among four species (A–D) for two resources (Resource 1 and Resource 2). The grey dash line is the line on which the Zero Net Growth Isocline (ZNGIs) are placed randomly. The ZNGIs and the consumption vectors are the continuous and dotted lines, respectively. The circles represent the microhabitats given by the 0.99 probability contour of the bivariate distribution. The number of species coexisting in each microhabitat is also specified near each circle. The grey stars are the cross points of two ZNGIs from which the consumption vectors are generated. The black dots define the diameter of the circle comprising all the potential species.

Fig. S2. COD removal (A), pH (B), concentrations of FA (C) and FNA (D) observed in the four bioreactors over time.

Fig. S3. Rarefaction curves based on the OTUs observed in the AOB clone libraries.

Table S1. Parameters used to simulate the AOB diversity.

4

Need for speed – finding productive mutations using transcription factor-based biosensors, fluorescence-activated cell sorting and recombineering

Michael Bott, Institute of Bio- and Geosciences, IBG-1: Biotechnology, Forschungszentrum Jülich, Jülich D-52428, Germany.

The establishment of a bioeconomy requires technologies enabling the rapid development of microbial production strains and enzymes as highly efficient biocatalysts for the conversion of substrates derived from renewable carbon sources into a multitude of chemicals. For strain development, two major strategies are available. The classical approach uses several rounds of random mutagenesis and screening for the desired producer clones and targets the entire genome irrespective of knowledge on function. The rational approach of metabolic and genetic engineering, on the other hand, is based on current knowledge and was strongly enhanced by the development of 'omics' technologies and novel tools for genetic engineering, leading to the establishment of systems biology and synthetic biology (Keasling, 2010). Despite the huge success of metabolic engineering, many of the industrial production strains are still based on the classical approach alone or a mixture of both approaches because these strains outperform the ones based exclusively on rational engineering. This is simply due to the fact that our knowledge of even the best studied microorganisms such as *Escherichia coli* or *Saccharomyces cerevisiae* is still far from complete, as obviously shown by the fact that the functions of hundreds of genes in these species are still unknown (because of the lack of a universal strategy to identify these functions). Furthermore, despite the enormous advances in our understanding of enzymes, it is still difficult to predict which mutations in an enzyme will be favourable for a certain purpose. Consequently, random mutagenesis and screening still offer great chances to identify novel mutations that enhance production. Because of the possibility of cheap re-sequencing of clones obtained from screening, the productive mutations can in principle be identified, thus providing the basis for a molecular understanding of the effect of the mutation.

The major bottleneck in the random mutagenesis and screening approach is screening, as the majority of molecules of interest do not have an easily observable phenotype. In the case of amino acid producers, for example, the library of mutagenized clones has to be screened by chromatography or a colorimetric assay. Therefore, the number of variants that can be screened in a reasonable time is much lower than the number of variants that are obtained by random mutagenesis. In the past years, novel tools have been developed that are capable of overcoming the screening bottleneck. The key are genetically encoded biosensors for intracellular molecules. Several types of such sensors have been described, such as those based on Förster resonance energy transfer (Frommer et al., 2009) or on RNA switches (Michener and Smolke, 2012). For *in vivo* screening, however, biosensors based on transcriptional regulators (TRs) appear to be the most promising ones, particulary those systems that can be coupled to fluorescence-activated cell sorting (FACS) as an ultrahigh-throughput method allowing screening of more than 50 000 cells per second. A prominent example of such a TR-based biosensor that can be combined with FACS is LysG of *Corynebacterium glutamicum*, a LysR-type TR that activates expression of its target gene *lysE* (encoding a lysine exporter) in response to elevated cytoplasmic levels of L-lysine and a few other amino acids (Bellmann et al., 2001). LysG requires binding of L-lysine or L-arginine as inducer in order to activate expression of *lysE* and thus functions as an L-lysine or L-arginine sensor with a K_d in the low millimolar range. Using a plasmid with a reporter gene coding for an autofluorescent protein like eYFP under the control of the P_{lysE} promoter, the intracellular lysine concentration can be converted into a fluorescence signal and in a certain concentration range, the cell-specific fluorescence is proportional to the intracellular L-lysine concentration. Importantly, the increased intracellular concentrations correlate with increased extracellular concentrations (Binder et al., 2012).

A biosensor like the one based on LysG can be used to screen by FACS libraries generated by random genome mutagenesis, gene-directed mutagenesis or site-directed mutagenesis. For example, using the L-lysine sensor it was possible to screen with FACS in 30 min a *C. glutamicum* library of seven million individual cells obtained by random chemical mutagenesis, a task that would have required weeks or months to do it in a convential way. More than 100 L-lysine-producing cells were isolated from the library, and their analysis led to the identification of novel productive mutations in previously known target genes, but also in novel target genes, such as *murE* (Binder *et al.*, 2012). Besides screening of productive mutations in the genome, a targeted approach yielded productive mutations in the key enzymes of L-arginine, L-lysine and L-histidine biosynthesis using the LysG sensor/FACS combination (Schendzielorz *et al.*, 2014). For application in site-directed mutagenesis, sensor-based FACS screening can be combined with recombineering, in which site-directed mutations are introduced using single-stranded oligonucleotides and a recombinase. In this way, transformation of a *C. glutamicum* strain carrying a recombinase and the LysG-based sensor with a mixture of 19 oligonucleotides covering all possible amino acid exchanges at *murE*-G81 led to the FACS-based isolation of mutants carrying 12 different productive chromosomal codon exchanges at *murE*-G81 in a single experiment, associated with a spectrum of different lysine titres (Binder *et al.*, 2013).

The examples described above demonstrate the power of TR-sensor-based FACS screening for strain and enzyme development, and the number of examples using this combination is increasing rapidly (Mustafi *et al.*, 2012; Jha *et al.*, 2014; Siedler *et al.*, 2014a) and includes also a biosensor for detecting the NADPH/NADP⁺ ratio in *E. coli* cells (Siedler *et al.*, 2014b). This type of sensor is suitable for evolving NADPH-dependent dehydrogenases by FACS screening of mutant libraries independent of a specific enzyme assay. Major prerequisites for the use of TR-based FACS screenings are a host microorganism whose morphology is suitable for FACS analysis and the existence of TR-based biosensor for the target molecule. Although thousands of TRs have been identified in the course of genome annotation, for most of them neither their ligand specificity nor their target genes are known. Thus, the question arises how to build a sensor for a molecule for which no TR is known at present. At least two approaches have been reported that address this problem. One is based on altering the ligand specificity of a TR to the molecule of interest. A prominent example is AraC from *E. coli*, whose ligand specificity was changed from L-arabinose to e.g. mevalonate or triacetic acid lactone (Tang and Cirino, 2011; Tang *et al.*, 2013). Another approach makes use of metagenomic libraries and is termed SIGEX for substrate-induced gene expression (Uchiyama *et al.*, 2005; Uchiyama and Watanabe, 2008). It was developed for the isolation of novel catabolic operons and is based on the fact that such operons are usually induced by the substrate via a TR whose gene is located in close proximity to the catabolic genes. Metagenomic fragments are cloned into an operon trap

gfp-expression vector, and positive clones becoming fluorescent in the presence of the target molecule are selected by FACS. In this way, TRs for molecules of interest may be identified.

The production of a wide spectrum of chemicals from renewable carbon sources using microbial strains or isolated enzymes requires a shortening of the times required for their development and a maximization of their efficiency. As a complementary approach to metabolic engineering, the use of TR-based biosensors in combination with FACS and next-generation sequencing technologies is a promising route to rapidly identify the key productive mutations, which can then be combined in a suitable host or target protein to obtain superior producer strains or enzymes. Besides their contribution to strain and enzyme development, the recognition of these productive mutations will also increase our understanding of the physiology, metabolism and regulation of cells, and they offer the possibility to discover unexpected linkages in the metabolic network.

Acknowledgements

Work performed in the author's institute was supported by the German Federal Ministry of Education and Research (FlexFit project, support code 0315589A) and the Ministry of Innovation, Science, Research and Technology of North Rhine-Westphalia (BioNRW, Technology Platform Biocatalysis, RedoxCell, support code W0805wb001b).

References

Bellmann, A., Vrljic, M., Patek, M., Sahm, H., Krämer, R., and Eggeling, L. (2001) Expression control and specificity of the basic amino acid exporter LysE of *Corynebacterium glutamicum*. *Microbiology* **147**: 1765–1774.

Binder, S., Schendzielorz, G., Stäbler, N., Krumbach, K., Hoffmann, K., Bott, M., and Eggeling, L. (2012) A high-throughput approach to identify genomic variants of bacterial metabolite producers at the single-cell level. *Genome Biol* **13**: R40.

Binder, S., Siedler, S., Marienhagen, J., Bott, M., and Eggeling, L. (2013) Recombineering in *Corynebacterium glutamicum* combined with optical nanosensors: a general strategy for fast producer strain generation. *Nucleic Acids Res* **41**: 6360–6369.

Frommer, W.B., Davidson, M.W., and Campbell, R.E. (2009) Genetically encoded biosensors based on engineered fluorescent proteins. *Chem Soc Rev* **38**: 2833–2841.

Jha, R.K., Kern, T.L., Fox, D.T., and Strauss, C.E.M. (2014) Engineering an *Acinetobacter* regulon for biosensing and high-throughput enzyme screening in *E. coli* via flow cytometry. *Nucleic Acids Res* **42**: 8150–8160.

Keasling, J.D. (2010) Manufacturing molecules through metabolic engineering. *Science* **330**: 1355–1358.

Michener, J.K., and Smolke, C.D. (2012) High-throughput enzyme evolution in *Saccharomyces cerevisiae* using a synthetic RNA switch. *Metab Eng* **14**: 306–316.

Mustafi, N., Grünberger, A., Kohlheyer, D., Bott, M., and Frunzke, J. (2012) The development and application of a single-cell biosensor for the detection of L-methionine and branched-chain amino acids. *Metab Eng* **14:** 449–457.

Schendzielorz, G., Dippong, M., Grünberger, A., Kohlheyer, D., Yoshida, A., Binder, S., *et al.* (2014) Taking control over control: use of product sensing in single cells to remove flux control at key enzymes in biosynthesis pathways. *ACS Synth Biol* **3:** 21–29.

Siedler, S., Stahlhut, S.G., Malla, S., Maury, J., and Neves, A.R. (2014a) Novel biosensors based on flavonoid-responsive transcriptional regulators introduced into *Escherichia coli. Metab Eng* **21:** 2–8.

Siedler, S., Schendzielorz, G., Binder, S., Eggeling, L., Bringer, S., and Bott, M. (2014b) SoxR as a single-cell biosensor for NADPH-consuming enzymes in *Escherichia coli. ACS Synth Biol* **3:** 41–47.

Tang, S.Y., and Cirino, P.C. (2011) Design and application of a mevalonate-responsive regulatory protein. *Angew Chem Int Ed* **50:** 1084–1086.

Tang, S.Y., Qian, S., Akinterinwa, O., Frei, C.S., Gredell, J.A., and Cirino, P.C. (2013) Screening for enhanced triacetic acid lactone production by recombinant *Escherichia coli* expressing a designed triacetic acid lactone reporter. *J Am Chem Soc* **135:** 10099–10103.

Uchiyama, T., and Watanabe, K. (2008) Substrate-induced gene expression (SIGEX) screening of metagenome libraries. *Nat Protoc* **3:** 1202–1212.

Uchiyama, T., Abe, T., Ikemura, T., and Watanabe, K. (2005) Substrate-induced gene-expression screening of environmental metagenome libraries for isolation of catabolic genes. *Nat Biotechnol* **23:** 88–93.

5

A refined technique for extraction of extracellular matrices from bacterial biofilms and its applicability

Akio Chiba,[1†] Shinya Sugimoto,[1†*] Fumiya Sato,[2]
Seiji Hori[2] and Yoshimitsu Mizunoe[1]
Departments of [1]Bacteriology and
[2]Infectious Disease and Control, The Jikei University
School of Medicine, 3–25-8, Nishi-Shimbashi,
Minato-ku, Tokyo 105-8461, Japan.

Summary

Biofilm-forming bacteria embedded in polymeric extracellular matrices (ECMs) that consist of polysaccharides, proteins and/or extracellular DNAs (eDNAs) acquire high resistance to antimicrobial agents and host immune systems. To understand molecular mechanisms of biofilm formation and maintenance and to develop therapeutic counter-measures against chronic biofilm-associated infections, reliable methods to isolate ECMs are inevitable. In this study, we refined the ECM extraction method recently reported and evaluated its applicability. Using three *Staphylococcus aureus* biofilms in which proteins, polysaccharides or eDNAs are major contributors to their integrity, ECMs were extracted using salts and detergents. We found that extraction with 1.5 M sodium chloride (NaCl) could be optimum for not only ECM proteins but also polysaccharides and eDNAs. In addition, long-time incubation was not necessary for efficient ECM isolation. Lithium chloride (LiCl) was comparative to NaCl but is more expensive. In contrast to SDS, NaCl hardly caused leakage of intracellular proteins and did not affect viability of bacterial cells within biofilms. Furthermore, this method is applicable to other bacteria such as Gram-positive *Staphylococcus epidermidis* and Gram-negative *Escherichia coli* and *Pseudomonas aeruginosa*. Thus, this refined method is very simple,

*For correspondence. E-mail ssugimoto@jikei.ac.jp

Funding Information This work was supported by a Grant-in-Aid for Young Scientists (B) to S.S. from JSPS, by a Grant-in-Aid for Scientific Research (B) to Y.M. from JSPS and by a grant to Y.M. from MEXT-Supported Program for the Strategic Research Foundation at Private Universities, 2012–2016.

rapid, low cost and non-invasive and could be used for a broad range of applications.

Introduction

Biofilms are intricate communities of microorganisms embedded in a self-produced matrix of extracellular polymer substances (EPS) and are adherent to an abiotic or biotic surface (Costerton *et al.*, 1999). The extracellular matrix (ECM) consists of proteins (O'Neill *et al.*, 2008), polysaccharides (O'Gara, 2007) and/or extracellular DNAs (eDNAs) (Mann *et al.*, 2009). Extracellular matrix has diverse functions to maintain the structural integrity of the biofilm and to adapt to surrounding environments (Flemming and Wingender, 2010). A deleterious property of ECM for human beings is to confer resistance to anti-microbial agents (Davies, 2003) and host immune systems (Archer *et al.*, 2011), which often becomes problematic in the clinical settings. Therefore, once biofilms are established on infected tissues or medical devices such as catheters and orthopaedic implants, it becomes difficult to eradicate them by chemotherapeutic strategies and biofilm-associated infections (e.g. catheter-related blood stream infections, prosthetic joint infections and artificial valve infections) become intractable and chronic (Del Pozo and Patel, 2009). In these cases, surgical intervention is often required to remove infected tissues or medical devices (Thwaites *et al.*, 2011).

A large variety of bacteria exhibit the capacity to form biofilms in varying degrees (Dalton and March, 1998). The biofilm phenotypes differ between strains of a single species; for example, various strains of *Staphylococcus aureus* which is a major cause of biofilm-associated infections produce several types of biofilm with different ECM components. These biofilm phenotypes are simply classified according to their susceptibilities to ECM-degrading enzymes, because the formation and maintenance of biofilms definitively depend on the production and quality of ECMs. Conventionally, proteases (e.g. proteinase K), glycolytic enzymes (e.g. dispersin B) (Kaplan *et al.*, 2004) and deoxyribonucleases (e.g. DNase I) are used for identifying proteinaceous, polysaccharide and DNA biofilms respectively (Sato *et al.*, unpublished). If multiple components in an ECM are crucial for biofilm integrity, a combination of these enzymes, rather than single ones, is more effective in inhibiting the formation of biofilm and promoting its dispersal. In order to understand basic principles of the biofilm lifestyle at a molecular level, identification of ECM components is a primary important task. Although much effort to isolate ECM components have been reported (Liang *et al.*, 1992; Tabouret *et al.*, 1992; Tapia *et al.*, 2009; Wu and Xi, 2009), efficient ECM isolation is still challenging in terms of cost performance, simplicity and/or applicability to various types of components and bacteria.

Recently, we reported a simple method for extracting ECMs from *S. aureus* biofilms using a high concentration of sodium chloride (NaCl) at 1 M (Sugimoto *et al.*, 2013). A principle of this method may be similar to that of ion-exchange chromatography. Usually, bacterial surface are considered to be negatively charged due to the electrical state of outer components such as outer membranes (phospholipids and lipopolysaccharides) in Gram-negative bacteria and cell walls (peptidoglycans and teichoic acids) in Gram-positive ones. However, some positively charged molecules (membrane embedded proteins and cell wall-anchored proteins) are also displayed on bacterial surfaces. Therefore, there are negatively or positively charged loci on the surfaces. Extracellular matrix components (proteins, polysaccharides and eDNAs) are also thought to be positively or negatively charged depending upon individual characteristics under biofilm conditions used. Although Van der Waals force and hydrophobic interactions may not be ignored, ionic interactions seem to be an important driving force for the adherence of ECM constituents to bacterial surfaces (Dunne and Burd, 1992; Frølund *et al.*, 1996; Jucker *et al.*, 1996; Dunne, 2002). Higher ionic strength of Na^+ and Cl^- ions might thus trigger the release of ECM components from bacterial cells (Fig. S1).

In this study, we refined our NaCl-based ECM extraction method using *S. aureus* biofilm models and compared with other methods previously reported elsewhere. We found that, using 1.5 M NaCl solution, ECM can be extracted from *S. aureus* cells more easily, inexpensively, rapidly and/or non-invasively compared with the other method using other chemicals such as sodium dodecyl sulfate (SDS) and lithium chloride (LiCl). Furthermore, our method was applicable to not only other Gram-positive *Staphylococcus epidermidis* but also Gram-negative *Escherichia coli* and *Pseudomonas aeruginosa*.

Results and discussion

Optimization of ECM extraction conditions

To optimize reagents for ECM extraction, we selected a biofilm model of clinically isolated methicillin-resistant *S. aureus* strain, MR23, since this strain produces a robust biofilm in which major components of the ECM are proteins including extracellular adherence protein (Eap) (Sugimoto *et al.*, 2013). Of note, we used conical tube

biofilms to ensure reproducibility and simplicity as shown in Fig. S1. In this method, planktonic and biofilm cells are incorporated into the pellet fraction at the first centrifugation step. If only biofilm cells are desired, planktonic cells can be removed by discarding the culture medium and gentle washing with certain buffers before the centrifugation. However, there is no significant difference in the profiles between the isolated ECM of biofilm cells and that of planktonic cells under the tested conditions (Fig. S2). We first extracted ECMs using various concentrations of NaCl and analysed them by SDS-PAGE (Fig. 1A). Several remarkable bands were detectable when ECMs were extracted with NaCl concentrations of 0.5 M and above. As previously reported (Sugimoto *et al.*, 2013), the prominent band with molecular mass of 70 kDa corresponds to Eap that is a specific ECM protein of *S. aureus* (Hussain *et al.*, 2008) and forms ECM architectures (Sugimoto *et al.*, 2013). Therefore, we used Eap as an ECM marker of the MR23 biofilm. The band intensity of Eap increased in a dose-dependent manner, ranging from 0.5 to 1.5 M NaCl but decreased at 2 M and above. The result of protein quantification by Bradford method also showed a maximum peak at around 1–1.5 M NaCl (Fig. 1B). These results strongly suggest that 1.5 M is a proper concentration for ECM extraction. Henceforth, 1.5 M NaCl was used for further analyses. Next, we examined incubation time for the ECM isolation. There was no significant difference between the indicated incubation times at 25°C (Fig. 1C), revealing that the addition of 1.5 M NaCl could immediately lead to a detachment of ECMs from bacterial cells. Indirect immunofluorescence microscopy using anti-Eap polyclonal primary and Cy3-labeled secondary antibodies also demonstrated that Eap localized surface and extracellular milieu, and the signals disappeared almost completely after the 1.5 M NaCl treatment (Fig. 1D). Dispersed signals could be observed in the isolated ECM fraction probably due to the presence of solubilized Eap proteins.

We next compared ECM extraction efficiency of NaCl with those of other salts [potassium chloride (KCl) and LiCl] and detergents (SDS, NP-40, Triton X-100 and Tween 20). The reasons why we selected these reagents are: (i) KCl is approximately equivalent to NaCl in terms of cost and conventional utility, (ii) LiCl was used for ECM extraction as reported previously (Liang *et al.*, 1992) (iii) and some detergents, especially SDS, were generally used for protein solubilization (Tabouret *et al.*, 1992). Potassium chloride was inconvenient, since the addition of high concentration of KCl into SDS sample buffer or running buffer led to precipitation of the ECM samples, making the analysis difficult (data not shown). In contrast, LiCl was found as efficient as NaCl in extracting ECMs as judged by SDS-PAGE (Fig. 2A), but LiCl is more expensive than NaCl. The detergents used in this study, except

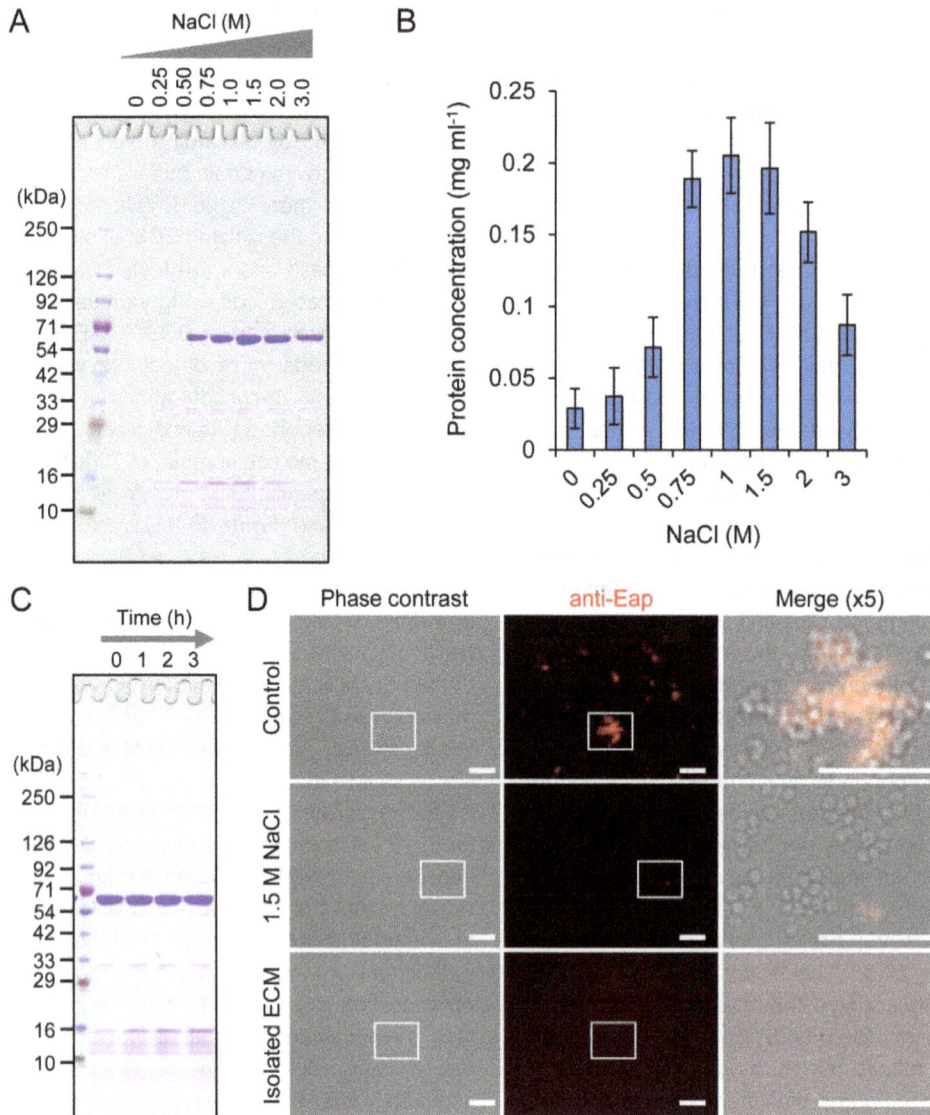

Fig. 1. Sodium chloride is available for isolating ECM proteins from biofilms.
A. Extracellular matrices of *S. aureus* MR23 were extracted with various concentrations of NaCl and were applied to SDS-PAGE with CBB staining.
B. Protein concentrations in the extracted ECMs were quantified by Bradford method. The means and standard deviations of triplicate determinations are represented.
C. Extracellular matrices were extracted after the indicated incubation periods in the presence of 1.5 M NaCl and were then dissolved by SDS-PAGE.
D. MR23 biofilm cells treated with or without 1.5 M NaCl, and the extracted ECM were probed with anti-Eap primary and Cy3-labeled secondary antibodies and were observed with a fluorescence and phase-contrast microscope. Higher magnification (fivefold) images of the white rectangles in the phase contrast and fluorescence images are merged. Bars indicate 10 μm. The positions of molecular mass markers in kilodaltons (kDa) are shown at the left of each panel (A and C).

for SDS, were not meaningful for ECM extraction at all the concentrations tested (Fig. S3). On the other hand, the addition of SDS resulted in the recovery of a large variety of proteins including Eap (Fig. 2A). To address the issue that cytoplasmic and membrane proteins may leak to the extracellular milieu due to altered membrane permeability and integrity, we evaluated the effects of these chemicals on the viability of biofilm cells by live/dead staining and colony-forming units (CFU) counting. Notably, a large number of cells were alive but only a small population of cells were dead even in the case of non-treated cells, indicating that a minority of dead cells coexist with live cells in the biofilm as judged by live/dead staining (Fig. 2B). Although NaCl treatment did not affect the cell viability, SDS-treated cells were almost dead. Colony-forming unit counting support these result.

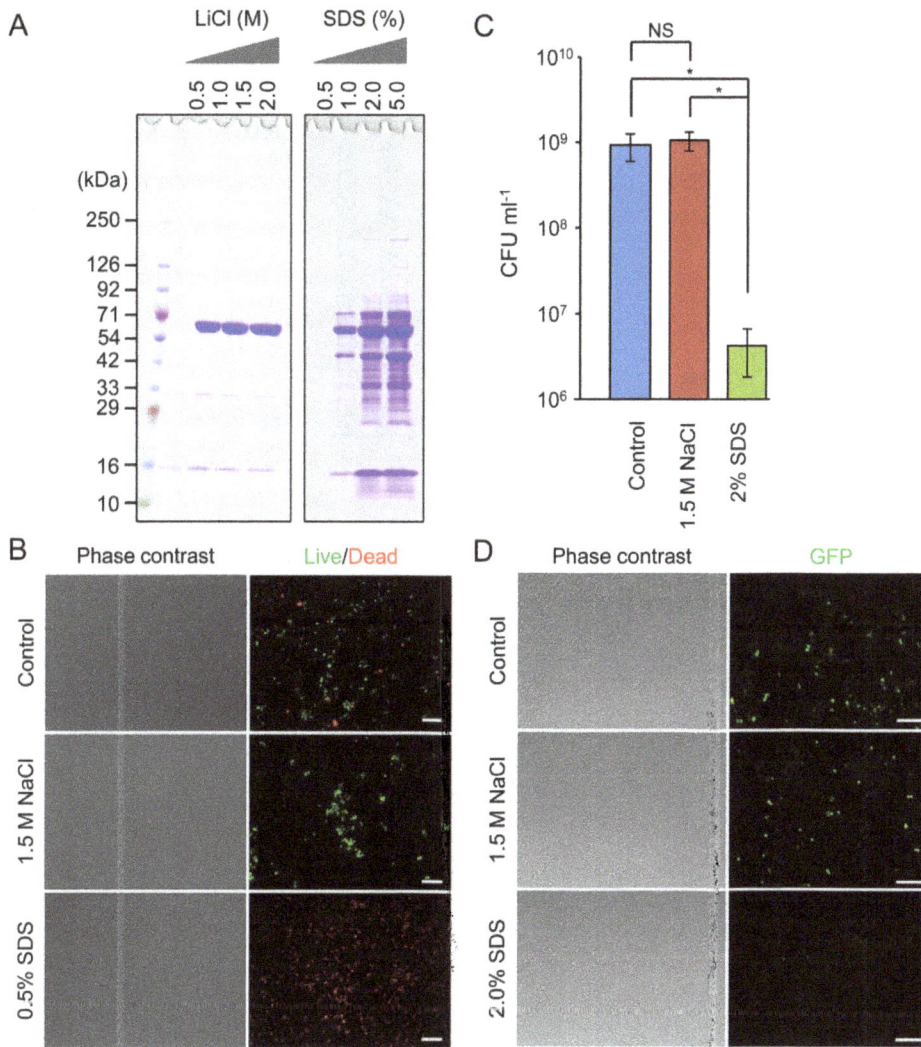

Fig. 2. Assessment of cytotoxicity of NaCl and SDS.
A. Extracellular matrices of *S. aureus* MR23 were extracted by the addition of the indicated concentrations of LiCl or SDS and were applied to SDS-PAGE with CBB staining. A molecular mass marker was also loaded to the left lane.
B. Live/Dead staining images of MR23 biofilm cells after the treatment with 1.5 M NaCl or 0.5% (w/v) SDS are shown. A solution of 0.9% NaCl was used as a control. Phase contrast and fluorescence images are shown. Live and dead cells are stained in green and red respectively.
C. Colony-forming unit of biofilm cells before and after the treatment with 1.5 M NaCl or 2% (w/v) SDS were measured. The means and standard deviations of triplicate determinations are represented.
D. *Staphylococcus aureus* MR23 pP1GFP cells were treated with 1.5 M NaCl, 2% (w/v) SDS or PBS (as control) was observed with a fluorescence and phase-contrast microscope. Bars indicate 10 μm (B and D).

Colony-forming unit of these SDS-treated cells was significantly reduced compared with those of control and NaCl-treated ones (*, $P < 0.01$) (Fig. 2C). In addition, fluorescence of cytoplasmic green fluorescent protein (GFP) in recombinant MR23 pP1GFP cells (Sugimoto *et al.*, 2013) disappeared only when the cells were treated with SDS (Fig. 2D), suggesting that cytoplasmic GFP leaked out from the cells. Taken altogether, judging from efficiency, cost performance and non-invasiveness, NaCl is the best among the chemicals tested in this study.

The isolated ECMs will be used for further biochemical analyses to identify individual components and to

examine their functional roles in biofilm development. If high concentrations of NaCl interfere with further analyses, they can be easily removed from the ECM fraction by the additional procedure such as dialysis and gel filtration chromatography. In our recent study, removal of NaCl was not necessary for identification of ECM proteins by a top-down proteome approach (Sugimoto *et al.*, 2013). Although it is often difficult to purify a demanded ECM constituent apart from other components, by combining our ECM isolation method with a conventional chromatographic technique such as nickel-affinity column (Sugimoto *et al.*, 2013) or fibrinogen-affinity column

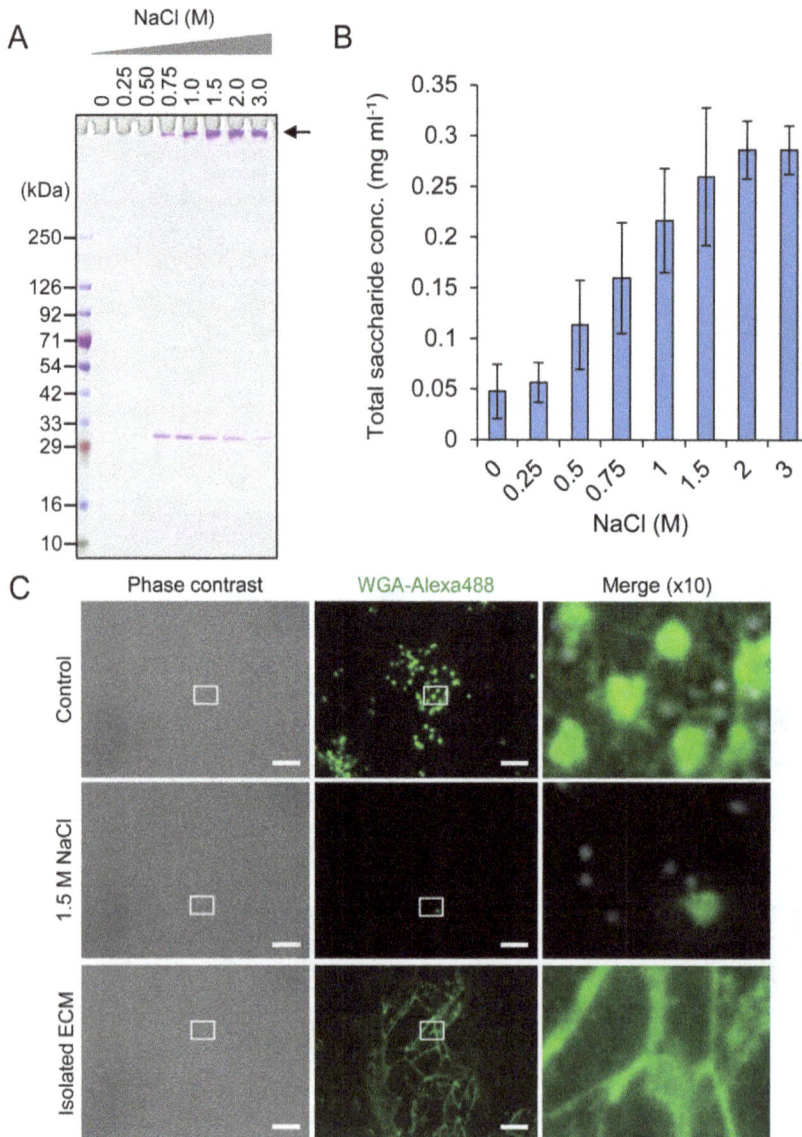

Fig. 3. Sodium chloride is valuable in extracting ECM polysaccharides from biofilms.
A. Extracellular matrices of *S. aureus* MR10 were extracted with various concentrations of NaCl and were dissolved by SDS-PAGE with CBB staining. The bands stacked at the top of the gel (shown as an arrow) are hallmark of polysaccharides as recently reported (Sugimoto *et al.*, 2013). The positions of molecular mass markers in kilodaltons (kDa) are represented at the left.
B. Total saccharide concentrations in the extracted ECM were quantified by phenol sulfuric acid method. The means and standard deviations of triplicate determinations are represented.
C. MR10 biofilm cells with and without 1.5 M NaCl treatment and the extracted ECMs were stained with green fluorescence probe-labeled lectin (WGA-Alex488) and were observed with a fluorescence and phase-contrast microscope. Higher magnification (10-fold) images of the white rectangles in the phase contrast and fluorescence images are merged. Bars are 10 μm.

chromatography (Palma *et al.*, 1999) (Fig. S4), we succeeded in purifying recombinant Eap protein and native one, supporting the merit of our method, too.

Applicability of NaCl for extraction of ECMs containing polysaccharides and eDNAs

Here, we investigated the applicability of our newly refined method for extraction of polysaccharide components in ECMs. Polysaccharide intercellular adhesin (PIA), also known as poly N-acetylglucosamine (PNAG), is a major ECM constituent in certain *S. aureus* strains (Kaplan *et al.*, 2004). The *S. aureus* MR10, clinically isolated in the Jikei hospital, produces a biofilm sensitive to dispersin B, but neither proteinase K nor DNase I (Sato *et al.*, unpublished), suggesting that this strain produces a

polysaccharide biofilm. We therefore used this strain in the present study and extracted ECMs from MR10 cells using various concentrations of NaCl. The profile of polysaccharides and proteins were analysed by SDS-PAGE (Fig. 3A). The band that is stacked at the top of the polyacrylamide gel and degraded by dispersin B is a feature of insoluble polysaccharides as recently reported (Sugimoto *et al.*, 2013). The present study also showed that the stacked bands of the ECMs isolated from MR10 biofilms were degraded by dispersin B (Fig. S5). The polysaccharides bands were detected in the ECM samples extracted with NaCl concentrations of 0.75 M and above, and the intensity was not apparently changed at higher concentrations 1.5 M and above. The profile of total sugar concentration in the ECM determined by phenol sulfuric acid method (Fig. 3B) was parallel with the result of SDS-PAGE (Fig. 3A). Furthermore, the other bands with molecular mass of approximately 31 kDa were also detected in these ECM (Fig. 3A), and these bands were degraded by proteinase K but not dispersin B and DNase I (Fig. S5), denoting that the ECM of MR10 is composed of PIAs and protein(s). Judging from the result that the band intensity of the 31 kDa protein(s) was decreased at the NaCl concentrations of more than 1.5 M (Fig. 3A), 1.5 M NaCl is enough to extract polysaccharides and better to isolate polysaccharides and proteins simultaneously. Since diverse sugar monomers and linkages exist in bacterial extracellular polysaccharides (Flemming et al., 2007), carbohydrate chemical analyses are still challenging. Characterization and biochemical analysis of the polysaccharides in the ECM samples are of great interest and will be our future direction to understand molecular mechanisms of biofilm development.

To confirm whether polysaccharides were detached by NaCl treatment, they were stained with Alexa488-labeled wheat germ agglutinin (WGA-Alex488) (Fig. 3C). The fluorescent signals of polysaccharides were detected on surfaces and interspaces of the cells. Interestingly, networks of filamentous structures were also observed. After ECM extraction with 1.5 M NaCl, these signals disappeared almost completely. In addition, filamentous networks in the extracted ECM fraction bound to WGA-Alexa488. These results represent that 1.5 M NaCl is applicable to dispersal of ECM polysaccharides from S. aureus cells.

To clarify whether NaCl can be used for extraction of ECMs composed of not only proteins and polysaccharides but also eDNAs, we selected a biofilm model of S. aureus MS10 strain which produces a DNase I-sensitive biofilm (Sato et al., unpublished). Extracellular matrices of MS10 biofilms were also extracted by the addition of various concentrations of NaCl and were analysed by agarose gel electrophoresis. In the 1.5% agarose gel, prominent bands with a huge molecular mass were detected in the ECM samples extracted with NaCl at 0.5 M and above (Fig. 4A). In addition, eDNAs were also detected in the ECM factions of MR23 and MR10 (Fig. S6). According to the migration of DNAs in the agarose gel, they seem to be whole genomic DNAs released from the cells probably due to autolysis. Since the isolated ECMs of MS10 contained a small proportion of proteins as in the case of those from MR10 (Fig. S5), we purified DNA in the extracted ECMs using a DNA purification kit and determined the concentration of DNA by measuring the absorbance at 260 nm. The yield of DNA increased with the dose of NaCl and reached a plateau at 1.5 M or 2 M (Fig. 4B). This profile was apparently equivalent to the data obtained by agarose gel electrophoresis (Fig. 4A). Furthermore, we stained eDNAs with propidium iodide (PI) and observed them under fluorescence microscopy. In addition to intracellular DNAs within dead cells, PI-positive signals of eDNAs were observed in the control sample, but the signals of eDNAs became invisible almost completely after the extraction of ECM with 1.5 M NaCl. In addition, clusters of DNA were observed in the extracted ECM fraction (Fig. 4C).

These results together indicate that 1.5 M NaCl could be used for isolation of diverse types of ECMs from S. aureus biofilms.

Applicability of NaCl for extraction of ECMs from other bacteria

Finally, we tested whether the ECM-isolation method sophisticated in this study is applicable to other Gram-positive and Gram-negative bacteria. Gram-positive bacterium S. epidermidis is also a major cause of biofilm diseases in clinical settings, and one of the major components of its ECM is PIA whose biosynthesis is regulated by ica gene products (O'Gara, 2007). Using the same procedure with S. aureus, ECM was isolated from S. epidermidis SE04, a clinically isolated strain that forms a PIA-positive biofilm (Sato et al., unpublished). As in the case of MR10, the bands stacked at the top of the polyacrylamide gel were detected, and they were completely degraded by dispersin B (Fig. 5A). In addition, proteinaceous components were also detected in the gel, indicating the presence of certain proteins that may contribute to the formation and maintenance of S. epidermidis biofilms.

Previously, isolation of TasA amyloid fibres from Bacillus subtilis biofilms by using 1 M NaCl has been reported (Romero et al., 2010), suggesting that high concentrations of NaCl is useful for extracting not only natively folded or denatured proteins but also ordered protein aggregates. Here, we use an E. coli biofilm model in which curli amyloid fibres are major components of the ECM and prominently contribute to the structural integrity of the biofilm (Maeyama et al., 2004). The curli-dependent biofilm that formed when a wild-type YMel strain was grown in YESCA plates at 25°C for 3 days was collected and used for ECM extraction. Here, we selected CsgA, a major structural component of curli, as a maker of the ECM in E. coli. The ECM extracted with 1.5 M NaCl was treated with formic acid to depolymerize curli amyloid fibres into CsgA monomers (Chapman et al., 2002) and Western blotting using an anti-CsgA antiserum (Fig. 5B). In parallel, curli was purified by conventional homogenization and centrifugation procedures (Chapman et al., 2002) and was used as a positive control. The ECM extracted from the isogenic csgA mutant of YMel (YMel-1) was also used as a negative control (Kikuchi et al., 2005). We detected specific bands corresponding to CsgA monomers (17 kDa band) in the ECM fraction of YMel (Fig. 5B).

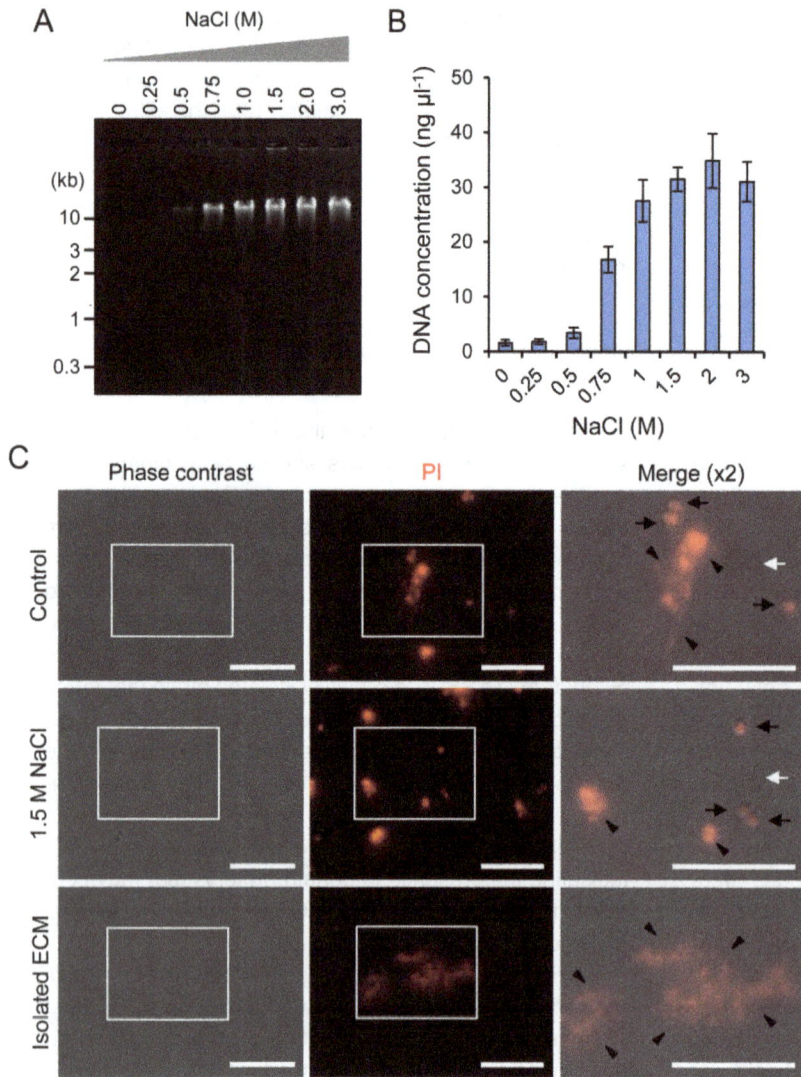

Fig. 4. Extracellular DNAs are detached from biofilms by the addition of NaCl and harvested quantitatively.
A. Extracellular matrices from *S. aureus* MS10 biofilms were extracted with the indicated concentrations of NaCl and were subjected to agarose gel electrophoresis. The gel was stained with ethidium bromide. The positions of molecular mass markers in kilobase pairs (kb) are shown at the left.
B. Extracellular DNAs in the extracted ECM were purified to remove contaminated proteins and quantified by measuring absorbance at 260 nm. The means and standard deviations of triplicate determinations are represented.
C. MS10 biofilm cells treated with or without 1.5 M NaCl and the extracted ECM were stained with PI and were observed with a fluorescence and phase-contrast microscope. Higher magnification (twofold) images of the white rectangles in the phase contrast and fluorescence images are merged. Black arrows, white arrows and black arrowheads represent bacterial cells stained with PI (dead cells), non-stained cells (variable cells) and eDNAs respectively. Bars indicate 10 μm.

In *P. aeruginosa,* eDNA is an integral structural component in biofilms (Whitchurch *et al.*, 2002). We tested whether eDNA could be extracted from colony biofilms of *P. aeruginosa* PAO1 strain. As shown in Fig. 5C, eDNA in the isolated ECM was detected by agarose gel electrophoresis, and the signal disappeared when treated with DNase I.

It is well known that *staphylococci* have salinity tolerance. We wondered whether 1.5 M NaCl may cause cell death of *E. coli* and *P. aeruginosa* cells. Therefore, we assessed cell damages caused by 1.5 M NaCl in these Gram-negative bacteria. As shown in Fig. S7, CFUs of these bacteria were not significantly reduced after the treatment with 1.5 M NaCl compared with those of PBS, suggesting that the ECM extraction method does not cause severe damages leading to cell death. Taken altogether, these findings support the notion that high concentrations of NaCl are applicable to isolation of ECMs from various microorganisms including not only Gram-positive bacteria but also Gram-negative ones.

Conclusion

We refined ECM extraction method for various bacterial biofilms. The merits of this method are its simplicity and applicability. This method may be useful for typing of biofilms produced by diverse bacteria. Typing of biofilms at the strain level is also an important issue, since biofilm phenotypes differ between strains. Typing of biofilms has conventionally been conducted by testing their susceptibility to ECM-degrading enzymes. Our results imply that biofilms are not composed of only single component, even though they are susceptible to an ECM-degrading enzyme. Characterization of the individual ECM constituents rather than susceptibility to ECM-degrading enzymes may be more important for understanding how biofilms

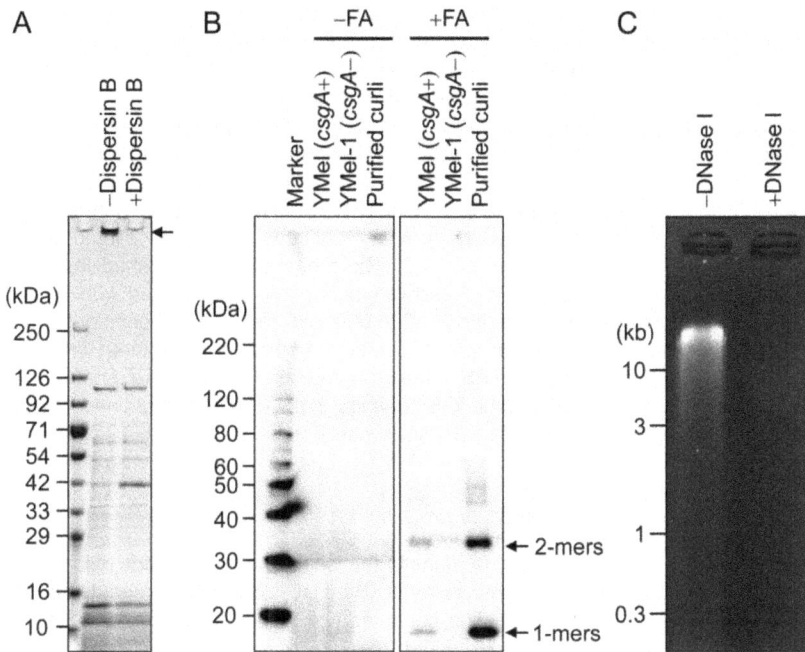

Fig. 5. Applicability of the ECM extraction method to *S. epidermidis*, *E. coli* and *P. aeruginosa* biofilms.
A. Extracellular matrices extracted from *S. epidermidis* SE04 by the addition of 1.5 M NaCl were treated with or without dispersin B and were then applied to SDS-APGE. The gel was stained with CBB. An arrow indicates polysaccharides.
B. Extracellular matrices of *E. coli* YMel and its isogenic *csgA* mutant YMel-1 were isolated with 1.5 M NaCl. Curli amyloid fibres in the ECM fraction were treated with or without formic acid. Purified curli was also used as a positive control. The proteins were analysed by Western blotting using anti-CsgA antibody. Arrows indicate monomeric and dimeric CsgA. FA, formic acid.
C. Extracellular matrices of *P. aeruginosa* PAO1 were extracted with 1.5 M NaCl and were subjected to agarose gel electrophoresis. The extracted ECMs were treated with or without DNase I. The gel was stained with ethidium bromide. The positions of molecular mass markers in kilodaltons (kDa) (A and B) and kilobase pairs (kb) (C) are shown at the left of each panel respectively.

are developed and maintained and for taxonomical and epidemiological investigations.

Experimental procedures

Bacterial strains and culture media

Bacterial strains used in this study were listed in Table S1 in the supplemental material. *Staphylococcus aureus* and *S. epidermidis* were grown at 37°C in brain heart infusion (BHI) medium (Becton Dickinson, Franklin Lakes, NJ, USA) or BHI medium supplemented with 1% (w/v) glucose (BHIG). *Escherichia coli* strains were grown in Luria-Bertani (LB) medium (Merk, Darmstadt, Germany) or YESCA plate composed of 1% (w/v) casamino acid (Becton Dickinson), 0.1% (w/v) yeast extract (Becton Dickinson) and 2% (w/v) agar plate. *Pseudomonas aeruginosa* was cultured in LB medium and LB plate containing 2% agar.

To reduce non-specific immunoglobulin G (IgG) binding in immunofluorescence microscopy, the *srtA* gene, which encodes sortase A (Mazmanian *et al.*, 2001), was deleted by in-frame deletion using pKOR1 (Bae and Schneewind, 2006). Because sortase A cleaves the LPXTG motif of cell wall-anchoring proteins including protein A, which is one of the major IgG binding proteins in *S. aureus* (Forsgren and Sjöquist, 1966). Briefly, approximately 500 bp upstream and downstream sequences of the *srtA* gene were amplified by PCR from MR23 genomic DNA using the following primer sets respectively: *attB1-srtA*-F (5'-ggggacaagtttgtacaaaaa agcaggctttaataatcttattttcactcgttatctta-3') and *srtA*-R (5'-att catccattagcgtaatagaacgttaaggctccttttataca-3'); *srtA*-F (5'-tgtataaaaggagccttaacgttctattacgctaatggatgaat-3') and *attB2-srtA*-R (5'- ggggaccactttgtacaagaaagctgggttaaccatctattaaa tttaaaacctacatt-3'). These fragments were connected by splicing by overlap extension polymerase chain reaction (Horton *et al.*, 1989). The created PCR product was cloned

into pKOR1 by using Gateway BP Clonase II enzyme mix (Life Technologies, Palo Alto, CA, USA), and the resulting plasmid was named as pKOR1-ΔsrtA (Table S1). Using pKOR1-ΔsrtA, the *srtA* gene was deleted from the MR23 genomic DNA according to the procedure reported previously (Bae and Schneewind, 2006).

Reagents for ECM extraction

Sodium chloride was purchased from Wako (Osaka, Japan). Lithium chloride and SDS were purchased from Nacalai Tesque (Kyoto, Japan). All reagents were dissolved into distilled deionized water to the indicated concentrations, and the resulting solutions were supplemented with a protease inhibitor cocktail (Nacalai) to block degradation of proteins in the isolated ECM.

ECM-degrading enzymes

Dispersin B from *Aggregatibacter actinomycetemcomitans* (20 μg ml^{-1}, Kane Biotech Inc., Manitoba, Canada), Proteinase K from *Tritirachium album* (100 μg ml^{-1}, Sigma, St Louis, MO, USA) and DNase I (100 U ml^{-1}, Roche Diagnostics, Mannheim, Germany) were used for dispersal of preformed biofilms or degradation of ECM components.

Antibodies and fluorescent dyes

An anti-Eap polyclonal antibody was raised in a rabbit against purified Eap (Sugimoto *et al.*, 2013) (Scrum, Tokyo, Japan). The anti-Eap IgG was purified with a MAbTrap Kit (GE Healthcare, Buckinghamshire, UK) according to the manufacturer's instructions. A rabbit anti-CsgA polyclonal antibody was raised against the CsgA peptide (LDQWNGKNSEM TVKQFGGGN) (Loferer *et al.*, 1997) with the C-terminal carrier cysteine residue (Medical and Biological Laboratories, Aichi, Japan). Horseradish peroxidase (HRP)-conjugated goat anti-rabbit IgG and Cy3-labeled goat anti-rabbit IgG were purchased from Bio-Rad Laboratories (Hercules, CA, USA) and GE Healthcare respectively. WGA-Alexa488, SYTO9 and PI were purchased from Life Technologies.

Biofilm formation and ECM extraction

Staphylococcal biofilms were grown as follows: single colonies grown on BHI plates were inoculated into 3 ml of BHI medium and incubated overnight at 37°C with shaking. The overnight cultures were 1000-fold diluted in 10 ml of BHIG medium in a conical tube (15 ml, Becton Dickinson) and were incubated at 37°C for 24 h under static conditions. After the incubation, the conical tubes were centrifuged at 8000 g for 10 min at 25°C, and the supernatants were discarded completely. To extract ECM components, the residual pellets were suspended with the indicated regents (100 μl) at various concentrations. The suspensions were centrifuged at 5000 g for 10 min at 25°C, and the supernatants were transferred to a new test tube (1.5 ml) as ECM fractions (Fig. S1B). If required, the suspensions were incubated for the indicated periods at 25°C before centrifugation.

Escherichia coli and *P. aeruginosa* colony biofilms were cultivated on YESCA agar plates at 25°C for 72 h and on LB agar plates at 37°C for 24 h respectively. After the incubation, the colony biofilms were scraped with a scraper and suspended with 1 ml of 1.5 M NaCl solution. These suspensions were centrifuged at 5000 g for 10 min at 25°C without any incubation period. The supernatants were harvested as ECM fractions.

Electrophoreses for proteins, polysaccharides and eDNAs in the isolated ECM fractions

To analyse protein and polysaccharide components in the extracted ECM fractions, they were subjected to SDS-PAGE. Here, 5% to 20% (w/v) polyacrylamide gradient gels (Atto, Tokyo, Japan) were used to separate proteins with a broad range of molecular masses and polysaccharides which are stacked on the top of the gels. Five microlitres of ECM samples were applied per lane. After the electrophoresis, the gels were stained with Coomassie Brilliant Blue (CBB).

For DNA analysis, extracted ECMs were separated by 1.5% (w/v) agarose gel electrophoresis. Ten microlitres of the fractions were applied per lane. After the electrophoresis, the gels were stained with ethidium bromide according to a conventional procedure.

Western blotting

To assess whether ECM components of *E. coli* biofilms can be isolated by the addition of 1.5 M NaCl, CsgA, a major structural component of curli amyloid fibres, was detected by Western blotting, since YMel, an *E. coli* K-12 strain, produces a curli-dependent biofilm under the tested conditions (Maeyama *et al.*, 2004). The ECM fractions from YMel and its isogenic *csgA* mutant, YMel-1, were separated by SDS-PAGE as mentioned above. As a control, curli purified from YMel biofilms

was also used as a positive control. If required, curli amyloid fibres were depolymerized with 90% formic acid (Sigma) as previously reported (Chapman *et al.*, 2002). The proteins were separated by SDS-PAGE and transferred to a polyvinylidene difluoride membrane. The membrane was treated with blocking solution composed of 5% (w/v) skim milk and Tris buffered saline containing 0.1% Tween 20 (TBS-T) overnight at 25°C. After gentle washing with TBS-T, the membrane was probed with an anti-CsgA primary antibody (1000-fold diluted in TBS-T) for 1 h at 25°C. The membrane was washed three times with TBS-T and was subsequently incubated with the secondary antibody (200 000-fold diluted HRP-conjugated goat anti-rabbit IgG in TBS-T) for 1 h at 25°C. After three times washing with TBS-T, the protein-related signals were detected using ECL prime (GE healthcare) on a LAS 4000 luminescent image analyser (GE healthcare).

Quantification of proteins

We quantified proteins in the isolated ECM fractions using a Bradford protein assay kit (Bio-Rad) according to manufacturer's protocol. The concentration was measured at 590 nm with Infinite F200 Pro (Tecan, Männedorf, Switzerland). Bovine serum albumin (BSA) was used as a standard.

Quantification of total saccharide

Total saccharide concentrations in the ECMs were measured by phenol sulfuric acid method. Twenty microlitres of the isolated ECM was mixed with 20 μl of 5% phenol in the 96 wells plate. Then, 100 μl of sulfuric acid was added and incubated for 10 min at 25°C. The concentration was measured at 492 nm with Infinite F200 Pro (Tecan). Glucose was used as a standard.

Quantification of DNA

To quantify DNA in the isolated ECM fractions, we purified the DNA using a Promega SV Wizard DNA purification system (Promega, Madison, WI, USA) according to manufacturer's protocol. The concentration of the purified DNA was measured with NanoDrop 2000 (Thermo Fisher Scientific, Waltham, MA, USA).

Fluorescence microscopy

To observe the distribution of Eap, a major ECM component in *S. aureus* MR23, immunofluorescence microscopy was performed using anti-Eap primary and Cy3-labeled secondary antibodies. Small aliquots (10 μl) of non-treated MR23 biofilm cells, 1.5 M NaCl-treated ones and the extracted ECM were spotted on a slide glass and fixed with 10% (w/v) paraformaldehyde for 10 min at 25°C. One hundred microlitres of blocking solution [PBS containing 0.5% (w/v) BSA and 5% (v/v) goat serum] was placed on the slide and removed after overnight at 25°C. The primary antibody (100-fold diluted anti-Eap antibody in the blocking solution, 100 μl) was placed on the slide and left for 1 h at 25°C. The sample on the slide was then washed three times with PBS. Remaining solution was removed with paper pieces. Then, in a dark room, the secondary antibody (200-fold diluted Cy3-labeled goat anti-rabbit IgG in the blocking solution, 100 μl) was placed on the slide and removed after 1 h by washing five times with PBS. The immunostained sample was covered with one drop of a mounting medium (Vector Laboratories, Burlingame, CA, USA) and a cover glass. The immunostained cells on the slide were observed with a fluorescence and phase-contrast microscope.

To visualize ECM polysaccharides, biofilm cells of MR10 were stained with WGA-Alexa488. Aliquots (10 μl) of the non-treated biofilm cells, 1.5 M NaCl-treated ones, and the extracted ECM were placed on a slide glass and air-dried for 30 min at 25°C. After twice washing with PBS, the cells were covered with WGA-Alexa488 (5 μg ml^{-1}) and removed after 20 min at 37°C. The cells were washed two times with PBS and covered with one drop of the mounting medium and a cover glass. The lectin-stained samples on the slide were observed with a fluorescence and phase-contrast microscope.

To detect eDNAs, the biofilm cells of MS10 were stained by PI (Life Technologies). One hundred microlitres of non-treated biofilm cells, 1.5 M NaCl-treated ones and the extracted ECM were mixed with 0.3 μl of PI solution in the test tube and incubated for 15 min at 25°C in the dark room. Aliquots of the solution (5 μl) were placed on a slide glass and covered with a cover glass. The stained samples on the slide were observed with a fluorescence and phase-contrast microscope.

Cytotoxicity assay

Viability of MR23 biofilm cells were evaluated by counting CFU after the treatment with 1.5 M NaCl, 2% (w/v) SDS or 0.9% (w/v) NaCl (control). To distinguish dead cells from live ones, they were stained with PI and SYTO9 (Life Technologies), respectively, according to manufacturer instruction and were observed with a fluorescence and phase-contrast microscope. The leakage of cytoplasmic proteins was confirmed by observing fluorescence of GFP as a marker of cytoplasmic protein. Here, we used a recently constructed recombinant *S. aureus* strain, MR23 pP1GFP (Table S1) (Sugimoto *et al.*, 2013). The biofilm cells were treated with 1.5 M NaCl, 2.0% SDS or PBS (control) and were subsequently observed under a fluorescence and phase-contrast microscope.

Acknowledgements

We acknowledge all the members in the Department of Bacteriology in The Jikei University School of Medicine for valuable discussions and suggestions. The authors appreciate Dr. Bae for providing pKOR1.

Conflict of Interest

Seiji Hori has received speaker's honoraria from Daiichi Sankyo Co., Ltd. (Tokyo, Japan).

References

Archer, N.K., Mazaitis, M.J., Costerton, J.W., Leid, J.G., Powers, M.E., and Shirtliff, M.E. (2011) *Staphylococcus*

aureus biofilms: properties, regulation, and roles in human disease. *Virulence* **2:** 445–459.

Bae, T., and Schneewind, O. (2006) Allelic replacement in *Staphylococcus aureus* with inducible counter-selection. *Plasmid* **55:** 58–63.

Chapman, M.R., Robinson, L.S., Pinkner, J.S., Roth, R., Heuser, J., Hammar, M., *et al.* (2002) Role of *Escherichia coli* curli operons in directing amyloid fiber formation. *Science* **295:** 851–855.

Costerton, J.W., Stewart, P.S., and Greenberg, E.P. (1999) Bacterial biofilms: a common cause of persistent infections. *Science* **284:** 1318–1322.

Dalton, H.M., and March, P.E. (1998) Molecular genetics of bacterial attachment and biofouling. *Curr Opin Biotechnol* **9:** 252–255.

Davies, D. (2003) Understanding biofilm resistance to antibacterial agents. *Nat Rev Drug Discov* **2:** 114–122.

Del Pozo, J.L., and Patel, R. (2009) Clinical practice. Infection associated with prosthetic joints. *N Engl J Med* **361:** 787–794.

Dunne, W.M., Jr (2002) Bacterial adhesion: seen any good biofilms lately? *Clin Microbiol Rev* **15:** 155–166.

Dunne, W.M., Jr, and Burd, E.M. (1992) The effects of magnesium, calcium, EDTA, and pH on the *in vitro* adhesion of *Staphylococcus epidermidis* to plastic. *Microbiol Immunol* **36:** 1019–1027.

Flemming, H.C., and Wingender, J. (2010) The biofilm matrix. *Nat Rev Microbiol* **8:** 623–633.

Flemming, H.C., Neu, T.R., and Wozniak, D.J. (2007) The EPS matrix: the 'house of biofilm cells. *J Bacteriol* **189:** 7945–7947.

Forsgren, A., and Sjöquist, J. (1966) 'Protein A' from *S. aureus*: I. Pseudo-immune reaction with human γ-globulin. *J Immunol* **97:** 822–827.

Frølund, B., Palmgren, R., Keiding, K., and Nielsen, P.H. (1996) Extraction of extracellular polymers from activated sludge using a cation exchange resin. *Water Res* **30:** 1749–1758.

Horton, R.M., Hunt, H.D., Ho, S.N., Pullen, J.K., and Pease, L.R. (1989) Engineering hybrid genes without the use of restriction enzymes: gene splicing by overlap extension. *Gene* **77:** 61–68.

Hussain, M., von Eiff, C., Sinha, B., Joost, I., Herrmann, M., Peters, G., and Becker, K. (2008) *eap* gene as novel target for specific identification of *Staphylococcus aureus*. *J Clin Microbiol* **46:** 470–476.

Jucker, B.A., Harms, H., and Zehnder, A.J. (1996) Adhesion of the positively charged bacterium *Stenotrophomonas* (*Xanthomonas*) *maltophilia* 70401 to glass and Teflon. *J Bacteriol* **178:** 5472–5479.

Kaplan, J.B., Velliyagounder, K., Ragunath, C., Rohde, H., Mack, D., Knobloch, J.K., and Ramasubbu, N. (2004) Genes involved in the synthesis and degradation of matrix polysaccharide in *Actinobacillus actinomycetemcomitans* and *Actinobacillus pleuropneumoniae* biofilms. *J Bacteriol* **186:** 8213–8220.

Kikuchi, T., Mizunoe, Y., Takade, A., Naito, S., and Yoshida, S. (2005) Curli fibers are required for development of biofilm architecture in *Escherichia coli* K-12 and enhance bacterial adherence to human uroepithelial cells. *Microbiol Immunol* **49:** 875–884.

Liang, O.D., Ascencio, F., Fransson, L.A., and Wadström, T. (1992) Binding of heparan sulfate to *Staphylococcus aureus*. *Infect Immun* **60:** 899–906.

Loferer, H., Hammar, M., and Normark, S. (1997) Availability of the fibre subunit CsgA and the nucleator protein CsgB during assembly of fibronectin-binding curli is limited by the intracellular concentration of the novel lipoprotein CsgG. *Mol Microbiol* **26:** 11–23.

Maeyama, R., Mizunoe, Y., Anderson, J.M., Tanaka, M., and Matsuda, T. (2004) Confocal imaging of biofilm formation process using fluoroprobed *Escherichia coli* and fluoro-stained exopolysaccharide. *J Biomed Mater Res A* **70:** 274–282.

Mann, E.E., Rice, K.C., Boles, B.R., Endres, J.L., Ranjit, D., Chandramohan, L., *et al.* (2009) Modulation of eDNA release and degradation affects *Staphylococcus aureus* biofilm maturation. *PLoS ONE* **4:** e5822.

Mazmanian, S.K., Ton-That, H., and Schneewind, O. (2001) Sortase-catalysed anchoring of surface proteins to the cell wall of *Staphylococcus aureus*. *Mol Microbiol* **40:** 1049–1057.

O'Gara, J.P. (2007) *ica* and beyond: biofilm mechanisms and regulation in *Staphylococcus epidermidis* and *Staphylococcus aureus*. *FEMS Microbiol Lett* **270:** 179–188.

O'Neill, E., Pozzi, C., Houston, P., Humphreys, H., Robinson, D.A., Loughman, A., *et al.* (2008) A novel *Staphylococcus aureus* biofilm phenotype mediated by the fibronectin-binding proteins, FnBPA and FnBPB. *J Bacteriol* **190:** 3835–3850.

Palma, M., Haggar, A., and Flock, J.I. (1999) Adherence of *Staphylococcus aureus* is enhanced by an endogenous secreted protein with broad binding activity. *J Bacteriol* **181:** 2840–2845.

Romero, D., Aguilar, C., Losick, R., and Kolter, R. (2010) Amyloid fibers provide structural integrity to *Bacillus subtilis* biofilms. *Proc Natl Acad Sci USA* **107:** 2230–2234.

Sugimoto, S., Iwamoto, T., Takada, K., Okuda, K., Tajima, A., Iwase, T., and Mizunoe, Y. (2013) *Staphylococcus epidermidis* Esp degrades specific proteins associated with *Staphylococcus aureus* biofilm formation and host-pathogen interaction. *J Bacteriol* **195:** 1645–1655.

Tabouret, M., de Rycke, J., and Dubray, G. (1992) Analysis of surface proteins of Listeria in relation to species, serovar and pathogenicity. *J Gen Microbiol* **138:** 743–753.

Tapia, J.M., Muñoz, J.A., González, F., Blázquez, M.L., Malki, M., and Ballester, A. (2009) Extraction of extracellular polymeric substances from the acidophilic bacterium *Acidiphilium 3.2Sup(5)*. *Water Sci Technol* **59:** 1959–1967.

Thwaites, G.E., Edgeworth, J.D., Gkrania-Klotsas, E., Kirby, A., Tilley, R., Török, M.E., *et al.* (2011) Clinical management of *Staphylococcus aureus* bacteraemia. *Lancet Infect Dis* **11:** 208–222.

Whitchurch, C.B., Tolker-Nielsen, T., Ragas, P.C., and Mattick, J.S. (2002) Extracellular DNA required for bacterial biofilm formation. *Science* **295:** 1487.

Wu, J., and Xi, C. (2009) Evaluation of different methods for extracting extracellular DNA from the biofilm matrix. *Appl Environ Microbiol* **75:** 5390–5395.

Supporting information

Additional Supporting Information may be found in the online version of this article at the publisher's web-site:

Fig. S1. The principle of the method employed in extracting bacterial extracellular matrices (ECMs) composed of proteins, polysaccharides and/or eDNAs.
A. Graphical image of bacterial cells embedded in ECMs. ECM components including proteins, polysaccharides and eDNAs can bind to bacterial cells or each other via electrostatic attractive forces, ionic attractive forces, hydrophobic interactions and van der Waals interactions (Flemming and Wingender, 2010). Increasing concentrations of salts and detergents (e.g. NaCl and SDS) potentially disrupt these interactions by coating cell surfaces and ECM constituents and therefore lead to detachment of ECMs from the cells. This image is an example using 1.5 M NaCl.
B. The procedure for preparation of biofilm and isolation of ECMs. Biofilms are grown in an appropriate medium (e.g. BHIG: brain heart infusion medium containing 1% glucose) in a conical tube. The bacterial cells covered with a self-produced ECM are sedimented by centrifugation at 8000 g for 10 min at 25°C. The cells and the ECM are suspended in an appropriate solution (e.g. 1.5 M NaCl). The suspension is immediately centrifuged at 5000 g for 10 min at 25°C, and the ECM components in the supernatant detached from bacterial cells are transferred to a new tube.

Fig. S2. SDS-PAGE profiles of planktonic and biofilm cells. Biofilms are grown in a conical tube as described in Fig. S1. Before the first centrifugation, cultured medium in which many planktonic cells existed was harvested and transferred to another conical tube. The residual bacterial cells attached to the tube were used as biofilm cells. Extracellular matrices extracted from planktonic and biofilm cells with 1.5 M NaCl and were analysed by SDS-PAGE. There is no significant difference in the profile of proteins between them.

Fig. S3. Non-ionic detergents are not effective in detachment of ECMs from biofilm cells. Using the indicated percentages (w/v) of Triton X100, Tween 20 and Nonidet 40 (NP-40), ECM proteins of *S. aureus* MR23 were extracted according to the procedure shown in Fig. S1 and were subjected to SDS-PAGE. The gel was stained with CBB. The positions of molecular mass markers in kDa are shown at the left.

Fig. S4. Purification of Eap from the isolated ECM of *S. aureus* MR23.
A. Native Eap was purified form the isolated ECM of *S. aureus* MR23 with a fibrinogen-affinity column. The isolated ECM (input), flow through, wash and elution fractions with PBS or acetic acid were subjected to SDS-PAGE.
B. The concentrated fractions of flow through, wash and elution were also analysed by SDS-PAGE. The wash fraction contained highly purified Eap, and in addition, the elution fraction contained both Eap and Efb with some contaminated proteins.
The positions of molecular mass markers in kDa are shown at the left (A and B).

Fig. S5. Susceptibilities of ECMs to Proteinase K, DNase I and Dispersin B. ECMs were extracted from *S. aureus* MR10 and MS10 biofilms and were treated with Proteinase

K (100 µg ml⁻¹), DNase I (100 U ml⁻¹) and Dispersin B (20 µg ml⁻¹) for 2 h at 37°C. They were subsequently applied to SDS-PAGE with CBB staining. The positions of molecular mass markers in kDa are shown at the left. Proteins in the ECMs from MR10 and MS10 were degraded by Proteinase K. In the case of the ECM from MR10, bands stacked at the top of the gel disappeared after the treatment with Dispersin B, and thus they are judged as polysaccharides as recently reported (Sugimoto *et al.*, 2013). Interestingly, their profiles are significantly different from each other and from that of MR23, suggesting the applicability of this method to typing of biofilms.

Fig. S6. Analyses of eDNAs in the ECMs isolated from *S. aureus* MR23, MR10 and MS10. The ECMs isolated from *S. aureus* MR23, MR10 and MS10 biofilms and treated with the indicated enzymes as described in Fig. S4 and applied to agarose gel electrophoresis. Non-treated ECMs were also used as a control. The gels were stained with ethidium bromide. The positions of molecular mass markers in kb are shown at the left. Notably, the bands were sensitive to DNase I neither Proteinase K nor Dispersin B, indicating that they contains DNA molecules to varying degrees.

Fig. S7. Assessment of cell damages for *E. coli* and *P. aeruginosa*. To assess cell damages for *E. coli* and *P. aeruginosa* caused by extraction with NaCl, CFUs of biofilm cells were counted. Briefly, colony biofilms on the plate were scraped and suspended with PBS. Then, this suspension was separated into two tubes equally. These samples were centrifuged at 8000 g for 10 min at 25°C. After centrifugation, the supernatants were discarded. One of these pellets was suspended with 1.5 M NaCl solution, and the other one was suspended with PBS (as control). These suspensions were centrifuged at 5000 g for 10 min at 25°C. After centrifugation, the supernatants were discarded. The pellets were suspended with PBS and the CFU was counted.

A. Colony-forming unit of *E. coli* biofilm cells before and after the treatment with PBS and 1.5 M NaCl for YMel strain. The means and standard deviations of triplicate determinations are represented.

B. Colony-forming unit of *P. aeruginosa* biofilm cells before and after the treatment with PBS and 1.5 M NaCl for PAO1 strain. The means and standard deviations of triplicate determinations are represented.

Table S1. Bacterial strains and plasmids used in this study.

6

Extraction and purification of high-value metabolites from microalgae: essential lipids, astaxanthin and phycobiliproteins

Sara P. Cuellar-Bermudez, Iris Aguilar-Hernandez,
Diana L. Cardenas-Chavez, Nancy Ornelas-Soto,
Miguel A. Romero-Ogawa and Roberto
Parra-Saldivar*
*Cátedra de Bioprocesos Ambientales, Centro del Agua
Para América Latina y el Caribe, Instituto Tecnológico y
de Estudios Superiores de Monterrey, Monterrey,
Nuevo Leon 64849, Mexico.*

Summary

The marked trend and consumers growing interest in natural and healthy products have forced researches and industry to develop novel products with functional ingredients. Microalgae have been recognized as source of functional ingredients with positive health effects since these microorganisms produce polyunsaturated fatty acids, polysaccharides, natural pigments, essential minerals, vitamins, enzymes and bioactive peptides. For this reason, the manuscript reviews two of the main high-value metabolites which can be obtained from microalgae: pigments and essential lipids. Therefore, the extraction and purification methods for polyunsaturated fatty acids, astaxanthin, phycoerythrin and phycocyanin are described. Also, the effect that environmental growth conditions have in the production of these metabolites is described. This review summarizes the existing methods to extract and purify such metabolites in order to develop a feasible and sustainable algae industry.

*For correspondence. E-mail r.parra@itesm.mx

Funding Information The Catedra de Bioprocesos Ambientales and Centro del Agua para America Latina y el Caribe of Tecnológico de Monterrey, Campus Monterrey support and assistance provided during this investigation is gratefully acknowledged.

Introduction

Microalgae are eucaryotic photosynthethic microorganisms that use solar energy, nutrients and carbon dioxide (CO_2) to produce proteins, carbohydrates, lipids and other valuable organic compounds like carotenoids (Mendes et al., 2003; Batista et al., 2013). The chemical compounds synthesized by microalgae have several applications. The high-protein content of some algae species is one of the main reasons to consider them as a non-conventional source of proteins. For example, Chlorella and Spirulina were the first microalgae species to be commercialized as a health food in Japan, Taiwan and Mexico (Sánchez et al., 2003; Borowitzka, 2013). However, information on the nutritional value of the protein and the degree of availability of the amino acids should be stated (Spolaore et al., 2006). Carbohydrates like starch, glucose, sugars and other polysaccharides have high overall digestibility for food or feeds (Spolaore et al., 2006). As an example, macroalgal polysaccharides like agar, alginates and carrageenans are used in diverse fields of industry due to their rheological gelling or thickening properties (Pulz and Gross, 2004). Finally, lipids and fatty acids from microalgae including the omega-3 (ω3) and omega-6 (ω6) families, have gained particular interest due to the health benefits related to its consumption (Spolaore et al., 2006). Aquaculture is also one of the main applications of lipids and pigments produced by algae since these microorganisms are essential in the food chain. Therefore, different studies have been carried in order to identify potential algae species to be used in aquaculture (Gladue and Maxey, 1994; Borowitzka, 1997; Makri et al., 2011; Birkou et al., 2012).

Depending on the microalgae species, various high-value compounds can be extracted from the biomass (Table 1). These include pigments, polysaccharides, triaglycerides, fatty acids and vitamins, which are also commonly used as bulk commodities and specialty chemicals in different industrial sectors (e.g. pharmaceuticals, cosmetics, nutraceuticals, functional foods, aquaculture, biofuels). High-value products from microalgae

Table 1. Microalgae species of high-value compounds extraction and applications (Pulz and Gross, 2004; Spolaore *et al.*, 2006; Casal *et al.*, 2011; Guedes *et al.*, 2011; Batista *et al.*, 2013; Borowitzka, 2013; Sørensen *et al.*, 2013).

Species	Product	Application areas
Chlorella vulgaris	Biomass, pigments	Health food, food supplement
Chlorella spp.		
Chlorella ellipsodea		
Coccomyxa acidophila	Lutein, β-carotene	Pharmaceuticals, nutrition
Coelastrella striolata var. multistriata	Canthaxanthin, astaxanthin, β-carotene	Pharmaceuticals, nutrition, cosmetics
Crypthecodinium conhi	Docosahexaenoic acid	Pharmaceuticals, nutrition
Diacronema vlkianum	Fatty acids	Pharmaceuticals, nutrition
Dunaliella salina	Carotenoids, β-carotene	Health food, food supplement, feed
Galdiera suphuraria	Phycocyanin	Pharmaceuticals, nutrition
Haematococcus pluvialis	Carotenoids, astaxanthin, cantaxanthin, lutein	Health food, pharmaceuticals, feed additives
Isochrysis galbana	Fatty acids, carotenoids, fucoxanthin	Pharmaceuticals, nutrition, cosmetics, animal nutrition
Lyngbya majuscule	Immune modulators	Pharmaceuticals, nutrition
Muriellopsis sp.	Lutein	Pharmaceuticals, nutrition
Nannochloropsis gaditana	Eicosapentaenoic acid	Pharmaceuticals, nutrition
Nannochloropsis sp.		
Odontella aurita	Fatty acids	Pharmaceuticals, cosmetics, baby food
Parietochloris incise	Arachidonic acid	Nutritional supplement
Phaedactylum tricornutum	Lipids, eicosapentaenoic acid, fatty acids	Nutrition, fuel production
Porphyridium cruentum	Arachidonic acid, polysaccharides	Pharmaceuticals, cosmetics, nutrition
Scenedesmus almeriensis	Lutein, β-carotene	Pharmaceuticals, nutrition, cosmetics
Schizochytrium sp.	Docosahexaenoic acid	Pharmaceuticals, nutrition
Spirulina platensis	Phycocyanin, γ-Linolenic acid, biomass protein	Health food, cosmetics
Ulkenia spp.	Docosahexaenoic acid	Pharmaceuticals, nutrition

are usually produced within a biorefinery model, since the composition of the microalgal cell allows for extraction of different co-products. In addition, specialty chemicals have higher revenues than bulk chemicals like algal oil for biofuels (Sharma *et al.*, 2010; Borowitzka, 2013; Gerardo *et al.*, 2014). Different commercialization prices of algae products have been reported (Table 2). Some reports show a decrease in the marketing prices (except for omega-3). This can be explained by the fact that more companies are focusing in the production and commercialization of compounds from microalgae (Table 3). Nevertheless, prices are still high due to the process expenses associated to the extraction and purification of intracellular metabolites. In microalgal biotechnological processes, the downstream stage can account for 50–80% of total production costs, depending on the biochemical characteristics of the compound and the purity ratio that needs to be achieved. (Molina Grima *et al.*, 2003).

This article, based on literature framework, describes lipids and pigments as two representative classes of high-value compounds synthesized by algae. In the case of lipids, metabolic production, extraction and quantification methods are discussed. For pigments, the extraction and purification methods for astaxanthin, phycocyanin and phycoerytrin are described.

Lipids and polyunsaturated fatty acids

The main components of the algae lipid fraction are fatty acids (FA), waxes, sterols, hydrocarbons, ketones and pigments (carotenoids, chlorophylls, phycobilins) (Halim *et al.*, 2011). Depending on the species, total lipids in microalgae usually represent 20% to 50% of total biomass in dry weight (DW). However, other values have also been reported in a range from 1% to 70% (Spolaore *et al.*, 2006). Fatty acids can generally be classified into two categories based on the polarity of the molecular head

Table 2. Commercialization price of some high value compounds from algae.

Product	Price (€) Spolaore *et al.*, 2006[a]	Price (€) Brennan and Owende, 2010	Price (€) Markou and Nerantzis, 2013[a]	Price (€) Borowitzka, 2013[a]
Phycobiliproteins	—	11–50 mg^{-1}	—	
B-phycoerythrin	105 mg^{-1}	—	0.036 mg^{-1}	
C-phycocyanin				360–72 460 Kg^{-1}
β-Carotene	—	215–2150 Kg^{-1}	218–510 Kg^{-1}	
Astaxanthin	—	7150 Kg^{-1}	1,450–5075 Kg^{-1}	
DHA oil/Omega-3	—	0.043 Kg^{-1}	0.63–2.78 Kg^{-1}	78–116 Kg^{-1}

a. Price expressed in USD by the authors. Conversion factor of 1.38.

Table 3. Global companies based in developing process and commercialization of high-value compounds from algae.

Company name	Location	Company name	Location
Algae. Tec	Australia	AlgaFuel, S.A (A4F)	Portugal
Solarvest BioEnergy	Canada	Necton	Portugal
Canadian Pacific Algae	Canada	Green Sea Bio Systems s.l.	Spain
Solarium Biotechnology S.A.	Chile	AlgaEnergy	Spain
BlueBio	China	Fitoplancton Marino	Spain
EcoFuel Laboratories	Czech Republic	Simris	Sweden
Aleor	France	Taiwan Chlorella Manufacturing Company	Taiwan
Fermentalg	France	Vedan	Taiwan
Roquette	France	AlgaeLink N.V	The Netherlands
Alpa Biotech	France	AlgaeBiotech	The Netherlands
IBV Biotech IGV GmbH	Germany	LGem	The Netherlands
Subitec	Germany	Solazyme, Inc.	USA
Algomed	Germany	Aurora Algae	USA
BlueBioTech	Germany	Solix Biosystems	USA
Phytolutions	Germany	Synthetic Genomics	USA
Algae Health	Ireland	Cellena	USA
Seambiotic	Israel	Cyanotech	USA
Algatechnologies	Israel	Algaeon	USA
UniVerve Biofuel	Israel	Alltech Algae	USA
Parry Nutraceuticals	India	Green Star Products, Inc	USA
Sunchlorella	Japan	Bionavitas	USA
Fuji Chemicals	Japan	Heliae	USA
DAESANG	Korea	Kuehnle Agro Systems	USA
Algaetech International	Malaysia	Photon8	USA
June Pharmaceutical	Malaysia	Ternion BioIndustries	USA
Tecnología Ambiental BIOMEX	Mexico	Algae to Omega Holdings	USA
Algae Technology Solutions	Mexico	Sapphire Energy	USA
Aquaflow Binomics	New Zealand	Algenol	USA
Photonz	New Zealand		

group: (i) neutral lipids which comprise acylglycerols and free fatty acids (FFA) and (ii) polar lipids or amphipathic lipids which can be further subcategorized into phospholipids and glycolipids. Acylglycerols consist of fatty acids ester bonded to a glycerol backbone, and according to its number of fatty acids are categorized in triacylglycerides (TAG), diacylglycerols, monoacylglycerols (Halim *et al.*, 2011).

Lipid production by microalgae depends on the species and is affected by culture conditions such as nutrients, salinity, light intensity periods, temperature, pH and even the association with other microorganisms (Richmond, 2004; Guschina and Harwood, 2006). Nitrogen limitation is considered the most efficient strategy to increase the content of neutral lipids in algae, mainly formed by the triglycerides with high degree of saturation. However, a decrease in biomass productivity occurs. Breuer and colleagues (2012) reported an increase in the TAG accumulation as response of nitrogen limitation in *Chlorella vulgaris*, *Chlorella zofingiensis*, *Neochloris oleoabundans* and *Scenedesmus obliquus*. Total lipid content in *Ulva pertusa*, *Euglena gracilis* and *Botryococcus* species increased with nitrogen starvation (Floreto *et al.*, 1993; Regnault *et al.*, 1995; Yeesang and Cheirsilp, 2011). In addition, in contrast to the polar lipids of nitrogen-sufficient cells, neutral lipids in the form of triacylglycerols

become predominant in lipids from nitrogen-depleted cells (Richmond, 2004).

As mentioned, nitrogen starvation is one the main strategies to increase lipid and TAG production by algae. However, it has also been reported that nitrogen starvation favours biosynthesis of starch (Ramazanov and Ramazanov, 2006; Wang *et al.*, 2009). Phosphorous limitation causes the replacement of membrane phospholipids by non-phosphorus glycolipids representing and effective phosphate-conserving mechanism. However, Guschina and colleagues (2003) reported that algae under phosphorous limitation, maintain their phosphoglyceride synthesis since significant endogenous phosphorus stores in the algae were found by X-ray electron microscopy. In case of the light intensity effect, diverse studies suggest that high light intensity and therefore high temperature favour the accumulation of triglycerides with high saturation profile (Floreto *et al.*, 1993; Van Wagenen *et al.*, 2012). Meanwhile, low light intensities and low temperature promote the synthesis of polyunsaturated fatty acids (PUFA) (Guschina and Harwood, 2006). The effect of carbon dioxide concentration has been studied in *C. Kessleri*, low-CO_2 cultures showed high contents of α-linolenate fatty acid (Sato *et al.*, 2003). In contrast, in *C. reinhardtii* mutant cia-3, higher content of PUFA was found in cultures with high CO_2 concentration. Finally, pH can also affect the lipid metabolism. Low pH stress in *Chlamydomonas* sp. increased the total lipid content compared with higher pH values (Tatsuzawa *et al.*, 1996). However in *Chlorella* spp., alkaline pH resulted in triacylglycerides accumulation (Guckert and Cooksey, 1990).

Fatty acids biosynthesis carries out by the conversion of acetyl coenzyme A (acetyl-CoA) to malonyl-CoA, catalysed by the complex enzyme acetyl-CoA carboxylase. One of the main biochemical pathways for acetyl-CoA production comes from the 3-phosphoglycerate (3-PG), primary product of carbon dioxide fixation. Also, 3-PG is also the precursor for the glycerol backbone of TAG, however, 3-PG also participates into the starch biosynthesis pathway (Wang *et al.*, 2009). Therefore, studies have been carried in order to identify the biochemical pathways for polysaccharides and lipids production in algae (Bellou and Aggelis, 2012). Also, to increase lipid and TAG accumulation in cells, non-starch algae mutant strains are been studied and characterized (Ramazanov and Ramazanov, 2006; Wang *et al.*, 2009).

Omega-3 (ω-3) PUFA are a specific group of polyunsaturated fatty acids in which the first double bond is located between the third and fourth carbon atom counting from the methyl end of the fatty acid (Ryckebosch *et al.*, 2012). Fish oil is a major non-sustainable source for the commercial production of these fatty acids. Fish oil quality depends on the fish species, the season/climate and geographical location of catching sites and food quality consumed by the fish. In addition, fish oil is not suitable for vegetarians, and its odour makes it unattractive for consumption. Moreover, in some cases, there is a contamination danger by lipid-soluble environmental pollutants (Ryckebosch *et al.*, 2012).

In addition, PUFAs are subject of intensive research due to the important health benefits associated with their consumption (Wen and Chen, 2003; Sijtsma and de Swaaf, 2004). These include the α-Linolenic acid (ALA 18:3 ω-3), γ-Linolenic acid (GLA, 18:3 ω-6), eicosapentaenoic acid (EPA, 20:5 ω-3), arachidonic acid (ARA, 20:6 ω-6) docosapentaenoic acid (22:5 ω-3) and docosahexaenoic acid (DHA, 22:6 ω-3) (Fraeye *et al.*, 2012; Ryckebosch *et al.*, 2012). According to Adarme-Vega and colleagues (2012), these long chain ω-3 PUFA provide significant health benefits particularly in reducing cardiac diseases such as arrhythmia, stroke and high blood pressure. As well, they have beneficial effects against depression, rheumatoid arthritis, asthma and can be used for treatment of inflammatory diseases such as rheumatoid arthritis, Crohn's disease, ulcerative colitis, psoriasis, lupus and cystic fibrosis. Additionally, in pregnant women, the adequate intake of EPA and DHA is crucial for healthy development of the fetal brain. In addition, ARA and DHA are required for normal growth and brain functional development, while EPA is essential for the regulation of some biological functions and prevention of arrhythmia, atherosclerosis, cardiovascular disease and cancer (Pulz and Gross, 2004). In Table 4 are shown some microalgae species source of high-quality PUFAs. In addition, the extraction method and yield of the FA is listed.

The biosynthesis of EPA/DHA (Fig. 1) occurs through a series of reactions which can be divided in two different steps. First is the novo synthesis of oleic acid (18:1) from acetate. This is followed by conversion of oleic acid (18:1) to linoleic acid (LA, 18:2) and α-linolenic acid (18:3). Finally, after a number of subsequent stepwise desaturation and elongation steps, the ω-3 PUFA family including EPA and DHA are formed (Wen and Chen, 2003).

In microalgae, EPA is found in the *Bacillariophceae* (diatoms) *Chlorophyceae*, *Chrysophyceae*, *Cryptophyceae*, *Eustigamatophyceae* and *Prasinophyceae* classes (Singh *et al.*, 2005). The heterotrophic marine dinoflagellate *Crypthecodinium cohnii*, together with *Schizochytrium* and related genera, represent the major commercial and interesting source of DHA. *Crypthecodinium cohnii* can accumulate a high percentage of DHA (25–60% of the total fatty acids) as TAG with only trivial amounts of other fatty acids (Mendes *et al.*, 2007; Zvi and Colin, 2010). In addition, other algae species have been studied as source of PUFA. Chen and colleagues (2007) analysed the microalgae *Nitzschia laevis*, founding that 75.9% of total EPA was accumulated as 37.4% in TAG, 22.6% in monoacylglycerides and 15.9% in phosphatidylcholine. Bigogno and colleagues (2002) analysed fatty acid production by the green algae *Parietochloris incise* concluding that ARA comprised 33.6% of total fatty acids in the logarithmic phase and 42.5% in the stationary phase with a 27% of total fatty acids.

Extraction and quantification methods

Historically, the three most common processes for recovering oil from most plant seeds are hydraulic pressing, expeller pressing and solvent extraction. Solvent extraction was originated as a batch process in Europe in 1870 (Zvi and Colin, 2010). Currently, algae oil extraction procedures include mechanical pressing, homogenization, milling and solvent extraction. The most common solvents for lipid extraction are chloroform–methanol, hexane, hexane–isopropanol or other solvent mixtures slightly soluble in each other. Depending on polarity and/or solubility of the lipid content, the adequate solvent or mixture must be chosen for the extraction.

Ryckebosch and colleagues (2011) tested different solvents for lipid extraction in four algae species

Table 4. PUFAs extracted in different microalgae species.

Species	Product	Yield	Extraction/Purification method	Reference
Arthrospira platensis	Total lipids	13.2% DW	Chloroform–methanol 1:1 (%v/v)	Ryckebosch and colleagues (2011)
Chlorella vulgaris		19.9% DW		
Chlorella vulgaris Green cells	ALA	661 mg/100 g	Acid digestion of biomass with 4 N HCl.	Batista and colleagues (2013)
	EPA	19 mg/100 g	Soxhlet method with petroleum ether for 6 h	
	DHA	16 mg/100 g		
	GLA	112 mg/100 g		
Chlorella vulgaris Orange cells	ALA	3665 mg/100 g		
	EPA	39 mg/100 g		
	DHA	80 mg/100 g		
	GLA	23 mg/100 g		
Crypthecodinium cohnii	DHA	99.2% purity	Purification by saponification, winterization and urea complexation	Mendes and colleagues (2007)
Diacronema vlkianum	ALA	14 mg/100 g	Acid digestion of biomass with 4 N HCl.	Batista and colleagues (2013)
	EPA	3212 mg/100 g	Soxhlet method with petroleum ether for 6 h	
	DHA	836 mg/100 g		
	GLA	112 mg/100 g		
H. pluvialis	ALA	3981 mg/100 g		
	EPA	579 mg/100 g		
	GLA	472 mg/100 g		
Isochrysis galbana	ALA	421 mg/100 g		
	EPA	4875 mg/100 g		
	DHA	1156 mg/100 g		
Scenedesmus obliquus	Total Lipids	29.7% DW	Chloroform–methanol 1:1 (%v/v)	Ryckebosch and colleagues (2011)
Spirulina maxima	ALA	40 mg/100 g	Acid digestion of biomass with 4 N HCl. Soxhlet method with petroleum ether for 6 h	Batista and colleagues (2013)
	GLA	452 mg/100 g		
Nannochloropsis salina	Total Lipids	34.4% DW	Chloroform–methanol 1:1 (%v/v)	Ryckebosch and colleagues (2011)
Nannochloropsis gaditana	EPA	3.7 g/100 g DW	Dichloromethane-ethanol (1:1)	Ryckebosch and colleagues (2013)
Parietochloris incise	ARA	9.1% DW	Methanol (10%DMSO) at 40°C for 5 min. Diethyl ether, hexane and water 1:1:1 (v/v/v)	Bigogno and colleagues (2002)
Phaeodactylum tricornutum	EPA	9.3% DW of total fatty acids (39%)	Chloroform–methanol 1:1 (%v/v)	Ryckebosch and colleagues (2011)
Porphyridium cruentum	EPA	50.8% recovery, 97% purity	Purification by saponification and urea complexation	Guil-Guerrero and colleagues (2000)
Tetraselmis sp.	EPA	10.41% of phospholipid fraction	Chloroform-methanol 2:1 (%v/v). Extract was washed with 0.88% w/v KCl to remove non-lipids	Makri and colleagues (2011)

(*Nannochloropsis salina, S. obliquus, C. vulgaris* and *A Arthrospira platensis*) finding that chloroform–methanol 1:1 (%v/v) allows the highest lipid extraction and is thus the preferred solvent mixture for total lipids determination. Moreover, while this analysis is performed, lyophilized algae can be used without the need for biomass pretreatment. This refers to the addition of isopropanol or antioxidants to inactivate the lipases for further oxidation. Also, the authors concluded that no cell disruption method is necessary, and in general two series of solvent extractions must be carried out. However, other methods such as microwaves, supercritical fluid extraction, enzymatic extractions, ultrasonic-assisted extraction, pulsed electric field technology and osmotic shock are used to enhance the extraction of cellular lipids bodies by solvents (Mercer and Armenta, 2011; Zheng *et al.*, 2011; Goettel *et al.*, 2012).

Lipids also need to be separated from carbohydrates, proteins or salts to enhance extraction or analysis. Acid hydrolysis has been used to break the lipids bonds with other compounds, increasing the extraction yield. As an example, hydrochloric acid or glacial acetic acid are used as pretreatment or mixed with the solvent used in the Soxhlet extraction (Palmquist and Jenkins, 2003). The moisture content in a sample can affect the performance of the lipid extraction (Griffiths *et al.*, 2010).

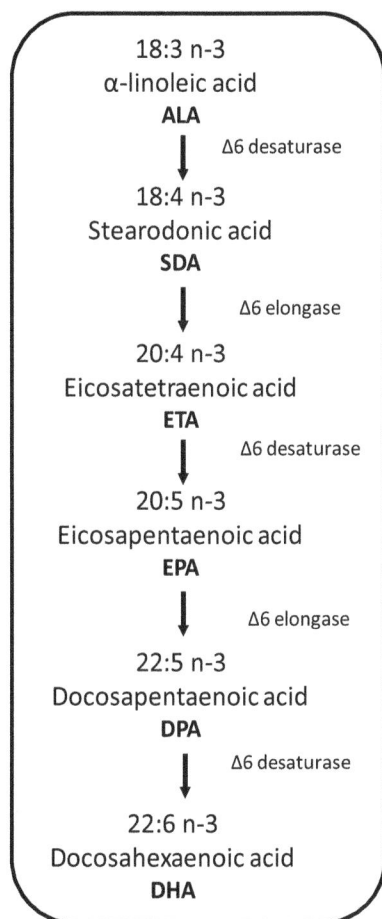

18:3 n-3
α-linoleic acid
ALA

↓ Δ6 desaturase

18:4 n-3
Stearodonic acid
SDA

↓ Δ6 elongase

20:4 n-3
Eicosatetraenoic acid
ETA

↓ Δ6 desaturase

20:5 n-3
Eicosapentaenoic acid
EPA

↓ Δ6 elongase

22:5 n-3
Docosapentaenoic acid
DPA

↓ Δ6 desaturase

22:6 n-3
Docosahexaenoic acid
DHA

Fig. 1. Pathway for the biosynthesis of omega-3 LC-PUFA (Wen and Chen, 2003; Adarme-Vega et al., 2012; Ryckebosch et al., 2012).

Goettel and colleagues (2012) studied the effect of pulsed electric field treatment as a cell disintegration method altering the structure of the cell membrane (permeability) in *Auxenochlorella prototecoides*. The results showed an increment in conductivity and a drop of pH of the treated suspensions due to the release of soluble cell material; however the lipid droplets remained inside the cells. Therefore, the authors proposed a two-stage process, first for water soluble cell materials extraction and subsequently for material soluble in organic solvents.

Most authors reporting lipid determination commonly use chromatography with flame-ionization detectors after transesterification of fatty acids (acylglycerols) with alcohol (methanol) to produce fatty acid methyl esters. However, some other techniques are currently used: fluorometry, colourimetry, Raman spectroscopy, gas chromatography with mass spectrometric detector, high performance liquid chromatography (HPLC) with pulsed amperometric detection, reverse phase HPLC, UV detection at 205 nm, evaporative light scattering detec-

tion, atmospheric pressure chemical ionization mass spectrometry, nuclear magnetic resonance, near infrared and Fourier transform infrared spectroscopy (Cooksey et al., 1987; Lee et al., 1998; Knothe, 2001; Meher et al., 2006; Elsey et al., 2007; Huang et al., 2010; Laurens and Wolfrum, 2010; Wawrik and Harriman, 2010; Cheng et al., 2011; Davey et al., 2012; De la Hoz Siegler et al., 2012).

Diverse technologies besides solvent extraction have been developed to achieve fast quantifications of the lipid content in algae cultures in order to determine the optimal time for harvesting (De la Hoz Siegler et al., 2012). These techniques are non-destructive, and just small samples volume is needed (Davey et al., 2012). Specifically, fluorescence methods are used for these purposes. However, fluorescence intensity is affected by several factors as staining agent concentration, cell concentration, staining temperature and staining duration (De la Hoz Siegler et al., 2012). Therefore, according to the algae species, the stain method should be optimized. De la Hoz Siegler and colleagues (2012) studied the effects that biomass and staining concentration, and mixing time have on the results variability of Nile Red (9-diethylamino-5H-benzo[α]-phenoxazine-5-one) fluorescence method for lipid determination in 4 algae species: *C. vulgaris*, *A. prototecoides*, *Scenedesmus dimorphus* and *S. obliquus*. As well, the addition of ethanol as carrier solvent was also evaluated. The results showed that staining time less than 20 min did not affect the measurement error, and 10 μg/ml of Nile Red solution was an appropriate concentration for each 10 μg/ml of algal suspension (5 g/L DW). Moreover, the sample standard deviation was reduced by 56% when all the reagents were mixed before distributing the sample into the microwells. In comparison, Chen and colleagues (2011) used microwaves in combination with dimethyl sulfoxide (DMSO) for an effective staining of lipid bodies. However, DMSO could carry other compounds increasing the error measurement (De la Hoz Siegler et al., 2012).

Purification methods

Docosahexaenoic acid/eicosapentaenoic acid crude oil is unfit for consumption because of its impurities, odour and taste, as well as cloudy or turbid appearance. Hence, it needs to be refined. This process can be achieved applying standard vegetable oil refining steps as degumming, caustic refining, bleaching and deodorization. However, as the oil is sensitive to oxidation, process conditions and speed of operation are critical. Specifically, Martek Biosciences Corp. has developed an optimized method of purification conditions based in parameters including odour, taste and oxidative stability. In addition, the deodorized oil is blended with high oleic antioxidants, mainly, ascorbyl palmitate and tocopherols (Martek Biosciences Corporation, 2010; Zvi and Colin, 2010).

Guil-Guerrero and colleagues (2000) purified EPA from *Porphyridium cruentum* by simultaneous oil saponification (potassium hydroxide and ethanol at 60°C for 1 h) and fatty acid extraction of the microalgal biomass. After this, PUFA were concentrated by the urea method (methanol/urea ratio 3:1 w/w, crystallization temperature 28°C) and the concentrate was transmethylated (with acetyl chloride and methanol). Finally, EPA methyl esters were separated from PUFA concentrate by an argentated silica gel column chromatography with recoveries of 50.8% of EPA with 97% of purity.

Mendes and colleagues (2007) developed a procedure to concentrate DHA from *C. cohnii* involving saponification and methylation in wet biomass for further winterization and urea complexation. The process yield was 99.2% of total fatty acids. The urea/fatty acid ratio used was 3.5, and the temperatures of crystallization were 4°C and 8°C. Winterization (– 18°C) was used to fraction TAG with different melting points that are present in edible oils allowing the solid portion to crystallize. Subsequently, filtration of the two phases was performed. At low temperatures, long-chain saturated fatty acids, which have higher melting points, crystallize out and the PUFA remain in the liquid form, while urea molecules readily form solid-phase complexes with saturated FFA. In addition, urea complexation allows the handling of large quantities of material in simple equipment, requiring inexpensive solvents such as methanol or hexane. Finally, the separation is more efficient than with other methods such as fractional crystallization or selective solvent extraction. Besides reducing costs, urea complexation protects the ω-3 PUFA from autoxidation.

Oil recovery and purification process of DHA from algae oil has also been described in the petition of Market Biosciences Corporation (Martek Biosciences Corporation, 2010). For *Schizochytrium sp.*, a protease enzyme breaks the proteins in the cell releasing the oil from the cells into the culture broth forming an oil and water emulsion. Isopropyl alcohol is then added to break the emulsion and separate the oil. Subsequently, the alcohol is recovered from the broth, evaporated and distilled for reusing in subsequently extractions. In contrast, *C. cohnii* cannot be hydrolysed by enzymes because of the cellulosic thecal layer. Consequently, hexane solvent extraction on dry biomass is required. Later, the cell walls are removed through centrifugation, and the oil is recovered after solvent evaporation. Oil purification includes oil heating followed by the addition of an acidulated solution (oleic or citric), and later sodium hydroxide is added. These pH changes cause the formation of 'soaps' and 'gums' that can be easily separated and removed. Oil is

reheated and centrifuged to separate the refined oil. Later, sand chelators as citric acid, silica or clay are used to remove remaining residual compounds. Finally, the treated oil may be chill filtered to remove high-melting point components (stearines and waxes) to achieve the desired level of clarity. Subsequently, the oil is heated and then cooled to crystallize triacylglycerols and waxes. Later, diatomaceous earth is added to the chilled oil, and the crystallized solids are removed by filtration. Deodorizer is used to remove peroxides and the remaining low molecular weight compounds such as carbonyls and aldehydes that may cause off odours and flavours. Finally, tocopherols, rosemary extracts and ascorbyl palmitate are added to provide oxidative stability and flavour to the refined oil.

Natural pigments

Among functional ingredients identified from marine algae, natural pigments (NPs) have received particular attention. Natural pigments have an important role in the photosynthetic and pigmentation metabolism of algae, and also exhibit several beneficial biological activities like antioxidant, anti-carcinogenic, anti-inflammatory, anti-obesity, anti-angiogenic and neuroprotective (Guedes *et al.*, 2011; Pangestuti and Kim, 2011).

The three basic classes of NPs found in marine algae are chlorophylls, carotenoids and phycobiliproteins. Table 5 summarizes the main microalgal species used as a source of natural pigments along with the extraction and purification methods currently used for each species.

Chlorophylls are greenish lipid-soluble photosynthetic pigments with a porphyrin ring in their structure. Chlorophylls are found in algae, higher plants and cyanobacteria. Carotenoids are linear polyenes that act as light energy harvesters and antioxidants that inactivate reactive oxygen species (ROS) formed by the exposure of cells to environmental stress (Ioannou and Roussis, 2009). Carotenoids are considered accessory pigments since they increase the light-harvesting properties of algae by passing the light excitation to chlorophylls (Larkum and Kühl, 2005). Carotenoids can be classified into carotenes, which are unsaturated hydrocarbons, and xanthophylls, which present one or more functional groups containing oxygen. Beta-carotene, lutein and violaxanthin are green algae primary carotenoids, distributed within the chloroplasts along with chlorophylls. The substitution by a large variety of oxygen containing groups and their combination carries the existence of more than 600 xanthophylls (Lemoine and Schoefs, 2010). Phycobiliproteins are brilliant-coloured and water-soluble antennae-protein pigments organized in supramolecular complexes, called phycobilisomes, which are assembled

Table 5. Extraction of natural pigments in microalgae species.

Species	Product	Extraction/Purification method	Yield/Extraction efficiency	Reference
Chlorococcum sp.	Astaxanthin	Solvent system with methanol (75%) and dichloromethane (25%). French Pressure Cell (110 MPa). Solution filtration (0.45 µm). Saponification in darkness (50 mg NaOH in 100 ml methanol).	7.09 mg/g DW	Ma and Chen (2001)
Haematococcus pluvialis	Astaxanthin	Acid digestion, HCL 2N, 70°C. Acetone extraction for 1 h	87% efficiency	Sarada and colleagues (2006)
		Dodecane mixing for 48 h. Saponification with methanolic NaOH (0.02M). Sedimentation in darkness at 4°C, 12 h.	85% efficiency	Kang and Sim (2007)
		SC-CO_2 at 55 MPa and 343°K.	77.9% efficiency	Machmudah and colleagues (2006)
		Hexane : acetone : ethyl alcohol (100:70:70 %v/v).	N/A	Domínguez-Bocanegra and colleagues (2004)
		DMSO (55°C), vortex 30 s	N/A	Orosa and colleagues (2005)
		SC-CO_2 at at 20 MPa, 60°C, 2 ml of ethanol for 1 h of extraction time	2.45 mg/g DW	Fujii (2012)
		SC-CO_2 at 20 MPa, 55°C and 13% (w/w) ethanol for 120 min of extraction time. CO_2 expanded ethanol (50% %w/w ethanol), 7 MPa, 45°C, 120 min of extraction time.	83% recovery 124.2% recovery	Reyes and colleagues (2014)
Synechococcus 833	Allophycocyanin C-phycocyanin Phycoerythrin	Incubation of sample for 2 h at 37°C, nitrogen cavitation cycles at 1500 psi for 10 min, centrifugation for 40 min at 18 000 rpm to remove cell debris.	85.2–87.9% DW	Viskari and Colyer (2003)
Limnothrix sp.	C-phycocyanin	Distilled water, activated carbon (1% w/v) and chitosan (0.01 g/L) for extraction. Ammonium sulfate (25%) was used for purification at 4°C, overnight. Precipitate was resuspended in 0.1 M PBS (pH 7.0) and tangential flow filtration system (30 kDa membrane pore) was used for pigment concentration.	18% DW	Gantar and colleagues (2012)
Spirulina platensis	C-phycocyanin	0.1 M PBS at pH of 6.8 and sonication at 28 KHz for extraction. Ultracentrifugation at 200 000 × g for purification	90% purity	Furuki and colleagues (2003)
	Allophycocyanin C-phycocyanin	100 mM phosphate buffer (pH 7.0) at a ratio of 1:100 (w/v) with continuous stirring at 300 rpm at room temperature for 4 h.	N/A	Chaiklahan and colleagues (2012)
Leptolyngbya sp. KC45	Phycoerythrin	Ammonium sulfate at 85% saturation. Purification by three consecutive chromatographic steps; hydroxyapatite column eluted with 100 mM phosphate buffer (pH 7) 0.2 M of NaCl, a Q-sepharose column and a Sephacryl S-200 HR resin.	A_{565}/A_{280} = 17.3 1.36% yield	Pumas and colleagues (2012)
Porphyridium cruentum	Phycoerythrin	Homogenization in 1 M acetic acid–sodium acetate buffer sonication for 10 min, ammonium sulfate precipitation (65% saturation) and dialysis, followed by ion exchange chromatography.	32.7% DW	Bermejo Román and colleagues (2002)
		Cell maceration with glass beads, simultaneous recovery and purification with a PEG-phosphate ATPS.	A_{565}/A_{280} = 2.8 76% recovery	Benavides and Rito-Palomares (2006)
Phormidium sp. A27DM	Phycoerythrin	Freeze-thaw cycles (−30°C and 4°C) in 1 M Tris Cl Buffer, two-step ammonium sulfate precipitation at 20% and 70% saturation and purification by gel permeation chromatography with a Sephadex G-150 matrix.	A_{565}/A_{280} = 3.9 62.6% yield	Parmar and colleagues (2011)

DW, dry weight.

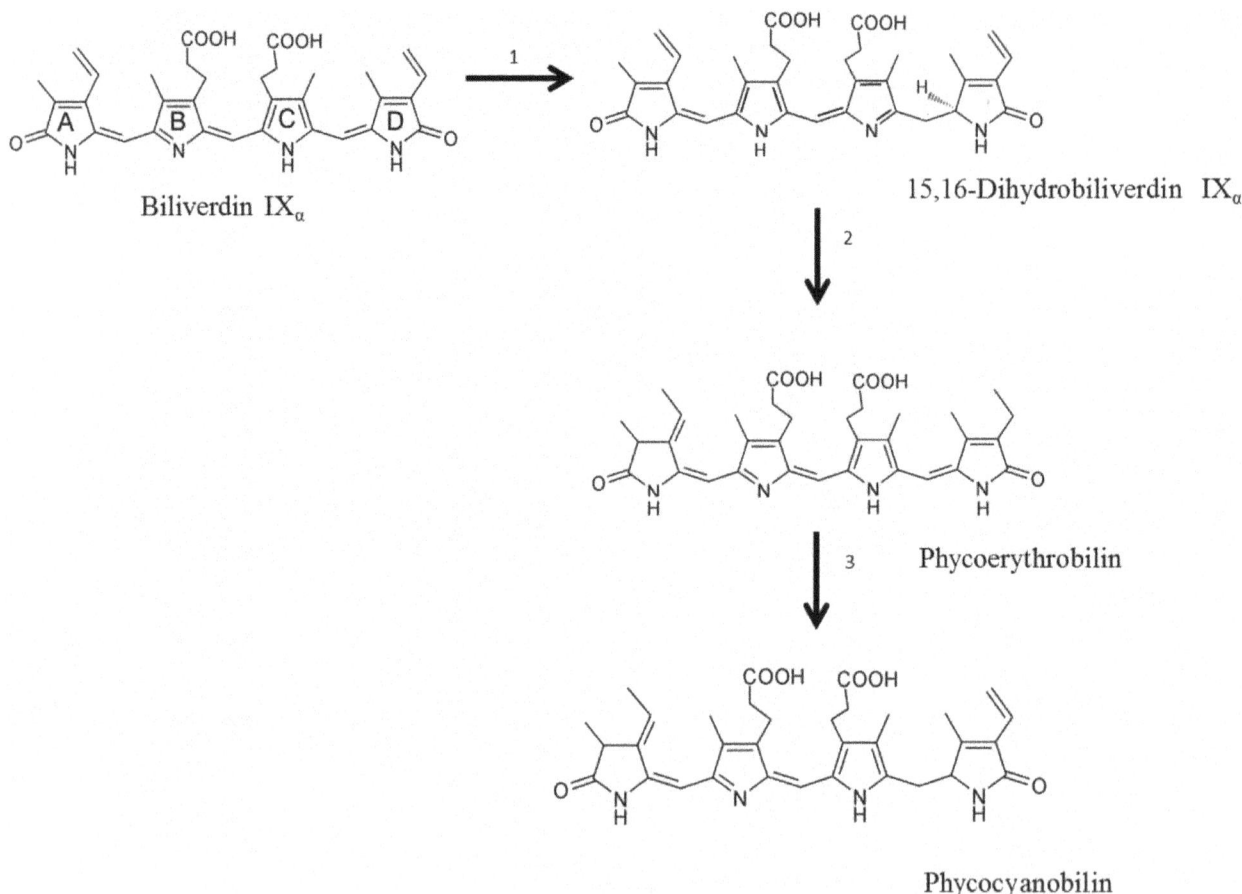

Fig. 2. Biosynthesis pathway for phycobilins formation from Biliverdin (Brown *et al.*, 1984; Lamparter *et al.*, 2002). Reactions are: (i) biliverdin 15,16-reductase, (ii) bilin 2,3-reductase (iii) phycobilin (15,16-metylene-to-18^2,18^3-ethyl) isomerase.

on the outer surface of the thylakoid membranes (Glazer, 1994; Hemlata and Fareha, 2011). The colours of phycobiliproteins originate mainly from covalently bound prosthetic groups that are open-chain tetrapyrrole chromophores bearing A, B, C and D rings named phycobilins (Sekar and Chandramohan, 2007). Phycobiliproteins absorb energy in portions of the visible spectrum (450–650 nm) and function as accessory pigments for photosynthetic light collection (Batista *et al.*, 2006). Phycobiliproteins are found in prokaryotic cyanobacteria and eukaryotic red algae (MacColl, 1998). In cyanobacteria and red algae, four main classes of phycobiliproteins are produced: allophycocyanin (APC, bluish green), phycocyanin (PC, blue), phycoerythrin (PE, purple), and phycoerythrocyanin (PEC, orange) (Bryant *et al.*, 1979; Sekar and Chandramohan, 2007). In many red algae, phycoerythrins are the most abundant phycobiliproteins, while phycocyanins are abundant in cyanobacteria (Bogorad, 1975; Glazer, 1994). Different biosynthesis pathways of phycobilins have been suggested including δ-Aminolevulinic acid, heme or Biliverdin as precursors (Brown *et al.*, 1984; Beale, 1993). Figure 2

shows the metabolic production of phycobilins from Biliverdin.

Astaxanthin

Metabolic production and applications. Astaxanthin (3,3′-dihydroxy-β,β′-carotene-4,4′-dione) is the main carotenoid found in aquatic animals (Hussein *et al.*, 2006). It serves as a precursor of vitamin A and is associated with cell reproduction and embryo development (Machmudah *et al.*, 2006). Studies on food, cosmetic and medical application of astaxanthin have been undertaken because its superior antioxidant activity compared with α, β-carotene, lutein, lycopene, cantaxanthin and vitamin E. As well in other biological functions is gaining widespread popularity as human dietary supplement (Machmudah *et al.*, 2006; Zhao *et al.*, 2006; Yuan *et al.*, 2011). The presence of this secondary carotenoid is equivalent to higher survival capacity of the cells since it enhances the cell resistance to oxidative stress generated by certain conditions of light, UV-B irradiation and nutrient conditions (Lemoine and Schoefs, 2010).

Astaxanthin can be synthetically produced or obtained from natural sources, e.g. microalgae, yeast or crustacean byproducts (Machmudah *et al.*, 2006). In addition, it has unique chemical properties based on its molecular structure. The presence of the hydroxyl (OH) and keto (C = O) moieties on each ionone ring explains some of its unique features, like the ability to be esterified, to have higher antioxidant activity and more polar nature than other carotenoids. In its free form, astaxanthin is considerably unstable and particularly susceptible to oxidation (Hussein *et al.*, 2006). However, free astaxathin is commercially more important than astaxanthin esters (Ma and Chen, 2001).

The green algae *Haematococcus pluvialis*, is one of the most important biological source of astaxanthin. Other microalgae species capable to accumulate secondary carotenoids are: *Botryococcus braunii, Chlamydomonas nivalis, Chlorella* sp., *Chlorococcum* sp., *Chloromonas nivalis, Coelatrella striolata* var. multistriata, *Dunaliella* sp., *Eremospherea viridis, Euglena* sp., *Neochloris wimmeri, Scenedesmus* sp., *S. obliquus, Scenedesmus komarekii, Scotiellopsis oocystiformis, Protosiphn botryoides, Tetracystis intermedim* and *Trachelomonas volvocina* (Lemoine and Schoefs, 2010). In contrast to the listed species, astaxanthin content in *H. pluvialis* represents 90% of total carotenoids. Therefore, extraction of this compound from other species represents a disadvantage in many markets due to purification steps and processing costs (Borowitzka, 2013).

Figure 3 shows the methabolic pathway of asxtanthin production from β-Carotene in *H. pluvialis*. Astaxanthin accumulation (2–3% w/w) in *H. pluvialis* has been observed only in encysted cells. This encystment has been reported to be induced by unfavourable growth conditions such as nitrogen and phosphorus starvation, heterotrofic media, salt stress or elevated temperature (Sarada *et al.*, 2002; 2006). In the absence of light, the use of suitable carbon source is crucial for attaining high biomass yield in algal culture. Moreover, nitrogen limitation in the presence of excess organic carbon substrates has increased astaxanthin production in mixotrophic cultures (Ip and Chen, 2005). The astaxanthin in encysted *H. pluvialis* cells consists of approximately 70% monoesters, 25% diesters and 5% free form (Lorenz and Cysewski, 2000).

Mature red cyst cells form a thick amorphous layer as a secondary wall inside the extracellular matrix, creating large interspace between the plasmalemma and the secondary wall (Montsant *et al.*, 2001; Hejazi and Wijffels, 2004; Wang *et al.*, 2004; Kang and Sim, 2007), which limits the bioavailability of astaxanthin from *H. pluvialis* (Kang and Sim, 2007).

Kobayashi and colleagues (1991) tested several carotenoids precursor in heterotrophic media. They con-

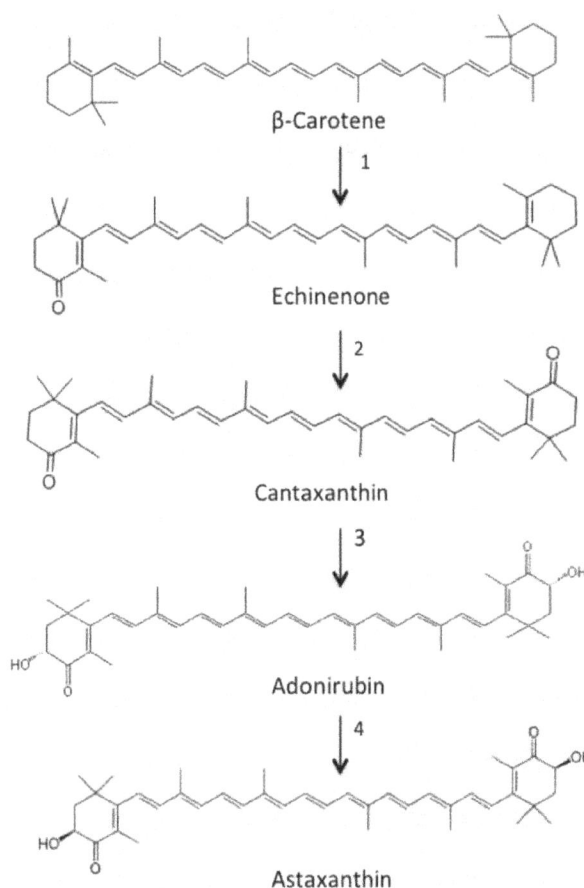

Fig. 3. Biosynthethic pathway for the formation of astaxanthin in the microalgae *H. pluvialis* (Lemoine and Schoefs, 2010). Enzymes are 1: 4,4′- ketolase, 2: 4,4′-ketolase, 3: 3′3-Hydroxylase, 4: 3′3-Hydroxylase.

clude that astaxanthin production in *H. pluvialis* at 4.5 Klux for 24 h continuous lighting with acetate (14.6 mM), yeast extract 0.5 g/L, L-asparagine 2.7 mM, trace minerals and metabolites as pyruvate (12 mM), mevalonate (3 mM), malonate (3 mM) or dimethylacrylate (6 mM) were effective in carotenoid formation. Tripathi and colleagues (1999) tested various culture media for astaxanthin production in *H. pluvialis* based on Bold's Basal Medium (BBM) (Kantz and Bold, 1969) and sodium acetate, L-asparagine and yeast extract (Kobayashi *et al.*, 1991). The results showed that BBM was the best for cell growth. However cells in heterotrophic medium (sodium acetate 1.99 g/L, L-Asparagine 0.4 g/L, Yeast extract 2 g/L) showed early encystment after 5 days with 1.86% w/w astaxanthin content at 1.5 Klux and 25°C in 40 days culture. Sarada and colleagues (2002) studied the influence of stress culture conditions on astaxanthin production in *H. pluvialis*. The results showed that BBM culture media at pH 7.0, salinity 0.5% (w/v) and sodium acetate (2 mM) for 12 days, allowed high astaxanthin production (1.2%w/w). Finally, further experiments based in different nitrogen sources indicated that calcium nitrate was the most effective in inducing astaxanthin formation, while sodium nitrate allowed higher biomass productivity.

Katsuda and colleagues (2004) cultivated *H. pluvialis* under Kobayashi and colleagues (1991) basal media and light emitting diodes (LEDs) at different wavelengths to study the effects of wavelength on the cells growth rate and astaxanthin accumulation. The results showed that high cell density and astaxanthin concentrations were obtained under illumination with LEDs emitting shorter wavelengths. The effectiveness of illumination at 380–470 nm for astaxanthin production was confirmed by the fact that the colour of the fermentation broths changed from green to red at 4–5 days. In addition, under illumination with the blue LEDs, the astaxanthin concentration increased to 16 μg/ml at light intensity of 8.0 μmol-photon/m²s while at 12.0 μmol-photon/m²s it increased to 25 μg/ml (astaxanthin content 6.5% of dry-cell weight). Lababpour and colleagues (2005) also reported the use of blue LEDs at 12.0 μmol-photon/m²s in *H. pluvialis* mixtotrophic culture to increase astaxanthin content in this algae, producing 70 μg/cm³ of astaxanthin in 400 h culture. Park and colleagues (2006) studied flashing light as light source for high astaxanthin production by *H. pluvialis*. The results indicated that light in short intense flashes induced astaxanthin synthesis more efficiently than giving the same amount of light energy in a continuous manner. Moreover, internally illuminated flashing light was found to be more efficient than externally illuminated flashing light. All these results indicate that light illumination induces morphological changes of *H. pluvialis*, enhancing the accumulation of astaxanthin.

Diverse authors have proposed immobilized biofilm cultivation of *H. pluvialis* in order to avoid excessive expenses of close photobioreactors (Wan *et al.*, 2014; Zhang *et al.*, 2014). Zhang *et al.* (2014) studied the effect of inoculum density, light intensity, nitrogen concentration and medium volume growth for astaxanthin accumulation. The optimized results showed that an inoculum density of 10 g/m² and 100 μmol/m²s of light intensity are needed. Then, nitrogen supply (in circulating media) of 30 L of BG-11 medium (Stanier *et al.*, 1971) with 1.8 mM of initial sodium nitrate strength for each square meter of cultivation surface should be supplied. Maximum astaxanthin productivity was 160 mg/m²d. According to the authors, this amount is higher than other productivities reported for indoor cultivation systems.

Astaxanthin is also produced by other microalgae species. Ma and Chen (2001) enhanced free *trans*-astaxanthin production on *Chlorococcum sp* from 3.6 mg/g cell DW to 5.72 mg/g cell DW when the culture was supplemented with hydrogen peroxide (H_2O_2) (0.1 mM) under mixotrophic (glucose 44 g/L) culture conditions at 22 μE/m²s for 7 days. After saponification,

7.09 mg of astaxanthin per g cell DW were achieved, with 81% as free *trans*-astaxanthin. Qin and colleagues (2008) studied the accumulation and metabolism of astaxanthin in *S. obliquus* in two-stage cultures. The results indicated that first culture stage (48 h) at 180 μmol/m²s and 30°C increased astaxanthin production (free and esters) with 44.6% of secondary carotenoids. In contrast, during the second stage (72 h) at 80 μmol/m²s and 25°C, the production of primary pigments as chlorophyll was enhanced. Ip and Chen (2005) analysed astaxanthin production of the green microalgae *Chlorella zofingiensis* in heterotrophic media with different C/N ratio with glucose concentrations (5–60 g/L) at 30°C and complete darkness. The alga exhibited the highest specific growth rate (0.031 h⁻¹) and the highest growth yield (0.44 g g⁻¹) at a glucose concentration of 20 g/L. However, the highest content and yield of astaxanthin were 1.01 mg/g biomass and 10.29 mg/L, respectively, at 50 g/L glucose. Therefore, glucose might be essential for providing the carbon skeleton for the formation of secondary carotenoids including astaxanthin formation in the carotenoids biosynthetic pathway. However, although nitrate is not directly involved in the biosynthetic pathway of carotenoids, it can modify the normal cellular metabolism such as protein synthesis, affecting the pigments formation in algae (Ip and Chen, 2005). In case of other nutrients, He and colleagues (2007) reported that sulfur starvation was more effective in high formation of astaxanthin compared with phosphorus or iron starvation. Therefore, culturing conditions under nitrogen and phosphate starvation, high illumination intensity, salt (NaCl) and hydrogen peroxide induce the production and accumulation of astaxanthin in algal cells.

Extraction and purification methods. Currently, different astaxanthin extraction methods are used in algae, including organic solvents, breakdown pretreatment process of encysted cells (cryogenic grinding and acid/base treatment), enzyme lysis (kitalase, cellulose and Abalone acetone powder mainly β-glucoronidase), mechanical disruption and spray drying (Sarada *et al.*, 2006; Kang and Sim, 2007).

Since astaxanthin is a xantophyll pigment, around 95% of astaxanthin accumulated in *H. pluvialis* cells is esterified. Thus, the extraction of astaxanthin can require a hydrolisis step in order to free the astaxanthin molecule. Sarada and colleagues (2006) tested the extractability of carotenoids from *H. pluvialis* cells. Results showed that a treatment with hydrochloric acid (2N) for 10 min at 70°C followed by acetone extraction for 1 h, extracted 87% (w/w) of astaxanthin in cells without affecting the astaxanthin composition. Kang and Sim (2007) developed a two-stage solvent procedure with dodecane and methanol to extract free astaxanthin from *H. pluvialis*. The culture broth was mixed with dodecane, and the mixture was sedimented 48 h. Later, the dodecane extract was separated from the cell debris and placed in another tank and mixed with NaOH in methanol (0.02 M) at a volume ratio of 1:1 (saponification of astaxanthin esters to free form). Later, the tank was kept in darkness at 4°C (12 h) for free astaxanthin extraction in the methanol phase. The results indicated a total recovery yield of free astaxanthin of over 85% DW. Zhao and colleagues (2006) studied the effect of microwaves and ultrasound on the stability of synthetic astaxanthin isomers, concluding that microwaves induce the conversion of other astaxanthin isomers, while ultrasound degrades this pigment into colourless compounds because of cavitation produced in the solvent by the propagation of an ultrasonic waves.

Supercritical fluids (SCF) are now widely used for extraction purposes since they are more efficient than the traditional liquid solvents. Supercritical fluids act like liquid solvents with selective dissolving powers. Supercritical CO_2 (SC-CO_2) is by far the most common supercritical fluid used in extraction of natural compounds and food processing (Machmudah et al., 2006). Supercritical CO_2 is non-flammable, non-toxic and relatively inert. Moreover, the addition of small amounts of other solvents (called entrainer or co-solvent) increase the solvent power of the SCF (Mendes et al., 2003). The extract obtained by SCF extraction is highly concentrated since the solvent is separated by process depressurization. Therefore, the extract is free of residues. The resulting CO_2 gas stream can be recycled, making SC-CO_2 extraction an environmental-friendly process (Machmudah et al., 2006).

Machmudah and colleagues (2006) tested different conditions to extract astaxanthin from H. pluvialis by SC-CO_2. The highest amount of total extract, astaxanthin extracted and astaxanthin content in the extract were 21.8%, 77.9% and 12.3%, respectively, at 55 MPa and 343 K. Results showed that the increment in temperature did not increase extraction yields. In contrast, higher pressure increased extraction yields. This is expected since CO_2 density increases at higher pressure, and therefore the solvent power to dissolve the substances increases. However, other compounds (lipids, proteins or polysaccharides) are also extracted leading more impurities in the extract. Total extract slightly increased at higher CO_2 flow rate (4 ml/min) with slight increment at higher flows. In contrast, Ethanol as entrainer 1.67% (v/v) at 40 MPa, 313 K and CO_2 flow rate of 3 ml/min increased 20% the astaxanthin extract. However, higher ethanol concentration (7.5% v/v) decreased the astaxanthin extracted, since high entrainer concentration decreases SCF density.

Fujii (2012) by SC-CO_2 and acid treatment improved the extraction efficiency of astaxanthin in Monoraphidium sp. GK12 cells. The author tested also the effect of ethanol as co-solvent in the extraction yield. Results showed that at 20 MPa, 60°C for 1 h extraction time, astaxanthin yield 2.02% mg/g-dry biomass. However, the addition of 2 ml of ethanol improved the extraction yield to 2.45 mg/g-dry biomass. In case of temperature and reaction times, the author concluded that biomass/solvent ratio of 1/20 at 30°C and 15 min are sufficient for astaxanthin extraction. Finally, when the extracts were treated with acid (sulphuric and chlorhydric acid. 0.01 N in the mixture), a large portion of the chlorophyll was removed ($\geq$ 80%).

Reyes et al., (2014) studied the effect that pressure, temperature and ethanol content have on the yield, astaxanthin content and antioxidant activity of H. pluvialis extracts during 120 min of SC-CO_2 extraction. Also, the authors studied the effect of temperature and ethanol content using CO_2-expanded ethanol (CXE) at 7 MPa. Optimized results showed that at 20 MPa, 55°C and 13% (w/w) ethanol content were the best conditions for SC-CO_2 extraction, with 282.5 mg/g, 53.48 mg/g and 83% (w/w) for extraction yield, astaxanthin content and astaxanthin recovery respectively. However, with CXE extraction (50% w/w ethanol content in CO_2 and 45°C), better results were obtained: 333.1 mg/g, 62.57 mg/g and 124.2% w/w of extraction yield, astaxanthin content and astaxanthin recovery respectively. Also, antioxidant activity in both extracts was the same. Therefore, the authors concluded that CXE extractions are much better than conventional extraction based in time difference, since just 2 h are needed for the extraction.

Phycoerythrin

Metabolic production and applications. Phycoerythrin (PE) is a red coloured phycobiliprotein with absorption maxima range at 565 nm. Phycoerythrin is found in the chloroplast of cyanobacteria and red algae. In aqueous solution, phycoerythrin exists as an hexameric disk-shaped complex of $(\alpha\beta)6\gamma$ subunits, with each hexamer carrying 34 bilins. There are two main classes of phycoerythrin found in red microalgae, B-phycoerythrin and R-phycoerythrin. R-phycoerythrin carries 25 phycoerythrobillins (PEBs) and nine phycourobillins (PUBs) while B-phycoerythrin carries 32 PEBs and two PUBs (Glazer, 1994).

Unlike other phycobiliproteins used as food and cosmetic colorant, phycoerythrin has advantageous physical properties that make it suitable for applications in clinical research and molecular biology. Phycoerythrin is highly fluorescent, exhibits high photostability, it has a fluorescent quantum yield independent of pH and a large Stokes shift that minimizes interferences from Rayleigh and Raman scatter (Sekar and Chandramohan, 2007; Pumas et al., 2012). Glazer (1994) pointed out that phycoerythrin is highly soluble in water and is negatively charged at physiological pH values, allowing minimal non-specific binding to cells, due to the fact that cells are also negatively charged. In addition, phycoerythrin can be used as label for biological molecules, as a reagent in fluorescence immunoassays, flow cytometry, fluorescence microscopy and diagnostics (Spolaore et al., 2006).

The microalgae species and culture conditions determine the variations in the final yield of the process. Specifically, phycoerythrin has been extracted and purified from the red algae P. cruentum (Tcheruov et al., 1993; Bermejo Román et al., 2002; Benavides and Rito-Palomares, 2004; Kathiresan et al., 2006), Phormidium sp. A27DM (Parmar et al., 2011) and the thermophile cyanobacterium Leptolyngbya sp. KC45 (Pumas et al., 2012). Pumas and colleagues (2012) screened cyanobacteria species from hot spring to obtain phycoerythrin with higher temperature stability than the phycoerythrin extracted from mesophile species like P. cruentum. Kathiresan and colleagues (2006) studied the effect of the macronutrients present in the culture media on the production of phycoerythrin by P. cruentum, finding that phycoerythrin production is mainly affected by phosphate, nitrate, chloride and sulfate concentration. Other authors have determined that the synthesis of phycobiliproteins is increased under light-limiting conditions (Bermejo Román et al., 2002).

Extraction and purification methods. Phycoerythrin must be highly purified in order to meet the standards of the pharmaceutical or molecular biology field. Purity is usually determined as the absorbance ratio of A_{565}/A_{280} which defines the relationship between the presence of phycoerythrin and other contaminating proteins. A purity ratio A_{565}/A_{280} greater than four corresponds to diagnostics and pharmaceutical grade phycoerythrin (Benavides and Rito-Palomares, 2004). Some authors use the A_{615}/A_{565} ratio to determine phycoerythrin purity in relation to phycocyanin, which is its closest contaminating protein.

Since phycoerythrin is an intracellular protein, the general purification process relies in three stages: protein extraction by cell disruption, primary recovery and purification. Disruption methods like sonication, mechanical maceration and lysozyme treatment have been successfully used to extract phycoerythrin from microalgae. Choosing the right cell disruption method has a significant impact in the recovery of the overall process. Benavides and Rito-Palomares (2006) assessed the protein efficiency release of two disruption methods, manual maceration and sonication, obtaining phycoerythrin release efficiency 5.5 times higher with sonication. Jubeau and colleagues (2012) evaluated the extraction of B-phycoerythrin from P. cruentum by high-pressure cell disruption varying the parameters of pressure (25–270 MPa), number of passages (1 to 3) and extraction mediums (culture media or distilled water). The authors found that contaminating proteins present in the cytoplasm are extracted at lower pressures (> 90 MPa), whereas B-phycoerythrin can be extracted at higher pressures since it is found inside the chloroplasts. Two passages achieved higher released of B-phycoerythrin, while three passages caused a loss of B-phycoerythrin due to the denaturation of the protein-pigment. Hence these authors propose a two-step selective extraction with a first passage at 50 MPa in culture medium followed by a second passage at 270 MPa in distilled water, achieving a 0.79 purity ratio.

Most authors carry out the primary recovery step by selective precipitation with ammonium sulfate. This solubility-based method is fast and fairly inexpensive. Bermejo Román and colleagues (2002) did a single precipitation step with ammonium sulfate at 65 % saturation. Parmar and colleagues (2011) recovered phycoerythrin from Phormidium spp. A27DM with a two-step ammonium sulfate precipitation, at 20% and 70% saturation. While Pumas and colleagues (2012) treated the cell homogenate of Leptolyngbya sp. KC45 with ammonium sulfate at 85% saturation.

Purification is a critical step of the downstream processing of phycoerythrin. Purification is typically achieved by chromatographic methods like ion exchange chromatography, hydroxyapatite chromatography, gel filtration and expanded bed absorption chromatography. Parmar and colleagues (2011) purified phycoerythrin from Phormidium sp. A27DM with a single-step gel permeation chromatography using a Sephadex G-150 matrix pre-equilibrated and eluted with a 10mM trisCl buffer (pH 8.1) at a flow rate of 60 ml h^{-1}. This protocol yielded a final purity ratio of 3.9. In contrast, the purity ratio achieved after ammonium sulfate precipitation was around 1.5.

Pumas and colleagues (2012) extracted and purified phycoerythrin from Leptolyngbya sp. KC45. Purification of the protein was carried in three chromatographic steps. The concentrated extract obtained after ammonium sulfate precipitation was treated with a hydroxyapatite column eluted with 100 mM phosphate buffer (pH 7) 0.2 M of NaCl. A yield of 37.35% and $A_{565}/A_{280} = 6.75$ was achieved. The eluted fractions were collected and applied on a Q-sepharose column, resulting on purity ratio of 15.48 but with a low process yield of 9.65%. After a third stage performed with a Sephacryl S-200 HR resin, high purity phycoerythrin was obtained ($A_{565}/A_{280} = 17.3$). However, extraction yield was low (1.36%). According to Kawsar and colleagues (2011) hydroxyapatite chromatography is not a reliable method since the separation ability depends on the quality of the particles, and the regeneration capacity is not good since the material might bind to other contaminating complexes.

Bermejo Román and colleagues (2002) purified phycoerythrin from *P. Cruentum* using an anionic chromatographic column of Diethylaminoethanol (DEAE) cellulose. Elution was performed as a discontinuous gradient of acetic acid-sodium acetate buffer (pH 5.5). The best results were achieved with flow rate of 100 ml h^{-1} with optimum protein loading of 7.3 mg. In addition, scale-up to a large preparative level was also tested. The capacity of the column was increased by modifying the cross-sectional area and maintaining the same superficial velocity. Final phycoerythrin recovery of 32.7% was achieved. The authors noted the importance of increasing the yield of the earlier stages since the recovery after cell disruption, and ammonium sulfate precipitation was only 45.4%.

Benavides and Rito-Palomares (2004) focused on an alternative method to concentrate and purify phycoerythrin with a polyethylene glycol (PEG)-phosphate aqueous two-phase system (ATPS) used as a single purification step. Aqueous two-phase system is an advantageous technique due to its biocompatibility and can easily be scaled. The authors found that it is possible to concentrate phycoerythrin in the PEG-rich top phase using a PEG 1450-phosphate system. The system constructed with a volume ratio (Vr) of 1, PEG 1450 of 24.9% (w/w), phosphate concentration of 12.6% (w/w) and pH value of 8 allowed the recovery of phycoerythrin with a 2.9 purity ratio. However, recovery varied depending on the parameters of the ATPS. The highest protein recovery was 73%, and the lowest recovery was 45%. Finally, purity and recovery rates were highly correlated to the system pH.

Bermejo and colleagues (2007) purified phycoerythrin from *P. Cruentum* by expanded bed adsorption chromatography (EBA) using a DEAE adsorbent. The authors focused on maximizing product recovery rather than purity, since the process is intended to replace low-resolution methods. A crude extract containing 0.033 mg phycoerythrin/ml was applied when the expanded bed reached a stable height. After processing the crude extract, the column was washed with a 50 mM acetic acid-sodium acetate buffer (pH 5.5). Finally, bound phycoerythrin was recovered by isocratic elution with a 250 mM acetic acid-sodium acetate buffer (pH 5.5). The results showed that a maximum recovery of 71% was achieved when the expanded bed volume was twice the settled bed volume (flow rate = 198 cm h^{-1}), resulting an application time of 108 min. Also, the process was scaled up from a 15 mm diameter column to a 60 mm diameter column reaching a recovery of 74% with a purity ratio higher than 3. As a result, the use of EBA chromatography allowed partial concentration of the product. Therefore, EBA chromatography works as a preparative method with little product loss and is suitable to large scale production (Nui *et al.*, 2010).

Phycocyanin

Metabolic production and applications. Phycobiliproteins are the major photosynthetic accessory pigments in cyanobacteria and red algae. However, most studies of phycocyanin (PC) have been done in *Spirulina sp.*, a cyanobacteria with high protein content. Phycobilisomes from *Spirulina sp.* consists of APC cores surrounded by c-phycocyanin (CPC) peripherally. c-Phycocyanin is the major phycobiliprotein in *Spirulina* and constitutes up to 20% of its DW. In solution, phycocyanin is found as a complex mix of monomers, trimers, hexamers and other oligomers, and its molecular weight, therefore, ranges from 44 to 260 kDa (Chaiklahan *et al.*, 2012). The reported molecular mass of CPC in different cyanobacteria ranges for the oligomer between 81 and 215 kDa and that for individual subunits (α,β) between 15.2 and 24.4 kDa. In most cases, the subunits are organized in trimmers (Patel *et al.*, 2005).

Phycocyanin is a colorant commonly used in food and cosmetics. In addition, it has gained importance probes for immunodiagnostics due to its fluorescence properties (Sarada *et al.*, 1999). Phycocyanin is an efficient scavenger of oxygen free radicals. Therefore, therapeutic use of PC appears to be promising since many diseases are related to an excessive formation of ROS (Romay *et al.*, 2003). However, the use of PC in food and other applications is limited due to its sensitivity to heat treatment, which results in precipitation and fading of the blue colour. Sodium azide and dithiothreitol are commonly used as preservatives for phycocyanin for analytical purposes, but they are toxic and thus cannot be used for food grade phycocyanin production. In contrast, sugars and polyhydric alcohols (safe for consumption) have been used to stabilize proteins. In addition, studies have reported that modification of the protein conformation itself can improve the stability of proteins (Chaiklahan *et al.*, 2012).

Extraction and purification methods. Allophycocyanin, a bluish green protein and CPC, a blue protein have the major absorption (λ_{max}) in the visible region of 650–655 nm and 610–620 nm, respectively, with emission light at 660 nm and 637 nm respectively (Bryant *et al.*, 1979; Sekar and Chandramohan, 2007). Determinations of these phycobiliproteins by spectrophotometry have been assessed by different authors (Furuki *et al.*, 2003; Chaiklahan *et al.*, 2012). The purity ratio of the phycocyanin extract is determined by the A_{620}/A_{280} ratio. High purity in the extract refers to high purity ratios (Chaiklahan *et al.*, 2012). Absorbance ratio ≥ 0.7 refers to food grade pigment, while reagent and analytical grade correspond to 3.9 and ≥ 4.0 respectively (Borowitzka, 2013).

Phycocyanin is water-soluble and can be easily extracted as a protein–pigment complex (Chaiklahan et al., 2012). General procedure includes a first extraction in buffer solutions (phosphate buffer) with sonication or ultrasound as cell disruption pretreatment. However, pretreatment times should not be long in order to avoid proteins destabilization. Later, proteins precipitation and recovery by ultracentrifugation or filtration are carried.

Sørensen and colleagues (2013) evaluated different extraction techniques for CPC extraction from the algae Galdieria sulphuraria. Also, all extractions were carried using PBS at pH of 7.2. The results showed contents of 25–30 mg/g of CPC for this algae. Ammonium sulfate concentration above 1.28 mol/L ensured only CPC precipitation with purity of 0.7. In case of ultrafiltration, more than 50% was lost at 100 kDa tangential flow filter, while 79% was retained by the 50 kDa filter. The authors also proposed to combine ammonium sulfate fractionation with the other methodologies tested (anion exchange chromatography, tangential flow filtration) in order to enhance purity of the recoveries (3.5–4.5).

Viskari and Colyer (2003), compared different cell disruption and extraction methods for phycobiliproteins (APC, CPC and PE) from Synechococcus 833. The best results were obtained using a buffer composed of 3% Chaps and 0.3% asolectin for protein solubilization combined with nitrogen cavitation for cell wall disruption. The extraction efficiencies varied between 85.2 and 87.9%.

Furuki and colleagues (2003) studied the effect of ultrasound on the extraction of phycocyanin from Spirulina platensis. Phycocyanin extraction was carried below 10°C, with 0.1 M phosphate buffer (pH 6.8). Later the solution was sonicated and centrifuged to remove cell debris. Results indicated that phycocyanin was not destroyed by ultrasonic irradiation as long as the irradiation time was appropriate in length. The purity of PC at 20 kHz was about 80% while at 28 kHz, the highest purity was obtained (90%).

Gantar and colleagues (2012) tested a procedure to extract, isolate and purify CPC from Limnothrix sp. concluding that this strain produces 18% CPC of total dry biomass. Purification of the extract was carried with chitosan (0.01 g/L) and activated carbon (1% w/v) followed of saturation with ammonium sulfate. Finally, extracts were concentrated by tangential flow filtration. The results showed that the introduction of the treatment with activated carbon and chitosan, significantly improved the purity of the pigment in the crude extract. This step increased the 620/280 ratio from 2.0 to 3.6 but also contributed to reduction of the pigment yield. Efficacy of CPC precipitation with different ammonium sulfate concentrations showed that an increment in the ammonium sulfate concentration from 20% to 25%, significantly increased the recovery of CPC. However, further increase of the

ammonium sulfate concentration reduced the purity of the pigment. The final step, which included tangential flow filtration, yielded 8% CPC of dry biomass with the purity index of 4.3. In comparison, Patel and colleagues (2005) and Zhang and Chen (1999) reported 50% ammonium sulfate solution for high purity extraction of CPC in Spirulina since lower concentration of ammonium sulfate extracted other proteins. Therefore, Gantar and colleagues (2012) concluded that there are differences in physicochemical properties of CPC among species. In addition, the authors reported for CPC isolated from Limnothrix 37-2-1, an oligomeric mass of ~50 kDa, with α and β subunits (organized in dimmers) of 13 and 11 kDa, one of the smallest molecular masses of CPC reported.

As mentioned, phycocyanin extract is not for a long term stable, since conditions as light, temperature and other microorganisms induce protein decomposition. Therefore, different studies have been assessed to increase phycocyanin's extract stability. Sarada and colleagues (1999) studied the stability of phycocyanin from Spirulina sp. for a period of 4 weeks at pH ranging from 2.5 to 13 using different buffers and different temperatures. The results indicated that phycocyanin was stable over a pH range of 5 to 7.5 at 25°C. However at lower temperature, phycocyanin was stable for longer periods.

Lawrenz and colleagues (2010) compared different extraction methods of phycocyanin in cell of the cryptophyte Rhodomonas salina and the cyanobacteria Synechococcus bacillaris. Also, the authors studied the effect of time storage in the degradation of phycobilins. Extractions by PBS at pH 6 were carried. Results showed that centrifugation of the crude extract ($10.870 \times g$) was slightly more effective than filtration by cellulose acetate membrane. Also, for cryptophyte and cyanobacteria samples, extraction times of 4 h and 96 h, respectively, were sufficient for total pigment recovery. Finally, samples can be stored at $-80°C$ during 6 months without degradation.

Chaiklahan and colleagues (2012) tested the stability of phycocyanin extracted from Spirulina sp. by different temperature and pH with the addition of certain preservatives (glucose, sucrose, D-sorbitol, sodium chloride, ascorbic acid, citric acid, sorbic acid and sodium azide). The best results were obtained at 47°C and pH value of 6.0 with relative concentration (C_R) value of 94%. Experiments under commercial high temperature–short time pasteurization conditions (72°C for 15 s) showed that phycocyanin remained almost intact. The C_R value of the solutions at pH 5.0, 6.0 and 7.0 was in 96–100% after incubation at 74°C for 1 min. In addition, phycocyanin solution was more stable at low temperature (4°C). Although the C_R value of the solutions at room temperature remained higher than 80% after incubation for 10 days, turbidity and odour of the solution was observed. In addition, the maximum stability at 50°C was observed at pH 6.0, while at 60°C the maximum stability was at pH 5.5. Also, glucose (20%), sucrose (20%) and sodium chloride (2.5%) were considered suitable for prolonging the stability of the phycocyanin extract. Antelo and colleagues (2008) reported that the extract of phycocyanin was longer stable at high temperature with low pH. Therefore, temperature and pH are inversely proportional with respect to the degradation of phycobiliproteins. In addition, at 50°C and 55°C, phycocyanin was more stable at pH 6.0 while at 57°C and 65°C, extract was more stable at pH 5.0. Finally, phycocyanin solution was longer stable at pH range of 5.5–6.0. In contrast, pH 5.0 leads to low stability.

Challenges and future perspectives

Most of the world companies commercializing products from algae started producing fuels for the transport sector (biodiesel, jet fuel). However, production process is not yet economically competitive with petroleum-based fuels. Therefore, economic incentives are needed besides low price feedstocks and low price processing. Facing this situation, these companies have invested in the production of other compounds with high price commercialization, pigments and fatty acids. Currently, these products have high demand and commercialization price. However, research is needed to reduce product losses during purification steps.

As mentioned, microalgal cultivation conditions have been exhaustively explored for strains that synthesize current high-value compounds (pigments, PUFAs, polysaccharides). The extraction and purification methods for high-quality lipids, carotenoids like astaxanthin, and phycobiliproteins like phycocyanin, and phycoerythrin have been also been studied at laboratory and small-scale level. However, the recovery of intracellular metabolites at large scale is still challenging since not every cell disruption, extraction or purification methods are scalable. In addition, these technologies can be energy intensive. Nevertheless, some technologies like bead mill or high-pressure homogenization can be viable for scale-up (Munir et al., 2013; Ruiz-Ruiz et al., 2013).

In protein purification, salt removal and buffer or water exchange are key steps. In this context, membrane technologies like ultrafiltration or diafiltration commonly used in other biotechnological processes have been extrapolated to microalgae products. In the case of solvent extraction methods, the solvent has to be accepted by regulatory agencies for animal or human consumption. Solvents must also be environmentally friendly. In this context, research has been focusing on supercritical CO_2 extraction to limit the use of other solvents like hexane, chloroform or acetone.

In the other hand, the biorefinery scheme is the key for the utilization of microalgal biomass, since high-value metabolites are viable products for researches and companies focusing on the production of biofuels. However, downstream process optimization is still a challenge faced in the large-scale production and commercialization of PUFAs and pigments. In most cases, final recoveries are low due to the number of steps required to achieve the purity levels specified by each industry. One way of reducing the number of purification steps is by process integration. Diminishing losses due to product degradation is also main concern that has to be addressed.

Currently, to increase production yields in these microorganisms, genetic modifications are being assessed. However, culture growth conditions (temperature, nutrients, light) must still be evaluated to increase the amount of specific compounds produced by the microorganisms. As well, studies in long-term stability of algae products should be evaluated. Pigments are easily degraded due to temperature, light or other microorganisms, while PUFAs oxidize by desaturation. Therefore, studies with additives or preservatives in extracts should be carried out.

Finally, economic and environmental studies regarding the production of high-value compounds from microalgae are needed. Recently, economic evaluation and life cycle analysis of some high-value compounds like phycoerythrin and astaxanthin have been carried out (Ruiz-Ruiz et al., 2013; Deenu et al., 2014; Pérez-López et al., 2014). In conclusion, efforts should focus in the reduction of product loss and equipment and energy costs associated with the extraction and purification steps. In addition, large-scale downstream processing must be further developed in order to achieve economically viable and environmentally friendly processes.

Conflict of interest

None declared.

References

Adarme-Vega, T.C., Lim, D.K.Y., Timmins, M., Vernen, F., Li, Y., and Schenk, P.M. (2012) Microalgal biofactories: a promising approach towards sustainable omega-3 fatty acid production. *Microb Cell Factories* **11**: 1–10.

Antelo, F.S., Costa, J.A.V., and Kalil, S.J. (2008) Thermal degradation kinetics of the phycocyanin from *Spirulina platensis*. *Biochem Eng J* **41**: 43–47.

Batista, A.P., Raymundo, A., Sousa, I., and Empis, J. (2006) Rheological characterization of coloured oil-in-water food emulsions with lutein and phycocyanin added to the oil and aqueous phases. *Food Hydrocoll* **20**: 44–52.

Batista, A.P., Gouveia, L., Bandarra, N.M., Franco, J.M., and Raymundo, A. (2013) Comparison of microalgal biomass profiles as novel functional ingredient for food products. *Algal Res* **2:** 164–173.

Beale, S.I. (1993) Biosynthesis of phycobilins. *Chem Rev* **93:** 785–802.

Bellou, S., and Aggelis, G. (2012) Biochemical activities in *Chlorella sp. and Nannochloropsis salina* during lipid and sugar synthesis in a lab-scale open pond simulating reactor. *J Biotechnol* **164:** 318–329.

Benavides, J., and Rito-Palomares, M. (2004) Bioprocess intensification: a potential aqueous two-phase process for the primary recovery of B-phycoerythrin from *Porphyridium cruentum. J Chromatogr B* **807:** 33–38.

Benavides, J., and Rito-Palomares, M. (2006) Simplified two-stage method to B-phycoerythrin recovery from *Porphyridium cruentum. J Chromatogr B* **844:** 39–44.

Bermejo, R., Ruiz, E., and Acien, F.G. (2007) Recovery of B-phycoerythrin using expanded bed adsorption chromatography: scale-up of the process. *Enzyme Microb Technol* **40:** 927–933.

Bermejo Román, R., Alvárez-Pez, J.M., Acién Fernández, F.G., and Molina Grima, E. (2002) Recovery of pure B-phycoerythrin from the microalga *Porphyridium cruentum. J Biotechnol* **93:** 73–85.

Bigogno, C., Khozin-Goldberg, I., Boussiba, S., Vonshak, A., and Cohen, Z. (2002) Lipid and fatty acid composition of the green oleaginous alga Parietochloris incisa, the richest plant source of arachidonic acid. *Phytochemistry* **60:** 497–503.

Birkou, M., Bokas, D., and Aggelis, G. (2012) Improving fatty acid composition of lipids synthesized by brachionus plicatilis in large scale experiments. *J Am Oil Chem Soc* **89:** 2047–2055.

Bogorad, L. (1975) Phycobiliproteins and complementary chromatic adaptation. *Annu Rev Plant Physiol* **26:** 369–401.

Borowitzka, M.A. (1997) Microalgae for aquaculture: opportunities and constraints. *J Appl Phycol* **9:** 393–401.

Borowitzka, M.A. (2013) High-value products from microalgae – their development and commercialisation. *J Appl Phycol* **25:** 743–756.

Brennan, L., and Owende, P. (2010) Biofuels from microalgae – a review of technologies for production, processing, and extractions of biofuels and co-products. *Renew Sustain Energy Rev* **14:** 557–577.

Breuer, G., Lamers, P.P., Martens, D.E., Draaisma, R.B., and Wijffels, R.H. (2012) The impact of nitrogen starvation on the dynamics of triacylglycerol accumulation in nine microalgae strains. *Bioresour Technol* **124:** 217–226.

Brown, S.B., Holroyd, J.A., and Vernon, D.I. (1984) Biosynthesis of phycobiliproteins. Incorporation of biliverdin into phycocyanin of the red alga *Cyanidium caldarium. Biochem J* **219:** 905–909.

Bryant, D., Guglielmi, G., Marsac, N., Castets, A.-M., and Cohen-Bazire, G. (1979) The structure of cyanobacterial phycobilisomes: a model. *Arch Microbiol* **123:** 113–127.

Casal, C., Cuaresma, M., Vega, J.M., and Vilchez, C. (2011) Enhanced productivity of a lutein-enriched novel acidophile microalga grown on urea. *Mar Drugs* **9:** 29–42.

Chaiklahan, R., Chirasuwan, N., and Bunnag, B. (2012) Stability of phycocyanin extracted from Spirulina sp.: influence of temperature, pH and preservatives. *Process Biochem* **47:** 659–664.

Chen, G.-Q., Jiang, Y., and Chen, F. (2007) Fatty acid and lipid class composition of the eicosapentaenoic acid-producing microalga, *Nitzschia laevis. Food Chem* **104:** 1580–1585.

Chen, W., Sommerfeld, M., and Hu, Q. (2011) Microwave-assisted Nile red method for in vivo quantification of neutral lipids in microalgae. *Bioresour Technol* **102:** 135–141.

Cheng, Y.-S., Zheng, Y., and VanderGheynst, J. (2011) Rapid quantitative analysis of lipids using a colorimetric method in a microplate format. *Lipids* **46:** 95–103.

Cooksey, K.E., Guckert, J.B., Williams, S.A., and Callis, P.R. (1987) Fluorometric determination of the neutral lipid content of microalgal cells using Nile Red. *J Microbiol Methods* **6:** 333–345.

Davey, P.T., Hiscox, W.C., Lucker, B.F., O'Fallon, J.V., Chen, S., and Helms, G.L. (2012) Rapid triacylglyceride detection and quantification in live micro-algal cultures via liquid state 1H NMR. *Algal Res* **1:** 166–175.

De la Hoz Siegler, H., Ayidzoe, W., Ben-Zvi, A., Burrell, R.E., and McCaffrey, W.C. (2012) Improving the reliability of fluorescence-based neutral lipid content measurements in microalgal cultures. *Algal Res* **1:** 176–184.

Deenu, A., Naruenartwongsakul, S., and Kim, S.M. (2014) Optimization and economic evaluation of ultrasound extraction of lutein from *Chlorella vulgaris. Biotechnol Bioprocess Eng* **18:** 1151–1162.

Domínguez-Bocanegra, A., Guerrero Legarreta, I., Martinez Jeronimo, F., and Tomasini Campocosio, A. (2004) Influence of environmental and nutritional factors in the production of astaxanthin from Haematococcus pluvialis. *Bioresour Technol* **92:** 209–214.

Elsey, D., Jameson, D., Raleigh, B., and Cooney, M.J. (2007) Fluorescent measurement of microalgal neutral lipids. *J Microbiol Methods* **68:** 639–642.

Floreto, E.A.T., Hirata, H., Ando, S., and Yamasaki, S. (1993) Effects of temperature, light intensity, salinity and source of nitrogen on the growth, total lipid and fatty acid composition of *Ulva pertusa Kjellman* (Chlorophyta). *Bot Mar* **36:** 149–158.

Fraeye, I., Bruneel, C., Lemahieu, C., Buyse, J., Muylaert, K., and Foubert, I. (2012) Dietary enrichment of eggs with omega-3 fatty acids: a review. *Food Res Int* **48:** 961–969.

Fujii, K. (2012) Process integration of supercritical carbon dioxide extraction and acid treatment for astaxanthin extraction from a vegetative microalga. *Food Bioprod Process* **90:** 762–766.

Furuki, T., Maeda, S., Imajo, S., Hiroi, T., Amaya, T., Hirokawa, T., *et al.* (2003) Rapid and selective extraction of phycocyanin from *Spirulina platensis* with ultrasonic cell disruption. *J Appl Phycol* **15:** 319–324.

Gantar, M., Simović, D., Djilas, S., Gonzalez, W.W., and Miksovska, J. (2012) Isolation, characterization and antioxidative activity of C-phycocyanin from Limnothrix sp. strain 37-2-1. *J Biotechnol* **159:** 21–26.

Gerardo, M.L., Oatley-Radcliffe, D.L., and Lovitt, R.W. (2014) Integration of membrane technology in microalgae biorefineries. *J Memb Sci* **464:** 86–99.

Gladue, R.M., and Maxey, J.E. (1994) Microalgal feeds for aquaculture. *J Appl Phycol* **6:** 131–141.

Glazer, A. (1994) Phycobiliproteins – a family of valuable, widely used fluorophores. *J Appl Phycol* **6:** 105–112.

Goettel, M., Eing, C., Gusbeth, C., Straessner, R., and Frey, W. (2012) Pulsed electric field assisted extraction of intracellular valuables from microalgae. *Algal Res* **2:** 401–408.

Griffiths, M.J., Hille, R.P., and Harrison, S.T.L. (2010) Selection of direct transesterification as the preferred method for assay of fatty acid content of microalgae. *Lipids* **45:** 1053–1060.

Guckert, J.B., and Cooksey, K.E. (1990) Triglyceride accumulation and fatty acid profile changes in chlorella (chlorophyta) during high pH-induced cell cycle inhibition. *J Phycol* **26:** 72–79.

Guedes, A.C., Amaro, H.M., and Malcata, F.X. (2011) Microalgae as sources of carotenoids. *Mar Drugs* **9:** 625–644.

Guil-Guerrero, J.L., Belarbi, E.-H., and Rebolloso-Fuentes, M.M. (2000) Eicosapentaenoic and arachidonic acids purification from the red microalga *Porphyridium cruentum*. *Bioseparation* **9:** 299–306.

Guschina, I.A., and Harwood, J.L. (2006) Lipids and lipid metabolism in eukaryotic algae. *Prog Lipid Res* **45:** 160–186.

Guschina, I.A., Dobson, G., and Harwood, J.L. (2003) Lipid metabolism in cultured lichen photobionts with different phosphorus status. *Phytochemistry* **64:** 209–217.

Halim, R., Gladman, B., Danquah, M.K., and Webley, P.A. (2011) Oil extraction from microalgae for biodiesel production. *Bioresour Technol* **102:** 178–185.

He, P., Duncan, J., and Barber, J. (2007) Astaxanthin accumulation in the green alga haematococcus pluvialis: effects of cultivation parameters. *J Integr Plant Biol* **49:** 447–451.

Hejazi, M.A., and Wijffels, R.H. (2004) Milking of microalgae. *Trends Biotechnol* **22:** 189–194.

Hemlata, G., and Fareha, B. (2011) Studies on Anabaena sp. nccu-9 with special reference to phycocyanin. *J Algal Biomass Util* **2:** 30–51.

Huang, Y.Y., Beal, C.M., Cai, W.W., Ruoff, R.S., and Terentjev, E.M. (2010) Micro-Raman spectroscopy of algae: composition analysis and fluorescence background behavior. *Biotechnol Bioeng* **105:** 889–898.

Hussein, G., Sankawa, U., Goto, H., Matsumoto, K., and Watanabe, H. (2006) Astaxanthin, a carotenoid with potential in human health and nutrition. *J Nat Prod* **69:** 443–449.

Ioannou, E., and Roussis, V. (2009) Natural products from seaweeds. In *Plant-Derived Natural Products*. Osbourn, A.E., and Lanzotti, V. (eds). New York, USA: Springer US, pp. 51–81.

Ip, P.-F., and Chen, F. (2005) Production of astaxanthin by the green microalga Chlorella zofingiensis in the dark. *Process Biochem* **40:** 733–738.

Jubeau, S., Marchal, L., Pruvost, J., Jaouen, P., Legrand, J., and Fleurence, J. (2012) High pressure disruption: a two-step treatment for selective extraction of intracellular components from the microalga *Porphyridium cruentum*. *J Appl Phycol* **25:** 983–989.

Kang, C.D., and Sim, S.J. (2007) Selective extraction of free astaxanthin from *Haematococcus* culture using a tandem organic solvent system. *Biotechnol Prog* **23:** 866–871.

Kantz, P.T., and Bold, H.C. (1969) *Morphological and Taxonomic Investigations of Nostoc and Anabaena in Culture*. Austin, TX, USA: University of Texas Press. pp. 67.

Kathiresan, S., Sarada, R., Bhattacharya, S., and Ravishankar, G.A. (2006) Culture media optimization for growth and phycoerythrin production from *Porphyridium purpureum*. *Biotechnol Bioeng* **96:** 456–463.

Katsuda, T., Lababpour, A., Shimahara, K., and Katoh, S. (2004) Astaxanthin production by *Haematococcus pluvialis* under illumination with LEDs. *Enzyme Microb Technol* **35:** 81–86.

Kawsar, S.M.A., Fujii, Y., Matsumoto, R., Yasumitsu, H., and Ozeki, Y. (2011) Protein R-phycoerythrin from marine red alga *Amphiroa anceps*: extraction, purification and characterization. *Phytol Balc* **17:** 347–354.

Knothe, G. (2001) Analytical methods used in the production and fuel quality assessment of biodiesel. *Trans ASAE* **44:** 193–200.

Kobayashi, M., Kakizono, T., and Nagai, S. (1991) Astaxanthin production by a green alga, *Haematococcus pluvialis* accompanied with morphological changes in acetate media. *J Ferment Bioeng* **71:** 335–339.

Lababpour, A., Shimahara, K., Hada, K., Kyoui, Y., Katsuda, T., and Katoh, S. (2005) Fed-batch culture under illumination with blue light emitting diodes (LEDs) for astaxanthin production by *Haematococcus pluvialis*. *J Biosci Bioeng* **100:** 339–342.

Lamparter, T., Michael, N., Mittmann, F., and Esteban, B. (2002) Phytochrome from *Agrobacterium tumefaciens* has unusual spectral properties and reveals an N-terminal chromophore attachment site. *Proc Natl Acad Sci USA* **99:** 11628–11633.

Larkum, A.W.D., and Kühl, M. (2005) Chlorophyll d: the puzzle resolved. *Trends Plant Sci* **10:** 355–357.

Laurens, L.M.L., and Wolfrum, E.J. (2010) Feasibility of spectroscopic characterization of algal lipids: chemometric correlation of NIR and FTIR spectra with exogenous lipids in algal biomass. *Bio Energy Res* **4:** 22–35.

Lawrenz, E., Fedewa, E.J., and Richardson, T.L. (2010) Extraction protocols for the quantification of phycobilins in aqueous phytoplankton extracts. *J Appl Phycol* **23:** 865–871.

Lee, S.J., Yoon, B.-D., and Oh, H.-M. (1998) Rapid method for the determination of lipid from the green alga *Botryococcus braunii*. *Biotechnol Tech* **12:** 553–556.

Lemoine, Y., and Schoefs, B. (2010) Secondary ketocarotenoid astaxanthin biosynthesis in algae: a multifunctional response to stress. *Photosynth Res* **106:** 155–177.

Lorenz, R.T., and Cysewski, G.R. (2000) Commercial potential for Haematococcus microalgae as a natural source of astaxanthin. *Trends Biotechnol* **18:** 160–167.

Ma, R.Y.-N., and Chen, F. (2001) Enhanced production of free trans-astaxanthin by oxidative stress in the cultures of the green microalga *Chlorococcum sp. Process Biochem* **36:** 1175–1179.

MacColl, R. (1998) Cyanobacterial phycobilisomes. *J Struct Biol* **124:** 311–334.

Machmudah, S., Shotipruk, A., Goto, M., Sasaki, M., and Hirose, T. (2006) Extraction of astaxanthin from

Haematococcus pluvialis using supercritical CO2 and ethanol as entrainer. *Ind Eng Chem Res* **45**: 3652–3657.

Makri, A., Bellou, S., Birkou, M., Papatrehas, K., Dolapsakis, N.P., Bokas, D., *et al.* (2011) Lipid synthesized by microalgae grown in laboratory- and industrial-scale bioreactors. *Eng Life Sci* **11**: 52–58.

Markou, G., and Nerantzis, E. (2013) Microalgae for high-value compounds and biofuels production: a review with focus on cultivation under stress conditions. *Biotechnol Adv* **31**: 1532–1542.

Martek Biosciences Corporation (2010) *Petition for the Addition of DHA Algal Oil to the National List of Allowed and Prohibited Substances*. Columbia, MD, USA: U.S. Department of Agriculture's Agricultural Marketing Service.

Meher, L.C., Vidya Sagar, D., and Naik, S.N. (2006) Technical aspects of biodiesel production by transesterification – a review. *Renew Sustain Energy Rev* **10**: 248–268.

Mendes, A., da Silva, T., and Reis, A. (2007) DHA concentration and purification from the marine heterotrophic microalga Crypthecodinium cohnii CCMP 316 by winterization and urea complexation. *Food Technol Biotechnol* **9862**: 38–44.

Mendes, R.L., Nobre, B.P., Cardoso, M.T., Pereira, A.P., and Palavra, A.F. (2003) Supercritical carbon dioxide extraction of compounds with pharmaceutical importance from microalgae. *Inorganica Chim Acta* **356**: 328–334.

Mercer, P., and Armenta, R.E. (2011) Developments in oil extraction from microalgae. *Eur J Lipid Sci Technol* **113**: 539–547.

Molina Grima, E., Belarbi, E.-H., Acién Fernández, F.G., Robles Medina, A., and Chisti, Y. (2003) Recovery of microalgal biomass and metabolites: process options and economics. *Biotechnol Adv* **20**: 491–515.

Montsant, A., Zarka, A., and Boussiba, S. (2001) Presence of a nonhydrolyzable biopolymer in the cell wall of vegetative cells and astaxanthin-rich cysts of *Haematococcus pluvialis* (Chlorophyceae). *Mar Biotechnol NYN* **3**: 515–521.

Munir, N., Sharif, N., Shagufta, N., Saleem, F., and Manzoor, F. (2013) Harvesting and processing of microalgae biomass fractions for biodiesel production (a review). *Sci Tech Dev* **32**: 235–243.

Nui, J., Chen, G., and Zhou, B. (2010) Purification of phycoerythrin from Porphyra yeoensis Ueda (Bangladesh, Rhodophyta) using expanded bed absorption. *J Appl Phycol* **22**: 25–31.

Orosa, M., Franqueira, D., Cid, A., and Abalde, J. (2005) Analysis and enhancement of astaxanthin accumulation in Haematococcus pluvialis. *Bioresour Technol* **96**: 373–378.

Palmquist, D.L., and Jenkins, T.C. (2003) Challenges with fats and fatty acid methods. *J Anim Sci* **81**: 3250–3254.

Pangestuti, R., and Kim, S.-K. (2011) Biological activities and health benefit effects of natural pigments derived from marine algae. *J Funct Foods* **3**: 255–266.

Park, T.H., Gu, M.B., Lee, J., Kim, Z.-H., Kim, S.-H., Lee, H.-S., and Lee, C.-G. (2006) Enhanced production of astaxanthin by flashing light using *Haematococcus pluvialis*. *Enzyme Microb Technol* **39**: 414–419.

Parmar, A., Singh, N.K., Kaushal, A., and Madamwar, D. (2011) Characterization of an intact phycoerythrin and its cleaved 14kDa functional subunit from marine cyanobacterium *Phormidium sp.* A27DM. *Process Biochem* **46**: 1793–1799.

Patel, A., Mishra, S., Pawar, R., and Ghosh, P.K. (2005) Purification and characterization of C-Phycocyanin from cyanobacterial species of marine and freshwater habitat. *Protein Expr Purif* **40**: 248–255.

Pérez-López, P., González-García, S., Jeffryes, C., Agathos, S.N., McHugh, E., Walsh, D., *et al.* (2014) Life cycle assessment of the production of the red antioxidant carotenoid astaxanthin by microalgae: from lab to pilot scale. *J Clean Prod* **64**: 332–344.

Pulz, M.O., and Gross, W. (2004) Valuable products from biotechnology of microalgae. *Appl Microb Biotech* **65**: 635–648.

Pumas, C., Peerapornpisal, Y., Vacharapiyasophon, P., Leelapornpisid, P., Boonchum, W., Ishii, M., and Khanongnuch, C. (2012) Purification and characterization of a thermostable phycoerythrin from hot spring cyanobacterium *Leptolyngbya* sp. KC45. *Int J Agric Biol* **14**: 121–125.

Qin, S., Liu, G.-X., and Hu, Z.-Y. (2008) The accumulation and metabolism of astaxanthin in *Scenedesmus obliquus* (Chlorophyceae). *Process Biochem* **43**: 795–802.

Ramazanov, A., and Ramazanov, Z. (2006) Isolation and characterization of a starchless mutant of Chlorella pyrenoidosa STL-PI with a high growth rate, and high protein and polyunsaturated fatty acid content. *Phycol Res* **54**: 255–259.

Regnault, A., Chervin, D., Chammai, A., Piton, F., Calvayrac, R., and Mazliak, P. (1995) Lipid composition of *Euglena gracilis* in relation to carbon-nitrogen balance. *Phytochemistry* **40**: 725–733.

Reyes, F.A., Mendiola, J.A., Ibañez, E., and del Valle, J.M. (2014) Astaxanthin extraction from *Haematococcus pluvialis* using CO_2-expanded ethanol. *J Supercrit Fluids* **92**: 75–83.

Richmond, A. (2004) *Handbook of Microalgal Culture*. Oxford, UK: Blackwell Publishing Ltd. pp. 83–90.

Romay, C., González, R., Ledón, N., Remirez, D., and Rimbau, V. (2003) C-phycocyanin: a biliprotein with antioxidant, anti-inflammatory and neuroprotective effects. *Curr Protein Pept Sci* **4**: 207–216.

Ruiz-Ruiz, F., Benavides, J., and Rito-Palomares, M. (2013) Scaling-up of a B-phycoerythrin production and purification bioprocess involving aqueous two-phase systems: practical experiences. *Process Biochem* **48**: 738–745.

Ryckebosch, E., Muylaert, K., and Foubert, I. (2011) Optimization of an analytical procedure for extraction of lipids from microalgae. *J Am Oil Chem Soc* **89**: 189–198.

Ryckebosch, E., Bruneel, C., Muylaert, K., and Foubert, I. (2012) Microalgae as an alternative source of omega-3 long chain polyunsaturated fatty acids. *Lipid Technol* **24**: 128–130.

Ryckebosch, E., Cuéllar-Bermúdez, S.P., Termote-Verhalle, R., Bruneel, C., Muylaert, K., Parra-Saldivar, R., and Foubert, I. (2013) Influence of extraction solvent system on the extractability of lipid components from the biomass of Nannochloropsis gaditana. *J Appl Phycol* **26**: 1501–1510.

Sánchez, M., Bernal-Castillo, J., Rozo, C., and Rodríguez, I. (2003) Spirulina (arthrospira): an edible microorganism: a review. *Univ Sci* **8:** 7–24.

Sarada, R., Pillai, M., and Ravishankar, G. (1999) Phycocyanin from *Spirulina sp*: influence of processing of biomass on phycocyanin yield, analysis of efficacy of extraction methods and stability studies on. *Process Biochem* **34:** 795–801.

Sarada, R., Tripathi, U., and Ravishankar, G. (2002) Influence of stress on astaxanthin production in *Haematococcus pluvialis* grown under different culture conditions. *Process Biochem* **37:** 623–627.

Sarada, R., Vidhyavathi, R., Usha, D., and Ravishankar, G.A. (2006) An efficient method for extraction of astaxanthin from green alga *Haematococcus pluvialis*. *J Agric Food Chem* **54:** 7585–7588.

Sato, N., Tsuzuki, M., and Kawaguchi, A. (2003) Glycerolipid synthesis in *Chlorella kessleri* 11 h. *Biochim Biophys Acta BBA – Mol Cell Biol Lipids* **1633:** 35–42.

Sekar, S., and Chandramohan, M. (2007) Phycobiliproteins as a commodity: trends in applied research, patents and commercialization. *J Appl Phycol* **20:** 113–136.

Sharma, N.K., Tiwari, S.P., Tripathi, K., and Rai, A.K. (2010) Sustainability and cyanobacteria (blue-green algae): facts and challenges. *J Appl Phycol* **23:** 1059–1081.

Sijtsma, L., and de Swaaf, M.E. (2004) Biotechnological production and applications of the omega-3 polyunsaturated fatty acid docosahexaenoic acid. *Appl Microbiol Biotechnol* **64:** 146–153.

Singh, S., Kate, B.N., and Banerjee, U.C. (2005) Bioactive compounds from cyanobacteria and microalgae: an overview. *Crit Rev Biotechnol* **25:** 73–95.

Sørensen, L., Hantke, A., and Eriksen, N.T. (2013) Purification of the photosynthetic pigment C-phycocyanin from heterotrophic Galdieria sulphuraria. *J Sci Food Agric* **93:** 2933–2938.

Spolaore, P., Joannis-Cassan, C., Duran, E., and Isambert, A. (2006) Commercial applications of microalgae. *J Biosci Bioeng* **101:** 87–96.

Stanier, R.Y., Kunisawa, R., Mandel, M., and Cohen-Bazire, G. (1971) Purification and properties of unicellular blue-green algae (order Chroococcales). *Bacteriol Rev* **35:** 171–205.

Tatsuzawa, H., Takizawa, E., Wada, M., and Yamamoto, Y. (1996) Fatty acid and lipid composition of the acidophilic green alga *Chlamydomonas sp*. *J Phycol* **32:** 598–601.

Tcheruov, A.A., Minkova, K.M., Georgiev, D.I., and Houbavenska, N.B. (1993) Method for B-phycoerythrin purification from *Porphyridium cruentum*. *Biotechnol Tech* **7:** 853–858.

Tripathi, U., Sarada, R., Rao, S.R., and Ravishankar, G.A. (1999) Production of astaxanthin in *Haematococcus pluvialis* cultured in various media. *Process Biochem* **37:** 623–627.

Van Wagenen, J., Miller, T.W., Hobbs, S., Hook, P., Crowe, B., and Huesemann, M. (2012) Effects of light and temperature on fatty acid production in nannochloropsis salina. *Energies* **5:** 731–740.

Viskari, P.J., and Colyer, C.L. (2003) Rapid extraction of phycobiliproteins from cultured cyanobacteria samples. *Anal Biochem* **319:** 263–271.

Wan, M., Hou, D., Li, Y., Fan, J., Huang, J., Liang, S., *et al.* (2014) The effective photoinduction of *Haematococcus pluvialis* for accumulating astaxanthin with attached cultivation. *Bioresour Technol* **163:** 26–32.

Wang, S.-B., Hu, Q., Sommerfeld, M., and Chen, F. (2004) Cell wall proteomics of the green alga *Haematococcus pluvialis* (Chlorophyceae). *Proteomics* **4:** 692–708.

Wang, Z.T., Ullrich, N., Joo, S., Waffenschmidt, S., and Goodenough, U. (2009) Algal lipid bodies: stress induction, purification, and biochemical characterization in wild-type and starchless Chlamydomonas reinhardtii. *Eukaryot Cell* **8:** 1856–1868.

Wawrik, B., and Harriman, B.H. (2010) Rapid, colorimetric quantification of lipid from algal cultures. *J Microbiol Methods* **80:** 262–266.

Wen, Z.-Y., and Chen, F. (2003) Heterotrophic production of eicosapentaenoic acid by microalgae. *Biotechnol Adv* **21:** 273–294.

Yeesang, C., and Cheirsilp, B. (2011) Effect of nitrogen, salt, and iron content in the growth medium and light intensity on lipid production by microalgae isolated from freshwater sources in Thailand. *Bioresour Technol* **102:** 3034–3040.

Yuan, J.-P., Peng, J., Yin, K., and Wang, J.-H. (2011) Potential health-promoting effects of astaxanthin: a high-value carotenoid mostly from microalgae. *Mol Nutr Food Res* **55:** 150–165.

Zhang, W., Wang, J., Wang, J., and Liu, T. (2014) Attached cultivation of *Haematococcus pluvialis* for astaxanthin production. *Bioresour Technol* **158:** 329–335.

Zhang, Y.-M., and Chen, F. (1999) A simple method for efficient separation and purification of c-phycocyanin and allophycocyanin from *Spirulina platensis*. *Biotechnol Tech* **13:** 601–603.

Zhao, L., Zhao, G., Chen, F., Wang, Z., Wu, J., and Hu, X. (2006) Different effects of microwave and ultrasound on the stability of (all-E)-astaxanthin. *J Agric Food Chem* **54:** 8346–8351.

Zheng, H., Yin, J., Gao, Z., Huang, H., Ji, X., and Dou, C. (2011) Disruption of *Chlorella vulgaris* cells for the release of biodiesel-producing lipids: a comparison of grinding, ultrasonication, bead milling, enzymatic lysis, and microwaves. *Appl Biochem Biotechnol* **164:** 1215–1224.

Zvi, C., and Colin, R. (2010) *Single Cell Oils, Microbial and Algal Oils*. Boulder, Urbana, IL, USA: AOCS Press.

7

Degradation of toluene by *ortho* cleavage enzymes in *Burkholderia fungorum* FLU100

Daniel Dobslaw and Karl-Heinrich Engesser*
Department of Biological Waste Air Purification, Institute of Sanitary Engineering, Water Quality and Solid Waste Management, University of Stuttgart, Bandtäle 2, Stuttgart, D-70569, Germany.

Summary

Burkholderia fungorum FLU100 simultaneously oxidized any mixture of toluene, benzene and mono-halogen benzenes to (3-substituted) catechols with a selectivity of nearly 100%. Further metabolism occurred via enzymes of *ortho* cleavage pathways with complete mineralization. During the transformation of 3-methylcatechol, 4-carboxymethyl-2-methylbut-2-en-4-olide (2-methyl-2-enelactone, 2-ML) accumulated transiently, being further mineralized only after a lag phase of 2 h in case of cells pre-grown on benzene or mono-halogen benzenes. No lag phase, however, occurred after growth on toluene. Cultures inhibited by chloramphenicol after growth on benzene or mono-halogen benzenes were unable to metabolize 2-ML supplied externally, even after prolonged incubation. A control culture grown with toluene did not show any lag phase and used 2-ML as a substrate. This means that 2-ML is an intermediate of toluene degradation and converted by specific enzymes. The conversion of 4-methylcatechol as a very minor by-product of toluene degradation in strain FLU100 resulted in the accumulation of 4-carboxymethyl-4-methylbut-2-en-4-olide (4-methyl-2-enelactone, 4-ML) as a dead-end product, excluding its nature as a possible intermediate. Thus, 3-methylcyclohexa-3,5-diene-1,2-diol, 3-methylcatechol, 2-methyl muconate and 2-ML were identified as central intermediates of productive *ortho* cleavage pathways for toluene metabolism in *B. fungorum* FLU100.

E-mail karl-h.engesser@iswa.uni-stuttgart.de

Funding Information This work was supported by grant EN 474/2-2 from the Deutsche Forschungsgemeinschaft (DFG).

Introduction

Because of widespread use of alkyl as well as halogen benzenes as reactants and solvents in chemical industry, they belong to the top ten of most frequently appearing contaminants in water and soil (UBA, 2013). Biological decontamination of these sites is highly interesting because of its low costs and 'low-tech' character (Christodoulatos *et al.*, 1996; Johnson and Odencrantz, 1999; Kao and Borden, 1999; SMUL Sachsen, 2000; Kao and Prosser, 2001; Kao *et al.*, 2006; Farhadian *et al.*, 2008). In contrast, degradation of cocktails of contaminants is still problematic (Corseuil *et al.*, 1998; Lovanh *et al.*, 2002), particularly with regard to mixtures of chloro- and methyl-substituted aromatics.

Chlorobenzene is the most widely used mono-halogen benzene. It is firstly converted to (chloro-)catechols as central intermediates via dioxygenation of the aromatic ring forming chlorocyclohexa-3,5-diene-1,2-diols (Reineke and Knackmuss, 1984; Beil *et al.*, 1997), via two successive monooxygenations producing chlorophenols as intermediates (Yen *et al.*, 1991), or via initial dehalogenation forming phenol (Zhang *et al.*, 2011).

Chlorocatechols are usually further transformed by intradiol (*ortho*) ring cleavage forming chloro-*cis/cis*-muconates (Dorn and Knackmuss, 1978; Schlömann, 1994; Reineke, 1998; Zerlin, 2004; Gröning *et al.*, 2012). Subsequent conversion is performed by a chloromuconate cycloisomerase directly forming diene-lactone (4-carboxymethylenebut-2-en-4-olides) with intermediary dehalogenation (Schmidt *et al.*, 1980; Kuhm *et al.*, 1990; Vollmer *et al.*, 1994; Vollmer and Schlömann, 1995; Reineke, 1998) or indirectly via chloromuconolactones and subsequent dehalogenation (Vollmer *et al.*, 1994; Moiseeva *et al.*, 2002; Skiba *et al.*, 2002; Nikodem *et al.*, 2003; Gröning *et al.*, 2012). The dienelactones are further cleaved hydrolytically to maleyl acetate being reduced in the next step to 3-oxoadipate, a central metabolite of the 3-oxoadipate pathway (Reineke, 1998). The degradation of fluorobenzene, bromobenzene and iodobenzene proceeds in a way similar to the degradation of chlorobenzene (Strunk, 2000; 2007; Carvalho *et al.*, 2006; 2007; 2009; Strunk *et al.*, 2006; Moreira *et al.*, 2013; Strunk and Engesser, 2013).

Toluene as a widely used alkyl-substituted aromatic compound is either converted to methylcatechols, catechol or 3,4-dihydroxybenzoate as central intermediates. Formation of methylcatechols – similar to chlorocatechols – takes place either via dioxygenation of the aromatic ring forming methylcyclohexa-3,5-diene-1,2-diole (Gibson *et al.*, 1974; Zylstra *et al.*, 1988; Warhurst *et al.*, 1994; Cho *et al.*, 2000; Kim *et al.*, 2002; Cafaro *et al.*, 2005; Chaikovskaya *et al.*, 2008) or via two successive monooxygenations with cresols as intermediates. Second, toluene can be transformed to catechol via side-chain oxidation forming benzyl alcohol, benzaldehyde and benzoate as intermediates (Worsey and Williams, 1975; Harayama, 1994). Finally, transformation of toluene to 3,4-dihydroxybenzoate via p-cresol and 4-hydroxybenzoate was also described (Richardson and Gibson, 1984; Shields *et al.*, 1989; Kaphammer *et al.*, 1990; Yen *et al.*, 1991). Subsequently, methylcatechols are usually mineralized by extradiol (*meta*) ring cleavage (Hou *et al.*, 1977; Taeger *et al.*, 1988; Reineke, 1998).

Whereas the mineralization of alkylated or halogenated aromatics as sole substrates is non-critical with most bacteria, mixtures of those compounds are hardly biodegradable, based preferentially on the incompatibility of individual pathways with potential formation of reactive intermediates in case of the *meta* cleavage pathway (Klecka and Gibson, 1981; Knackmuss, 1981; Bartels *et al.*, 1984; Rojo *et al.*, 1987; Pettigrew *et al.*, 1991). Only a few strains are able to deal with chloro-substituted catechols via *meta* cleavage pathway. Nonetheless, in most of these strains, inactivation of the *meta* pyrocatechase is only partially compensated by reactivation or pricey continuing de novo synthesis (Oldenhuis *et al.*, 1989; Haigler *et al.*, 1992; Arensdorf and Focht, 1994; 1995; Hollender *et al.*, 1994; 1997; Wieser *et al.*, 1994; Heiss *et al.*, 1997; Mars *et al.*, 1997; 1999; Franck-Mokross and Schmidt, 1998; Kaschabek *et al.*, 1998; Riegert *et al.*, 1998; 1999; Göbel *et al.*, 2004).

The second alternative to mineralize mixtures of alkyl- and halogen-substituted aromatics is to degrade not only the halogenated catechols, but also alkyl-substituted aromatics like toluene via *ortho* pathways. The *ortho* cleavage of methylcatechols as an individual reaction was described manifold in literature. For instance, 4-methylcatechol was converted to 4-carboxymethyl-4-methylbut-2-en-4-olide (4-methyl-2-enelactone, 4-methyl-muconolactone, 4-ML) being a dead-end metabolite in most strains possessing *ortho* cleavage pathways (Catelani *et al.*, 1971; Rojo *et al.*, 1987; Sovorova *et al.*, 2006; Marín *et al.*, 2010). However, *Cupriavidus necator* JMP134 (Pieper *et al.*, 1985; 1988; 1990; Bruce *et al.*, 1989), *Rhodococcus opacus* 1CP (Sovorova *et al.*, 2006) and *Pseudomonas knackmussii* B13 FR1(pFRC20p), a strain engineered genetically (Rojo *et al.*, 1987), were able to circumvent this bottle

neck by an isomerase enzyme, shifting the methyl group from 4-position to the 3-position in order to get 3-methylmuconolacton (4-carboxymethyl-3-methylbut-2-en-4-olide), which was a growth substrate for each strain.

However, the major obstacle for degradation of toluene via *ortho* cleavage reaction is the fact that mainly 3-methylcatechol is formed as an intermediate of the oxidation of the ring. This intermediate is further transformed to 2-methyl-*cis,cis*-muconic acid followed by formation of 4-carboxymethyl-2-methylbut-2-en-4-olide (2-methyl-2-enelactone, 2-methylmuconolactone, 2-ML), which has frequently been described to be a dead-end metabolite.

So far, only a few authors described a productive mineralization of 2-ML. Taeger *et al.* used 2-ML as a substrate for bacterial enrichments obtaining strains totally mineralizing 2-ML. However, exposing these strains to 3-methylcatechol, enzymes of the *meta* cleavage pathway were induced (Taeger *et al.*, 1988). Pettigrew *et al.* found a mutant of *Pseudomonas* sp. JS150, called JS6, which had a defect in the catechol-2,3-dioxygenase and thus was unable to cleave 3-methylcatechol by *meta* pathway (Pettigrew *et al.*, 1991). Furthermore, 3-methylcatechol was mineralized by a modified *ortho* pathway with 2-ML as intermediate. However, toluene was no inducer for this *ortho* pathway and the native strain preferred mineralization of toluene by *meta* cleavage pathway. Similar results were found with *Ralstonia* sp. PS12 (Lehning, 1998). Franck-Mokross and Schmidt reported about the strain *Pseudomonas* sp. D7-4 being able to degrade m-toluate productively via 3-methylcatechol and 2-ML by using modified *ortho* enzymatics (Franck-Mokross and Schmidt, 1998). However, 2-ML accumulated temporarily and was further mineralized after a lag phase lasting a few hours. Consistently, this strain was not able to mineralize toluene.

We previously described the strain *Burkholderia fungorum* FLU100 being able to mineralize all monohalogenated benzenes, benzene and toluene as pure substances by an *ortho* cleavage pathway (Strunk, 2000; 2007; Dobslaw, 2003; Strunk *et al.*, 2006; Strunk and Engesser, 2013). We postulated 2-ML as intermediate of the toluene degradation pathway.

In the present paper, we show data clearly demonstrating 2-ML as the principal intermediate in the productive degradation of toluene by modified *ortho* cleavage enzymes as well as the capability of strain FLU100 to degrade mixtures of benzene, toluene and monohalogen benzenes simultaneously. To our knowledge, this is the first description of a functional, non-engineered *ortho* pathway for total degradation of toluene.

Results and discussion

Simultaneous degradation of mixtures of aromatic compounds

Burkholderia fungorum FLU100 was able to degrade any mixture of the aromatic compounds benzene, toluene, fluorobenzene, chlorobenzene, bromobenzene as well as iodobenzene simultaneously presented as sole carbon source without observing a lag phase. Increasing the number of compounds, the substrate-specific transformation rates declined. However, the sum of the specific transformation rates stayed nearly constant during each test series. Thus, transformation of each substrate seems to be accomplished by the same initial dioxygenase previously described (Strunk and Engesser, 2013).

Extracts of cultivation media of FLU100 with 0.25% dimethylformamide (DMFA) and 2 to 3 mmol l^{-1} aromatic compounds as start concentrations showed transformation rates between 90 and 120 mg C l^{-1} h^{-1} OD^{-1}, and 180 mg C l^{-1} h^{-1} OD^{-1} at maximum in case of optimal induction conditions (Supporting Information Table S1).

Aqueous samples without DMFA, analysed by a high performance liquid chromatography – system with UV/visible light detector (HPLC-UV/VIS), revealed rates up to 280 mg C l^{-1} h^{-1} OD^{-1} in case of toluene as sole carbon source showing the toxic effect of DMFA. However, the substrate-specific transformation rates reflected differences in the affinity of the enzyme for different substrates (in mg C l^{-1} h^{-1} OD^{-1}: toluene, 280; benzene, 178; fluorobenzene, 127; chlorobenzene, 159).

As shown, the maximum specific transformation rate for toluene with FLU100 in case of variable concentrations of toluene was 280 mg C l^{-1} h^{-1} OD^{-1} and remained stable with declining toluene concentrations down to 60 mg toluene·l^{-1}. Thereby, a K_s value of 30 mg toluene·l^{-1} could be calculated.

Influence of pH value on mineralization behaviour of toluene

Transforming high amounts of toluene, benzene or monohalogen benzenes as sole substrates as well as in any mixture, supernatants and growing cells of FLU100 stained black. This is due to a well known pH-dependent polymerization of (substituted) catechols as soon as they are being accumulated. Hence, the rate of cleavage of catechols is a bottleneck in the degradation pathway for these aromatics impeding up-scaling for full-scale applications. The amount of polymerized products analysed photometrically was reduced by lowering the pH value of the cultivation medium from 7.15 to 5.0. Accordingly, corresponding HPLC analyses showed an increase in the concentration of 2-methylmuconate (Supporting Information Fig. S1) as well as a still unknown metabolite with higher lipophilicity

than 3-methylcatechol. Obviously, under this regime, higher amounts of methylcatechol could be enzymatically metabolized without suffering chemical autoxidation.

Substrate-specific oxygen uptake relationships

Strain FLU100 degrades a wide spectrum of aromatics, favouring a broad application in possible bioremediation. The oxidation potential of the initial enzymes was characterized by measuring the specific oxygen uptake rates for different carbon sources in dependence of fluorobenzene, toluene and benzene as pre-cultivation substrates, allowing to judge the number of initial enzymes by analysis of the relative maximum transformation rates. The corresponding results are shown in Supporting Information Tables S2 and S3. All results are standardized for toluene or catechol as 100% respectively. These data are giving the average result of several independent experiments each performed in double.

With the exception of the value for fluorobenzene in fluorobenzene-grown cells, the relative specific activities for each substrate were nearly independent of the pre-cultivation conditions. Thus, the same initial oxygenase seems to be induced independently from the nature of different benzene derivatives. As only low activities could be measured with the phenol derivatives tested as well as with benzylic alcohol and benzoate, the initial enzyme most probably is a dioxygenase directly oxidizing the aromatic ring. No activity for side chain oxidation, which is a possible reaction in case of toluene, was observed. Currently, we have no proof to explain higher activity for fluorobenzene after growth on fluorobenzene. The existence of a special transport system for fluorobenzene into the cells induced by fluorobenzene is the most probable explanation.

In contrast to the initial dioxygenase, oxygen uptake values of catechol cleavage enzymes strongly depend on the type of growth substrate. The absolute oxygen uptake of nearly 2100 U in case of cells pre-grown on fluorobenzene was 10-fold higher than that measured for toluene cells and 20-fold higher than that of cells pre-grown on benzene (see Supporting Information Table S3).

This intensified expression of catechol cleavage enzymes in case of fluorobenzene may reflect and compensate the lower relative activities for ring cleavage of 3-fluorocatechol in order to avoid its accumulation and autoxidation.

In general, muconates are described as inducers of catechol *ortho* cleavage pathways (Feist and Hegeman, 1969). According to literature, 2-halomuconates are stronger inducers for the modified chlorocatechol pathway than 2-methylmuconate or unsubstituted muconate (Reineke, 1998). However, comparison of the relative activities for muconates in cells pre-grown on fluorobenzene or toluene, respectively, showed high similarity. In contrast, cells grown on benzene were found to strongly deviate from this pattern, indicating the presence of an additional *ortho* pathway, being too specific to tolerate bulky substituents in the substrates. Thus, toluene is mainly transformed by the modified *ortho* pathway with chlorocatechol-1,2-dioxygenase as the initial enzyme, followed by a broad-specificity enzyme like chloromuconate cycloisomerase.

Proof of dioxygenase nature of the initial enzyme by accumulation of its products

The existence of an initial dioxygenase enzyme was also proofed by a transposon mutant of strain FLU100, called FLU100 P2R5 (Strunk, 2007), where the transposon element was inserted within the benzene dihydrodiol-dehydrogenase encoding gene sequence. During turnover of toluene, the corresponding 3-methylcyclohexa-3,5-diene-1,2-diol accumulated in the supernatant. This diendiol can be re-aromatized also chemically by heat treatment for 5 min under acidic conditions yielding the corresponding o-cresol quantitatively. For all mono-halogen benzenes, toluene and benzene, the corresponding diendiols were transformed to phenolic compounds, which were identified and verified by commercial standards.

Second, the same diendiol structures were produced by *Escherichia coli* pST04, an analogue to strain pSTE44, which contained a lac operon-regulated tetrachlorobenzene dioxygenase. This strain as well as the products of transformation was formerly described in literature in more detail (Pollmann *et al.*, 2003). This clearly establishes the nature of 3-methylcyclohexa-3,5-diene-1,2-diol to be the principal intermediate of the oxidation of toluene in FLU100 after growth with toluene.

Furthermore, using FLU100 P2R5, we were able to produce and isolate these diendiol structures for all six aromatic substrates (for diendiol structures of benzene, toluene and fluorobenzene, see Supporting Information Fig. S2) and additionally used them as substrates in conversion experiments. The process of conversion was followed by HPLC analyses. During conversion of the 3-methylcyclohexa-3,5-diene-1,2-diol, for example, a temporary accumulation of further intermediates of toluene degradation, namely 3-methylcatechol, 2-methyl-*cis,cis*-muconic acid, as well as 2-ML was observed, again proposing the way of toluene metabolism to follow the route for halo-benzene degradation up to the level of methyl lactone.

The conversion rate of the diendiol compound in comparison with conversion rates of 3-methylcatechol, 2-methylmuconic acid and 2-ML was extremely high. In case of 3-methylcyclohexa-3,5-diene-1,2-diol, nearly 0.6 mmol l^{-1} of the substrate, equivalent to 50% of the concentration supplied, was biologically converted to the corresponding catechol within 5 min (Fig. 1). Because of time delays of HPLC analyses, the conversion rate of diendiol between the second and third measurements was transferred to Table 1, showing absolute conversion rates for all relevant substrates and intermediates for cells pre-grown on toluene.

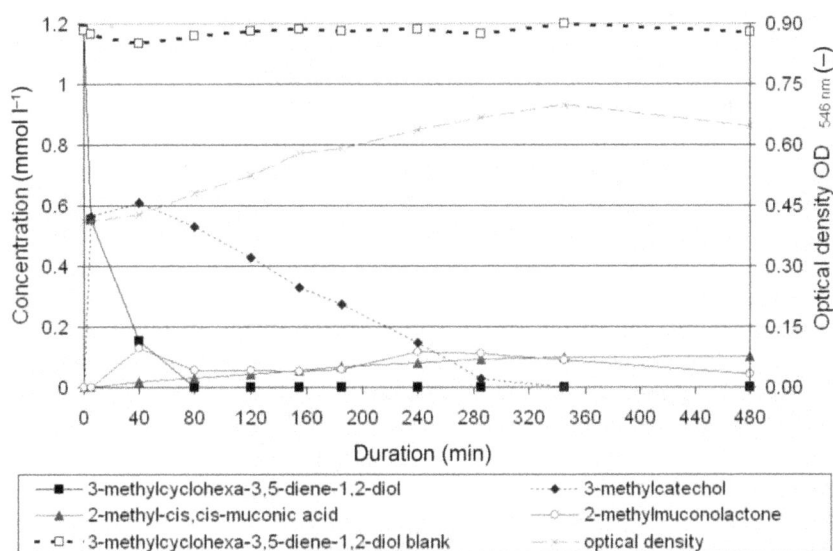

Fig. 1. Concentration of 3-methylcyclohexa-3,5-diene-1,2-diol as substrate and its intermediates during conversion by strain FLU100 pre-grown on toluene. Additionally, values of optical density measured at 546 nm wavelength are given.

Table 1. Maximum substrate specific conversion rates of cultures of FLU100 pre-grown on toluene (column, concentration > 1 mmol l^{-1}) in mmol l^{-1} h^{-1} OD^{-1}.

	Substrate				
Intermediate	Toluene	3-methyldiendiol	3-methylcatechol	2-methylmuconolactone	Succinate
Toluene	3.323				
3-methyldiendiol		1.636			
3-methylcatechol		0.251	0.306		
2-methylmuconic acid		0.219	0.236		
2-methylmuconolactone		0.195	0.219	0.796	
Succinate					3.332

Conversion of produced intermediates was measured after total conversion of the substrate (line, concentration <1 mmol l^{-1}). In case of toluene as substrate, no intermediate conversion rates were detected, because the solvent used was not adequate for intermediate measurement.

Identification of 3-methylcatechol and 2-methyl-cis/cis-muconic acid as further intermediates of the toluene degradation pathway

Corresponding catechols were formed out of the diendiol intermediates by dehydrogenation using viable cells (Fig. 1). Identification was performed using a gas chromatograph with a mass spectrometer as detector (GC-MS) and analysing the concentrated organic phase after a threefold extraction with ethyl acetate as well as methylation of the organic phase with diazomethane. For toluene, 3-methylcatechol (before methylation) and 2-methoxy-3-methylphenol (after methylation) were identified by using commercially available standard compounds (Supporting Information Table S4).

The catechols of all six substrates were cleaved by an *ortho* cleavage pathway and no *meta* cleavage activity was observed. Measurement of absolute and relative conversion activities of whole cells as well as crude extracts of strain FLU100 exhibited an activity pattern with low similarity to the conversion pattern in *P. knackmussii* B13 (Dorn and Knackmuss, 1978). In contrast, the activity pattern of FLU100 revealed high similarity to the activity pattern typical for chlorocatechol-1,2-dioxygenase of B13 readopted by the authors (Dobslaw, 2003).

Even though strain FLU100 is able to metabolize 4-substituted catechols, the main, nearly exclusively formed transformation products of the benzene derivatives examined in this study are the corresponding 3-substituted catechols. These catechols are further cleaved to the corresponding (2-substituted) *cis-cis*-muconic acids. These muconic acids were identified by comparison of retention times and wavelength spectra with reference substances produced by *E. coli* Klon 4, carrying plasmid pBBRCI, from the corresponding catechols. This mutant expressed an chlorocatechol-1,2-dioxygenase regulated by a lac operon and was previously described (Perez-Pantoja *et al.*, 2003).

The intermediate nature of 2-methylmuconolactone in the toluene degradation pathway of FLU100

While 2-halogenated muconic acids can be transformed to *trans*-dienlactone during lactone ring formation/elimination of halide by muconate-cycloisomerases (see for example Vollmer *et al.*, 1994; Vollmer and Schlömann, 1995; Reineke, 1998; Moiseeva *et al.*, 2002), 2-methylmuconic acid as intermediate of toluene cannot be transformed this way, because the methyl group is no leaving group at all. We observed that the cycloisomerase transformed 2-methylmuconic acid to 2-methylmuconolactone (2-ML), which was further mineralized. There was no indication for the existence of an exocyclic 5-methylmuconolactone, a possible product of alternative cycloisomerization of 2-methylmuconic acid.

In contrast to strain *Pseudomonas sp.* D7-4 (Franck-Mokross and Schmidt, 1998; Taeger *et al.*, 1988;) and other strains obtained by enrichment on 2-ML and 3-methylbenzoate (Taeger *et al.*, 1988), which were also able to mineralize 2-ML, strain FLU100 was able to grow on toluene as sole source of carbon and energy.

However, transforming 3-methylcatechol by cells of strain FLU100 pre-grown on benzene or mono-halogen benzenes, nearly 55–65% of the provided 3-methylcatechol temporarily accumulated as 2-ML. Further transformation was observed after a lag phase of 2–3 h (Fig. 2). By contrast, no or only small concentrations of 2-ML accumulated in case of toluene-induced/growing cells (Fig. 3). This means that key enzymes for degradation of 2-ML were only induced after adaptation on toluene, but not on benzene or mono-halogen benzenes.

To prove this interpretation and to avoid induction of new enzymes, the same experiments were done after addition of chloramphenicol (CAP) as an inhibitor of protein synthesis. Cells induced with benzene or mono-halogen benzenes showed similar concentrations of 2-ML accumulating in the supernatant compared with non-inhibited cells. However, no further transformation

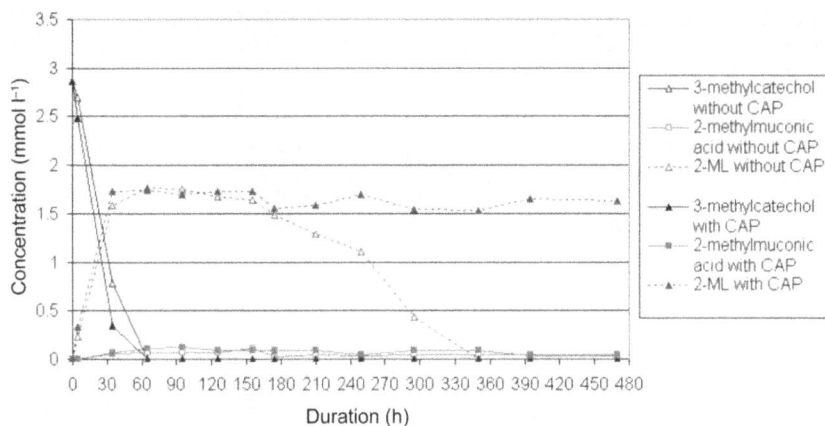

Fig. 2. Concentration of 3-methylcatechol as substrate and its transformation products during conversion by strain FLU100 pre-grown on benzene (OD_{546nm} = 1.5; double tested). One row contained chloramphenicol (CAP) as cytostatic drug avoiding induction of new enzymes. 2-methylmuconolactone (2-ML) was only transformed in absence of CAP.

appeared, even after prolonged incubation (Fig. 2). In contrast, no accumulation of 2-ML was detected when cells used were previously induced with toluene (Fig. 3).

In a second experiment, an aqueous solution of 2-ML was produced by complete transformation of 3-methycatechol by cells of FLU100 grown on benzene, followed by subsequent separation of the cells. The final concentration of 2-ML was adjusted to about 1.7 mmol l^{-1}. This solution was split and added to two liquid cultures of strain FLU100 pre-grown on toluene. One of these cultures has been supplemented with CAP. Both cultures transformed 2-ML showing no lag phase at transformation rates of 0.35 mmol l^{-1} h^{-1} OD^{-1} (culture without CAP) and 0.17 mmol l^{-1} h^{-1} OD^{-1} (with CAP) respectively. The transformation rate without CAP was twofold higher, because the formation of new enzymes was possible (Fig. 3). The results of both experimental concepts confirmed the

induction of 2-ML-specific transforming enzymes in case of toluene-metabolizing cells of FLU100.

2-ML as metabolite was identified by HPLC-MS and GC-MS analyses and identity was proofed by comparison with data of Knackmuss and colleagues (1976) and Pieper and colleagues (1985), given in Table 2.

A theoretical alternative dioxygenation route for toluene in 3,4-position, i.e. at *meta* and *para* position forming 4-methylcyclohexa-3,5-diene-1,2-diol, 4-methylcatechol, 3-methyl-*cis,cis*-muconic acid and 4-ML as intermediates, was found to be non-productive. Metabolism of 4-methylcatechol, added to whole cells of FLU100, lead to stoichiometric accumulation of 4-ML as a dead-end metabolite (data not shown). 4-ML itself as well as 4-methylmuconolactone methyl ester as reaction product after methylation were identified by GC-MS analyses (see Table 2).

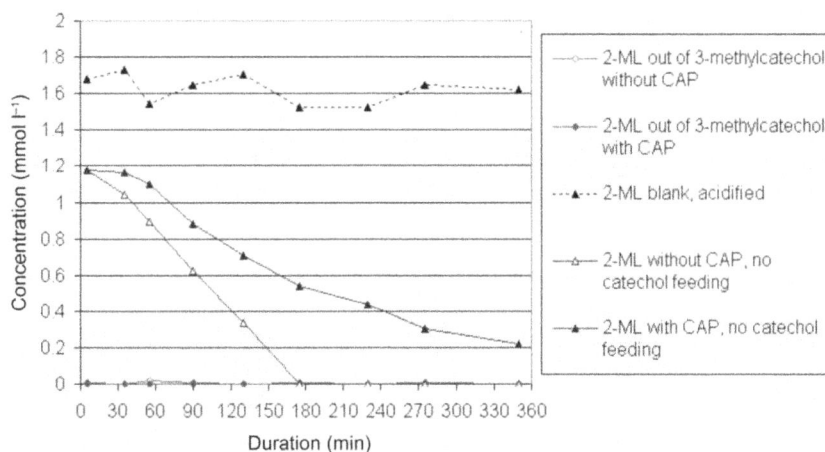

Fig. 3. Concentration of 2-methylmuconolactone (2-ML) as intermediate of the conversion of 3-methylcatechol as well as sole substrate by whole cells of strain FLU100 pre-grown on toluene. To the second flask of each test row, chloramphenicol (CAP) as cytostatic drug was added to avoid induction of new enzymes. The optical density was OD_{546nm} = 1.6. In all cases, 2-ML was transformed showing induction of relevant enzymes. Additionally, the concentration of 2-ML in a cell-free supernatant at a pH value of 1–2 as a blank row is presented.

Table 2. Comparison of detected GC-MS fragments and corresponding intensities of the identified metabolites 2-methylmuconolactone (2-ML, column 2) and 4-methymuconolactone (4-ML, column 5) of strain FLU100 with data of 2-methylmuconolactone (column 3), 3-methylmuconolactone (3-ML, column 4) and 4-methylmuconolactone (column 6) given by Knackmuss and colleagues (1976) and Pieper and colleagues (1985).

m/z	Intermediate 2-ML of strain FLU100; intensity (%)	Knackmuss et al., 1976 2-ML; intensity (%)	Pieper et al., 1985 3-ML; intensity (%)	Intermediate 4-ML of strain FLU100; intensity (%)	Knackmuss et al., 1976 4-ML; intensity (%)
156	1.5	3.0	7.9	8.8	17.0
141	0.0	0.0	n.d.	8.4	13.0
139	1.4	>0.0	n.d.	n.d.	n.d.
138	n.d.	n.d.	12.4	n.d.	n.d.
128	n.d.	n.d.	n.d.	3.6	n.d.
127	1.7	n.d.	n.d.	n.d.	n.d.
111	n.d.	n.d.	11.6	15.2	n.d.
110	100.0	100.0	82.8	15.2	22.0
99	0.8	n.d.	n.d.	36.0	n.d.
97	31.6	30.0	51.1	100.0	100.0
96	n.d.	n.d.	22.4	n.d.	n.d.
69	48.8	n.d.	100.0	41.6	n.d.
68	n.d.	n.d.	37.5	n.d.	n.d.
43	4.0	n.d.	n.d.	36.0	n.d.
41	34.4	n.d.	n.d.	n.d.	n.d.
λ_{max}	208 nm	210 nm	n.d.	208 nm	210 nm

The fragment with the highest intensity is normalized to 100% and other fragments are given as relative intensities. Additionally, the wave length at maximum absorption is given in the last line.
n.d. = not detected or intensity not given in literature.

Intermediates of 2-ML degradation

To identify further intermediates of 2-ML degradation pathway, analyses were performed by use of HPLCs with an evaporative light scattering detector (HPLC-ELSD) or a mass spectrometer (HPLC-MS). However, neither metabolites of 2-ML degradation were reproducibly detectable nor identification of detected compounds was possible. 2-methyl succinate as hypothetical intermediate of the 2-ML degradation pathway, however, was used as a substrate in conversion experiments. Whereas the reference substrate succinate was transformed without showing any lag phase, methyl substituted succinate was not converted at all by whole cells of FLU100 (see Supporting Information Table S5). Therefore, revelation of the complete degradation pathway downstream of the intermediate 2-ML was not successful. Maximum conversion rates of the identified intermediates of toluene degradation by FLU100 pre-grown on toluene are given in Table 1. The correlating degradation pathway is shown in Fig. 4.

Concluding remarks

Burkholderia fungorum strain FLU100 was shown to mineralize toluene alone and simultaneously in any mixture with benzene and mono-halogen benzenes, even including fluorobenzene. The initial attack on all these substrates by a dioxygenase attacking the aromatic ring yielded into the corresponding 3-substituted cyclohexa-3,5-diene-1,2-diols, being metabolized via a dehydrogenase to the corresponding 3-(substituted) catechols. These catechols were nearly exclusively cleaved by a chlorocatechol-1,2-dioxygenase to the muconic acid derivatives. Based on a lack of *meta* cleavage activity, toluene is atypically cleaved by this *ortho* enzyme resulting in the formation of 2-methylmuconic acid, followed by 2-methyl-2-ene-lactone (2-ML) as the next steps. In contrast to literature, the latter intermediate is not a dead-end metabolite. The further degradation of 2-ML by so far unknown enzymes, induced by growth with toluene, remains yet to be elucidated.

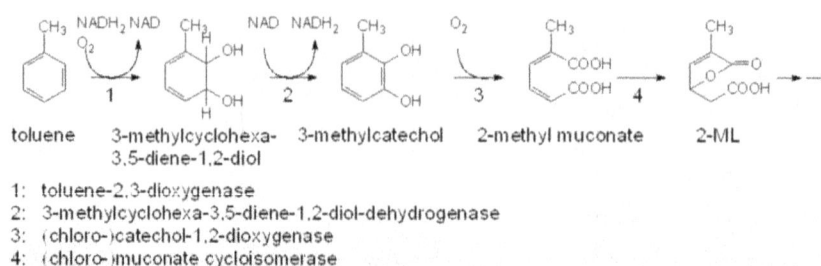

1: toluene-2,3-dioxygenase
2: 3-methylcyclohexa-3,5-diene-1,2-diol-dehydrogenase
3: (chloro-)catechol-1,2-dioxygenase
4: (chloro-)muconate cycloisomerase

Fig. 4. Postulated pathway of toluene degradation by strain *Burkholderia fungorum* FLU100 via 3–methylcyclohexa-3,5-diene-1,2-diol, 3-methylcatechol, 2-methylmuconic acid and 2-methyl-2-ene-lactone.

Experimental procedures

Cultivation conditions of FLU100 (wild type and mutants)

Wild type of strain FLU100 was cultivated at 30°C in 500 ml glass flasks containing 200 ml of a mineral salt medium previously described (Dobslaw, 2003). About 20 µl of toluene, benzene or mono-halogen benzenes was directly added to this medium and flasks were incubated for 48–72 h. After adding 10–15 µL of substrate, cultures were further incubated overnight and harvested by centrifugation. The pellet was re-suspended in fresh medium.

The mutant P2R5 of strain FLU100 was obtained via transposon mutagenesis by mating the recipient FLU100 with the donor strain *E. coli* pCro2a (Onaca *et al.*, 2007) and screening of nearly 16 800 mutants. The plasmid carries an ampicillin resistance gene for false-positive screening, whereas the Tn5 insert carries a kanamycin resistance gene and an *ori*-sequence. Mutants were screened by spreading cells on mineral medium plates containing 10 mmol l^{-1} glucose and 100 mg l^{-1} kanamycin. Colonies grown were transferred into liquid cultures (20 ml volume) with the same medium. The accumulation of intermediates of aromatic degradation was checked by adding 10 µl of toluene or another aromatic compound and 200 µl of a 1 M glucose solution after 48 h. Formation of intermediates was verified the next day via HPLC analyses.

For quantification of remaining concentrations of aromatics in the liquid medium via extraction and subsequent GC analyses, mixtures of aromatic compounds were supplied in DMFA with final concentrations of 2 to 3 mmol l^{-1} aromatics and 0.25% DMFA in the cultivation medium. Extraction was performed using dichloromethane.

In case of oxygen uptake experiments, the pellet was re-suspended in fresh mineral medium and culture was shaken at 30°C overnight without supplying substrates. As preparation, the culture was aerated, samples of 3 ml of suspension were filled into a 30°C tempered stirrer and oxygen uptake was recorded. For measurement of the substrate-specific oxygen uptake, 30 µL of a substrate solution (c = 0.1 mol L^{-1}) was added.

In the event of crude extract experiments, the centrifuged pellets of strain FLU100, *P. knackmussii* B13 or *E. coli* Klon 4, were washed twice with saline solution and re-suspended in 15 ml of a Tris buffer solution (8 g l^{-1} Tris, pH 8.0). Cells were cracked by pressure release in two to three runs in a French pressure cell with 10 000 psi. The extract was centrifuged for 10 min at 14 000 rpm at 4°C and cell-free supernatant was stored on ice. At the conversion tests of catechols, 100 µL of crude extract was added to 1 ml of Tris buffer pH 8.0, and 900 µl of deionized water and 26 µl of ethylenediaminetetraacetic acid (c = 0.1 mol L^{-1}) in a 4 ml fused silica cuvette. After aeration for 1 min, 20 µL of the tested catechol (c = 0.1 mol L^{-1}) was added and the change in extinction at 260 nm was determined. In case of high turnover, the samples were measured by HPLC analyses at 210 nm and 260 nm wave lengths. Quantification of conversion rates (µmol l^{-1} min^{-1} mg protein^{-1}) was performed via photometric measurement at maximum extinction wave lengths using extinction coefficients given by (Dorn and Knackmuss 1978). The

concentration of protein was measured by the method of Bradford.

Conversion tests of catechols in whole cells of strain FLU100 were performed using cultures with optical densities of $OD_{546nm} = 1.5$–2.5. Cultures were firstly pre-grown on an aromatic substrate, centrifuged and re-suspended in new mineral medium. Relevant catechols were added in final concentrations of 3 mmol l^{-1}. The concentrations of catechols and corresponding intermediates over time were observed by HPLC analyses.

2-ML as substrate for transformation tests was obtained via conversion of 3-methylcatechol by whole cells of strain FLU100 pre-grown on benzene. Directly after total transformation of 3-methylcatechol to 2-ML, reaction mixtures were centrifuged and the supernatant was directly used in subsequent conversion experiments as aqueous substrate solution. The solution can be stored at 4°C about 1 week. To avoid the induction of new enzymes during transformation of 2-ML, CAP as cytostatic drug was added (10 mg per 50 ml reaction volume). Actual concentrations of 2-ML in the reaction flasks were measured by HPLC analyses.

Cultivation conditions of strains of E. coli

All strains of *E. coli* were pre-grown on lysogeny broth media containing 10 g of tryptone, 5 g of yeast extract and 5 g of sodium chloride per liter of deionized water. The medium was autoclaved at 121°C for 30 min. With the exception of strain pCro2a, 2.5 mmol l^{-1} of lactose was added. For cultivation of pCro2a, 100 mg l^{-1} of ampicillin and 50 mg l^{-1} of kanamycin were added. In case of *E. coli* Klon 4, 100 mg l^{-1} of kanamycin and in case of *E. coli* pST04, 100 mg l^{-1} of ampicillin and 10 mmol l^{-1} of glucose were added. Cultures were cultivated at 37°C overnight. Afterwards, the pellets of pST04 and Klon 4 were harvested and re-suspended in new LB-media with the corresponding additives and additional 0.25–0.4 mmol l^{-1} IPTG and 5–10 mmol l^{-1} of aromatic substrate (pST04) or 1 mmol l^{-1} of catechol (Klon 4). After 4 h, a significant blue colour appeared showing expression of the lac-regulated gene sequences. The production of substituted cyclohexa-3,5-diene-1,2-diols was directly measured per HPLC analyses at 260 nm. For production of muconic acids, Klon 4 was analogously prepared like FLU100 for crude extract measurements.

Cultivation conditions of strains P. knackmussii B13

Strain B13 was cultivated at 30°C on a mineral medium containing benzoate or 3-chlorobenzoate as substrate. While a catechol-1,2-dioxygenase was induced in case of the former substrate, a chlorocatechol-1,2-dioxygenase was induced in the event of the latter substrate. The preparation was similar to that of crude extract measurements of FLU100.

Acknowledgements

The authors would like to gratefully acknowledge the extremely valuable contribution of D.H. Pieper, Helmholtz Centre for Infection Research in Braunschweig, for helpful comments as well as the supply of *Escherichia coli* pST04 and *E. coli* Klon 4. Special thanks to J. Altenbuchner, Institute of Industrial Genetics, University of Stuttgart for the supply of *E. coli* pCro2a. Finally we thank P. Fischer, Institute of Organic Chemistry, University of Stuttgart, for GC-MS and HPLC-MS analyses.

References

Arensdorf, J.J., and Focht, D.D. (1994) Formation of chlorocatechol *meta* cleavage products by a pseudomonad during metabolism of monochlorobiphenyl. *Appl Environ Microbiol* **60**: 2884–2889.

Arensdorf, J.J., and Focht, D.D. (1995) A *meta* cleavage pathway for 4-chlorobenzoate, an intermediate in the metabolism of 4-chlorobiphenyl by *Pseudomonas cepacia* P166. *Appl Environ Microbiol* **61**: 443–447.

Bartels, I., Knackmuss, H.-J., and Reineke, W. (1984) Suicide inactivation of catechol 2,3-dioxygenase from *Pseudomonas putida* mt-2 by 3-halocatechols. *Appl Environ Microbiol* **47**: 500–505.

Beil, S., Happe, B., Timmis, K.N., and Pieper, D.H. (1997) Genetic and biochemical characterization of the broad spectrum chlorobenzene dioxygenase from *Burkholderia* sp. strain PS12. Dechlorination of 1,2,4,5-tetrachlorobenzene. *Eur J Biochem* **247**: 190–199.

Bruce, N.C., Cain, R.B., Pieper, D.H., and Engesser, K.-H. (1989) Purification and characterization of 4-methylmuconolactone methyl-isomerase, a novel enzyme of the modified 3-oxoadipate pathway in nocardioform actinomycetes. *Biochem J* **262**: 303–312.

Cafaro, V., Notomista, E., Capasso, P., and Di Donato, A. (2005) Regiospecificity of two multicomponent monooxygenases from *Pseudomonas stutzeri* OX1: molecular basis for catabolic adaptation of this microorganism to methylated aromatic compounds. *Appl Environ Microbiol* **71**: 4736–4743.

Carvalho, M.F., Ferreira, M.I., Moreira, I.S., Castro, P.M., and Janssen, D.B. (2006) Degradation of fluorobenzene by *Rhizobiales* strain F11 via *ortho* cleavage of 4-fluorocatechol and catechol. *Appl Environ Microbiol* **72**: 7413–7417.

Carvalho, M.F., Duque, A.F., Goncalves, I.C., and Castro, P.M. (2007) Adsorption of fluorobenzene onto granular activated carbon: isotherm and bioavailability studies. *Bioresour Technol* **98**: 3424–3430.

Carvalho, M.F., De Marco, P., Duque, A.F., Pacheco, C.C., Janssen, D.B., and Castro, P.M. (2009) *Labrys potucalensis*, a bacterial strain with the capacity to degrade fluorobenzene. *New Biotechnol* **25**: S68.

Catelani, D., Fiechhi, A., and Galli, E. (1971) (+)-γ-Carboxymethyl-γ-methyl-Δ$^\alpha$-butenolide: a 1,2-ring-fission product of 4-methylcatechol by *Pseudomonas desmolyticum*. *Biochem J* **121**: 89–92.

Chaikovskaya, O.N., Sokolova, I.V., Karetnikova, E.A., and Maiera, G.V. (2008) Fluorescent analysis of photoinduced biodegradation of cresol isomers. *J Appl Spectrosc* **2**: 261–267.

Cho, M.C., Kang, D.-O., Yoon, B.D., and Lee, K. (2000) Toluene degradation pathway from *Pseudomonas putida* F1: substrate specificity and gene induction by 1-substituted benzenes. *J Ind Microbiol Biotechnol* **25**: 163–170.

Christodoulatos, C., Korfiatis, G.P., Pal, N., and Koutsospyros, A. (1996) In situ groundwater treatment in a trench bio-sparge system. *Hazard Waste Hazard Mater* **13**: 223–236.

Corseuil, H.X., Hunt, C.S., Ferreira dos Santos, R.C., and Alvarez, P.J.J. (1998) The influence of the gasoline oxygenate ethanol on aerobic and anaerobic BTX biodegadation. *Water Res* **32**: 2065–2072.

Dobslaw, D. (2003) Zur Analytik des Toluolabbaus in bakteriellen Biofilterisolaten: Nachweis eines neuartigen Ortho-Weges. Diplomarbeit. Universität Stuttgart.

Dorn, E., and Knackmuss, H.-J. (1978) Chemical structure and biodegradability of halogenated aromatic compounds – substituent effects on 1,2-dioxygenation of catechol. *Biochem J* **174**: 85–94.

Farhadian, M., Vachelard, C., Duchez, D., and Larroche, C. (2008) In situ bioremediation of monoaromatic pollutants in groundwater: a review. *Bioresour Technol* **99**: 5296–5308.

Feist, C.F., and Hegeman, G.D. (1969) Phenol and benzoate metabolism by *Pseudomonas putida*: regulation of tangential pathways. *J Bacteriol* **100**: 869–877.

Franck-Mokross, A.C., and Schmidt, E. (1998) Simultaneous degradation of chloro- and methylsubstituted aromatic compounds. Competition between *Pseudomonas* strains using the *ortho* and *meta* pathway or the *ortho* pathway exclusively. *Appl Microbiol Biotechnol* **50**: 233–240.

Gibson, D.T., Mahadavan, V., and Davey, J.F. (1974) Bacterial metabolism of *para*- and *meta*-xylene: oxidation of the aromatic ring. *J Bacteriol* **119**: 930–936.

Göbel, M., Kranz, O.H., Kaschabek, S.R., Schmidt, E., Pieper, D.H., and Reineke, W. (2004) Microorganisms degrading chlorobenzene via a *meta*-cleavage pathway harbour highly similar chlorocatechol 2,3-dioxygenase-encoding gene clusters. *Arch Microbiol* **182**: 147–156.

Gröning, J.A.D., Roth, C., Kaschabek, S.R., Sträter, N., and Schlömann, M. (2012) Recombinant expression of a unique chloromuconolactone dehalogenase ClCF from *Rhodococcus opacus* 1CP and identification of catalytically relevant residues by mutational analysis. *Arch Biochem Biophys* **526**: 69–77.

Haigler, B.E., Pettigrew, C.A., and Spain, J.C. (1992) Biodegradation of mixtures of substituted benzenes by *Pseudomonas* sp. strain JS150. *Appl Environ Microbiol* **58**: 2237–2244.

Harayama, S. (1994) Codon usage patterns suggest independent evolution of the two catabolic operons on toluene-degradative plasmid TOL pWW0 of *Pseudomonas putida*. *J Mol Evol* **38**: 328–335.

Heiss, G., Müller, C., Altenbuchner, J., and Stolz, A. (1997) Analysis of a new dimeric extradiol dioxygenase from a naphthalenesulfonate degrading Sphingomonad. *Microbiology* **143**: 1691–1699.

Hollender, J., Dott, W., and Hopp, J. (1994) Regulation of chloro- and methylphenol degradation in *Comamonas testosterone* JH5. *Appl Environ Microbiol* **60**: 2330–2338.

Hollender, J., Hopp, J., and Dott, W. (1997) Degradation of 4-chlorophenol via the *meta* cleavage pathway by *Comamonas testosteroni* JH5. *Appl Environ Microbiol* **63:** 4567–4572.

Hou, C.T., Patel, R., and Lillard, M.O. (1977) Extradiol cleavage of 3-methylcatechol by catechol 1,2-dioxygenase from various microorganisms. *Appl Environ Microbiol* **33:** 725–727.

Johnson, J.G., and Odencrantz, J.E. (1999) Management of a hydrocarbon plume using a permeable ORC™ barrier. In *Accelerated Bioremediation Using Slow Release Compounds*. Königsberg, S.S., and Norris, R.D. (eds). San Clemente: Regenesis Bioremediation Products, pp. 39–44.

Kao, C.M., and Borden, R.C. (1999) Enhanced aerobic bioremediation of a gasoline-contaminated aquifer by oxygen-releasing barriers. In *Accelerated Bioremediation using Slow Release Compounds*. Königsberg, S.S., and Norris, R.D. (eds). San Clemente: Regenesis Bioremediation Products, pp. 1–5.

Kao, C.M., and Prosser, J. (2001) Evaluation of natural attenuation rate at a gasoline spill site. *J Hazard Mater* **82:** 275–289.

Kao, C.M., Huang, W.Y., Chang, L.J., Chen, T.Y., Chien, H.Y., and Hou, F. (2006) Application of monitored natural attenuation to remediate a petroleum-hydrocarbon spill site. *Water Sci Technol* **53:** 321–328.

Kaphammer, B., Kukor, J.J., and Olsen, R.H. (1990) Cloning and characterization of a novel toluene degrading pathway from *Pseudomonas pickettii* PK01. In *Abstracts of the 90th Annual Meeting of the American Society for Microbiology*. Morello, J.A., and Domer, J.E. (eds). Washington, D.C.: American Society of Microbiology, p. 243. abstr. K-145.

Kaschabek, S.R., Kasberg, T., Müller, D., Mars, A.E., Janssen, D.B., and Reineke, W. (1998) Degradation of chloroaromatics: purification and characterization of a novel type of chorocatechol 2,3-dioxygenase of *Pseudomonas putida* GJ 31. *J Bacteriol* **180:** 296–302.

Kim, D., Kim, Y.-S., Kim, S.-K., Kim, S.-W., Zylstra, G.J., Kim, Y.M., and Kim, E. (2002) Monocyclic aromatic hydrocarbon degradation by *Rhodococcus* sp. strain DK17. *Appl Environ Microbiol* **68:** 3270–3278.

Klecka, G.M., and Gibson, D.T. (1981) Inhibition of catechol 2,3-dioxygenase from *Pseudomonas putida* by 3-chlorocatechol. *Appl Environ Microbiol* **41:** 1159–1165.

Knackmuss, H.-J. (1981) Degradation of halogenated and sufonated hydrocarbons. In *Microbial Degradation of Xenobiotics and Recalcitrant Compounds*. Leisinger, T.A., Cook, M., Hütter, R., and Nüetsch, J. (eds). London: Academic Press, pp. 189–212.

Knackmuss, H.-J., Hellwig, M., Lackner, H., and Otting, W. (1976) Cometabolism of 3-methylbenzoate and methylcatechols by a 3-chlorobenzoate utilizing *Pseudomonas*: accumulation of (+)-2,5-dihydro-4-methyl- and (+)-2,5-dihydro-2-methyl-5-oxo-furan-2-acetic acid. *Eur J Appl Microbiol* **2:** 267–276.

Kuhm, A.E., Schlömann, M., Knackmuss, H.-J., and Pieper, D.H. (1990) Purification and characterization of dichloromuconate cycloisomerase from *Alcaligenes eutrophus* JMP134. *Biochem J* **266:** 877–883.

Lehning, A. (1998) Untersuchungen zum Metabolismus von Chlortoluolen: Konstruktion Chlortoluol und Chlorbenzylalkohol verwertender Mikroorganismen. Dissertation. TU Braunschweig.

Lovanh, N., Hunt, C.S., and Alvarez, P.J. (2002) Effect of ethanol on BTEX biodegradation kinetics: aerobic continuous culture experiments. *Water Res* **36:** 3739–3746.

Marín, M., Pérez-Pantoja, D., Donoso, R., Wray, V., González, B., and Pieper, D.H. (2010) Modified 3-oxoadipate pathway for the biodegradation of methylaromatics in *Pseudomonas reinekei* MT1. *J Bacteriol* **192:** 1543–1552.

Mars, A.E., Kasberg, T., Kaschabek, S., van Agteren, M.H., Janssen, D.B., and Reineke, W. (1997) Microbial degradation of chloroaromatics: use of the *meta*-cleavage pathway for mineralization of chlorobenzene. *J Bacteriol* **179:** 4530–4537.

Mars, A.E., Kingma, J., Kaschabek, S.R., Reineke, W., and Janssen, D.B. (1999) Conversion of 3-chlorocatechol by various catechol 2,3-dioxygenases and sequence analysis of the chlorocatechol dioxygenase region of *Pseudomonas putida* GJ 31. *J Bacteriol* **181:** 1309–1318.

Moiseeva, O.V., Solyanikova, I.P., Kaschabek, S.R., Gröninger, J., Thiel, M., Golovleva, L.A., and Schlömann, M. (2002) A new modified *ortho* cleavage pathway of 3-chlorocatechol degradation by *Rhodococcus opacus* 1CP: genetic and biochemical evidence. *J Bacteriol* **184:** 5282–5292.

Moreira, I.S., Amorim, C.L., Carvalho, M.F., Ferreira, A.C., Afonso, C.M., and Castro, P.M. (2013) Effect of the metal iron, copper and silver on fluorobenzene biodegradation by *Labrys portucalensis*. *Biodegradation* **24:** 245–255.

Nikodem, P., Hecht, V., Schlömann, M., and Pieper, D.H. (2003) New bacterial pathway for 4- and 5-chlorosalicylate degradation via 4-chlorocatechol and maleylacetate in *Pseudomonas* sp. strain MT1. *J Bacteriol* **185:** 6790–6800.

Oldenhuis, R., Kuijk, L., Lammers, A., Janssen, D.B., and Witholt, B. (1989) Degradation of chlorinated and non-chlorinated aromatic solvents in soil suspensions by pure bacterial cultures. *Appl Microbiol Biotechnol* **30:** 211–217.

Onaca, C., Kieninger, M., Engesser, K.-H., and Altenbuchner, J. (2007) Degradation of alkyl methyl ketones by *Pseudomonas veronii* MEK700. *J Bacteriol* **189:** 3759–3767.

Perez-Pantoja, D., Ledger, T., Pieper, D.H., and Gonzalez, D. (2003) Efficient turnover of chlorocatechols is essential for growth of *Ralstonia eutropha* JMP134(pJP4) in 3-chlorobenzoic acid. *J Bacteriol* **185:** 1534–1542.

Pettigrew, C.A., Haigler, B.E., and Spain, J.C. (1991) Simultaneous biodegradation of chlorobenzene and toluene by a *Pseudomonas* strain. *Appl Environ Microbiol* **57:** 157–162.

Pieper, D.H., Engesser, K.-H., Don, R.H., Timmis, K.N., and Knackmuss, H.-J. (1985) Modified *ortho*-cleavage pathway in *Alcaligenes eutrophus* JMP134 for the degradation of 4-methylcatechol. *FEMS Microbiol Lett* **29:** 63–67.

Pieper, D.H., Reineke, W., Engesser, K.-H., and Knackmuss, H.-J. (1988) Metabolism of 2,4-dichlorophenoxyacetic acid, 4-chloro-2-methylphenoxyacetic acid and 2-methylphenoxyacetic acid by *Alcaligenes eutrophus* JMP134. *Arch Microbiol* **150:** 95–102.

Pieper, D.H., Stadler-Fritzsche, K., Knackmuss, H.-J., Engesser, K.-H., Bruce, N.C., and Cain, R.B. (1990) Purification and characterization of 4-methylmuconolactone methylisomerase, a novel enzyme of the modified 3-oxoadipate pathway in the gram-negative bacterium *Alcaligenes eutrophus* JMP134. *Biochem J* **271**: 529–534.

Pollmann, K., Wray, V., Hecht, H.-J., and Pieper, D.H. (2003) Rational engineering of the regioselectivity of TecA tetrachlorobenzene dioxygenase for the transformation of chlorinated toluenes. *Microbiology* **149**: 903–913.

Reineke, W. (1998) Development of hybrid strains for the mineralization of chloroaromatics by patchwork assembly. *Annu Rev Microbiol* **52**: 287–331.

Reineke, W., and Knackmuss, H.J. (1984) Microbial metabolism of haloaromatics: isolation and properties of a chlorobenzene-degrading bacterium. *Appl Environ Microbiol* **47**: 395–402.

Richardson, K.L., and Gibson, D.T. (1984) A novel pathway for toluene oxidation in *Pseudomonas mendocina*. In *Abstracts of the 84th Annual Meeting of the American Society of Microbiology*. Neidhardt, F.C. (ed.). Washington, D.C: American Society of Microbiology, p. 156. abstr. K54.

Riegert, U., Heiss, G., Fischer, P., and Stolz, A. (1998) Distal cleavage of 3-chlorocatechol by an extradiol dioxygenase to 3-chloro-2-hydroxymuconic semialdehyde. *J Bacteriol* **180**: 2849–2853.

Riegert, U., Heiss, G., Kuhm, A.E., Müller, C., Contzen, M., Knackmuss, H.-J., and Stolz, A. (1999) Catalytic properties of the 3-chlorocatechol-oxidizing 2,3-dihydroxybiphenyl 1,2-dioxygenase from *Sphingomonas* sp. strain BN6. *J Bacteriol* **181**: 4812–4817.

Rojo, F., Pieper, D.H., Engesser, K.-H., Knackmuss, H.-J., and Timmis, K.N. (1987) Assemblage of *ortho* cleavage route for simultaneous degradation of chloro- and methylaromatics. *Science* **238**: 1395–1398.

Schlömann, M. (1994) Evolution of chlorocatechol catabolic pathways. *Biodegradation* **5**: 301–321.

Schmidt, E., Remberg, G., and Knackmuss, H.-J. (1980) Chemical structure and biodegradability of halogenated aromatic compounds: halogenated muconic acids as intermediates. *Biochem J* **192**: 331–337.

Shields, M.S., Montgomery, S.O., Chapman, P.J., Cuskey, S.M., and Pritchard, P.H. (1989) Novel pathway of toluene catabolism in the trichloroethylene-degrading bacterium G4. *Appl Environ Microbiol* **55**: 1624–1629.

Skiba, A., Hecht, V., and Pieper, D.H. (2002) Formation of protoanemonin from 2-chloro-cis/cis-muconate by the combined action of muconate cycloisomerase and muconolactone isomerise. *J Bacteriol* **184**: 5402–5409.

SMUL Sachsen: Sächsisches Staatsministerium für Umwelt und Landwirtschaft (2000). Materialien zur Altlastenbehandlung Nr. 2/2001.

Sovorova, M.M., Solyanikova, I.P., and Golovleva, L.A. (2006) Specificity of catechol *ortho*-cleavage during *para*-toluate degradation by *Rhodococcus opacus* 1CP. *Biochemistry (Mosc)* **71**: 1316–1323.

Strunk, N. (2000). Halogenaromatenabbau in der biologischen Abluftreinigung. Diplomarbeit. Universität Stuttgart.

Strunk, N. (2007). Der Abbau von Fluorbenzol und seinen Homologen durch Burkholderia fungorum FLU100 – Biologische Grundlagen und Anwendung in der Abluftreinigung. Dissertation. Universität Stuttgart.

Strunk, N., and Engesser, K.-H. (2013) Degradation of fluorobenzene and its central metabolites 3-fluorocatechol

and 2-fluoromuconate by *Burkholderia fungorum* FLU100. *Appl Microbiol Biotechnol* **97**: 5605–5614.

Strunk, N., Dobslaw, D., Pieper, D.H., and Engesser, K.-H. (2006) Isolation and characterization of strain FLU100, a bacterium with the rare capability to simultaneously degrade toluene and all monohalosubstituted benzenes including fluorobenzene. BIOSPECTRUM Tagungsband Jena, KC 13.

Taeger, K., Knackmuss, H.-J., and Schmidt, E. (1988) Biodegradability of mixtures of chloro- and methylsubstituted aromatics: simultaneous degradation of 3-chlorobenzoate and 3-methylbenzoate. *Appl Microbiol Biotechnol* **28**: 603–608.

UBA (2013). Bundesweite Übersicht zur Altlastenstatistik. Daten zur Umwelt. Umweltbundesamt.

Vollmer, M.D., and Schlömann, M. (1995) Conversion of 2-chloro-cis,cis-muconate and its metabolites 2-chloro- and 5-chloromuconolactone by chloromuconate cycloisomerase of pJP4 and pAC27. *J Bacteriol* **177**: 2938–2941.

Vollmer, M.D., Fischer, P., Knackmuss, H.-J., and Schlömann, M. (1994) Inability of muconate cycloisomerases to cause dehalogenation during conversion of 2-chloro-cis,cis-muconate. *J Bacteriol* **176**: 4366–4375.

Warhurst, A.M., Clarke, K.F., Hill, R.A., Holt, R.A., and Fewson, C.A. (1994) Metabolism of styrene by *Rhodococcus rhodochrous* NCIMB 13259. *Appl Environ Microbiol* **60**: 1137–1145.

Wieser, M., Eberspächer, J., Vogler, B., and Lingens, F. (1994) Metabolism of 4-chlorophenol by *Azotobacter* sp. GP1: structure of the *meta*-cleavage product of 4-chlorocatechol. *FEMS Microbiol Lett* **116**: 73–78.

Worsey, M.J., and Williams, P.A. (1975) Metabolism of toluene and xylene by *Pseudomonas putida* (arvilla) mt-2: evidence for a new function of the TOL plasmid. *J Bacteriol* **124**: 7–13.

Yen, K.-M., Karl, M.R., Blatt, L.M., Simon, M.J., Winter, R.B., Fausset, P.R., *et al.* (1991) Cloning and characterization of a *Pseudomonas mendocina* KR1 gene cluster encoding toluene-4-monooxygenase. *J Bacteriol* **173**: 5315–5327.

Zerlin, K.F.T. (2004) Abbau von Chloraromaten in *Pseudomonas putida* GJ 31 und seinen Abkömmlingen: Charakterisierung von Genclustern auf den Plasmiden. Dissertation. Universität Wuppertal.

Zhang, L.L., Leng, S.Q., Zhu, R.Y., and Chen, J.M. (2011) Degradation of chlorobenzene by strain *Ralstonia pickettii* L2 isolated from a biotrickling filter treating a chlorobenzene-contaminated gas stream. *Appl Microbiol Biotechnol* **91**: 407–415.

Zylstra, G.J., McCombie, W.R., Gibson, D.T., and Finette, B.A. (1988) Toluene degradation by *Pseudomonas putida* F1: genetic organization of the tod operon. *Appl Environ Microbiol* **54**: 1498–1503.

Supporting information

Additional Supporting Information may be found in the online version of this article at the publisher's web-site:

Fig. S1. Influence of the pH value of the medium on the concentration of 2-methylmuconic acid as intermediate during conversion of 1.55 mmol toluene l⁻¹ by strain FLU100 pre-grown on toluene. The chart for pH 5.0 fell out of the series because of reduced conversion rates of toluene as substrate.

Fig. S2. Characterization and identification of formed 'diendiol' structures of benzene, toluene and fluorobenzene by mutant strain FLU100 P2R5 via HPLC analyses (column: ProntoSIL™ SC-04 Eurobond C18 column, 125 mm, 4 mm, i.d. 5 µm; solvent: H_2O : CH_3OH : H_3PO_4 (85 w/v%) = 74.9% : 25% : 0.1%). The flow rate was maintained at 1 ml min^{-1}.

Table S1. Conversion rates of aromatics in single, binary and ternary mixtures under different cultivation conditions. The corresponding generation times are given in the last column. B = benzene; CB = chlorobenzene; FB = fluorobenzene; T = toluene.

Table S2. Specific oxygen uptake activities for the initial enzyme(s) of cells of FLU100 pre-grown on fluorobenzene, toluene or benzene. The value in parenthesis represents the number of independent batch cultures tested twice. The first value of each column describes the average value of oxygen uptake while the second one describes the variation. The absolute activity for fluorobenzene grown cells was 465 ± 77.2 units, 623.8 ± 243.7 units for toluene and 629.4 ± 190.1 units for benzene. One unit of oxygenase activity was defined as the conversion of 1 µg O_2 l^{-1} min^{-1} OD^{-1}.

Table S3. Specific oxygen uptake activity of the (chloro)catechol-1,2-dioxygenase of strain FLU100 after growth on fluorobenzene (FB), toluene (T) or benzene (B). The value in parenthesis represents the number of independent batch cultures tested twice. The first value of each column describes the average oxygen uptake while the second one describes the variation. The absolute activity of fluorobenzene grown cells was 2090.5 ± 395.1 units, 189.4 ± 124.5 units for toluene and 88.7 ± 29.0 units for benzene. One unit of oxygenase activity was defined as the conversion of 1 µg O_2 l^{-1} min^{-1} OD^{-1}.

Table S4. Identification of 3-methylcatechol (native form) and 2-methoxy-3-methylphenol (methylated form) as intermediate of toluene degradation of strain FLU100 by GC-MS analyses. The fragment with the highest intensity is normalized to 100% and other fragments are given as relative intensities.

Table S5. Conversion rates (in mmol l^{-1} h^{-1} OD_{546}^{-1}) and generation times (in hours) of cells of FLU100 pre-grown on toluene or glucose as reference. n.d. = not detected or intensity not given in literature.

8

Synthetic biology approaches to improve biocatalyst identification in metagenomic library screening

María-Eugenia Guazzaroni,[1] Rafael Silva-Rocha[2] and Richard John Ward[1]*

[1]Departamento de Química, FFCLRP,

[2]Departamento de Bioquímica e Imunologia, FMRP,

University of São Paulo, Ribeirão Preto, SP, Brazil.

Summary

There is a growing demand for enzymes with improved catalytic performance or tolerance to process-specific parameters, and biotechnology plays a crucial role in the development of biocatalysts for use in industry, agriculture, medicine and energy generation. Metagenomics takes advantage of the wealth of genetic and biochemical diversity present in the genomes of microorganisms found in environmental samples, and provides a set of new technologies directed towards screening for new catalytic activities from environmental samples with potential biotechnology applications. However, biased and low level of expression of heterologous proteins in *Escherichia coli* together with the use of non-optimal cloning vectors for the construction of metagenomic libraries generally results in an extremely low success rate for enzyme identification. The bottleneck arising from inefficient screening of enzymatic activities has been addressed from several perspectives; however, the limitations related to biased expression in heterologous hosts cannot be overcome by using a single approach, but rather requires the synergetic implementation of multiple methodologies. Here, we review some of the principal constraints regarding the discovery of new enzymes in metagenomic libraries and discuss how these might be resolved by using synthetic biology methods.

*For correspondence. E-mail rjward@ffclrp.usp.br

Funding Information The study was supported by CNPq (CNPq 472893/2013-0 and 307795/2009-8) and FAPESP (2010/18850-2). MEG was a beneficiary of a CNPq Young Talent Fellowship (CNPq 370630/2013-0). RSR was supported by a FAPESP Post-Doctoral Fellowship (FAPESP 2013/04125-2).

Introduction

Biotechnology takes advantage of processes that utilize biological components or living organisms, and plays an increasing role in industry, agriculture, medicine and energy generation. In the industrial context, many manufacturing processes that previously depended strictly on complex (and frequently harmful) chemical reactions have been superseded by much simpler and safer enzyme-based catalysis (Kirk *et al.*, 2002). The introduction of biotechnology in industrial processes generates not only a reduction in the final amount and toxicity of effluents, but can also considerably reduce costs (Herrera, 2004). The number of biotechnology applications has expanded in recent years, and this has created a growing demand for biocatalysts with superior performance or tolerance to extreme application-specific conditions (Lorenz *et al.*, 2002; Schloss and Handelsman, 2003). This is particularly true in those industries that produce bulk commodities such as detergents (Maurer, 2004). Similarly, fine-chemical industries require multiple biocatalysts in order to perform highly diverse transformations for the production of new compounds (Homann *et al.*, 2004).

The use of enzymes in industrial applications has grown considerably, and represents a market of approximately US$4 billion in 2011 (GIA, 2012), and a number of different categories of enzymes have been used in a wide variety of applications (Schoemaker *et al.*, 2003). For example, proteases have been used in detergents, in pharmaceutical and chemical synthesis industries to degrade proteins into amino acids (Gupta *et al.*, 2002). Lipases hydrolyse fats to fatty acids and glycerol, and are useful for effluent treatment, detergent production and synthesis of fine chemicals among others (Hasan *et al.*, 2006). Glycosyl hydrolases (GHs) catalyse the hydrolysis of carbohydrates to sugars and have found applications in many processes in the textile, pulp and paper, and food production industries (Kirk *et al.*, 2002). The food industry also takes advantage of amylases, enzymes that hydrolase starch into sugars (Kirk *et al.*, 2002).

Furthermore, cellulases are not only useful for fuel production but are also applied in food processing, chemical synthesis and textile industries (Bhat, 2000).

The impact of the use of enzymes in industrial processes has stimulated an increased interest both in identifying new variants with enhanced kinetic parameters and in modifying previously characterized enzymes to increase their suitability for industrial applications (Lorenz et al., 2002; Schloss and Handelsman, 2003). Parameters such as activity, efficacy, specificity and stability are used to characterize and select enzymes for different applications (Lorenz and Eck, 2005). Enzymes used in industry have been identified from different sources through a combination of two major strategies: (i) the identification of novel enzymes from cultured microorganisms and (ii) molecular evolution by DNA shuffling and rational design (Lynd et al., 2002; Percival Zhang et al., 2006; Krogh et al., 2010; Ward, 2011; Chen et al., 2012)

(see Fig. 1). However, enzymes suitable for a given biotechnology application need to work efficiently within specified parameters (Fig. 1), and since those currently used are frequently not optimized, the industrial processes have to be adjusted to accommodate these suboptimal catalysts (Warnecke and Hess, 2009). As a consequence, there is an increasing demand for new biocatalysts with improved properties for industrial applications, such as higher catalytic efficiency on insoluble substrates (as in the case of cellulases used in the production of second generation bioethanol), increased stability at elevated temperature and at defined pH, and higher tolerance to end-product inhibition (Ward, 2011; Singhania et al., 2013).

Microorganisms play a central role in biotechnology, not only as tools in molecular biology techniques, but also as the major source of biocatalysts for industrial applications (Fernandez-Arrojo et al., 2010). Bearing in mind that the

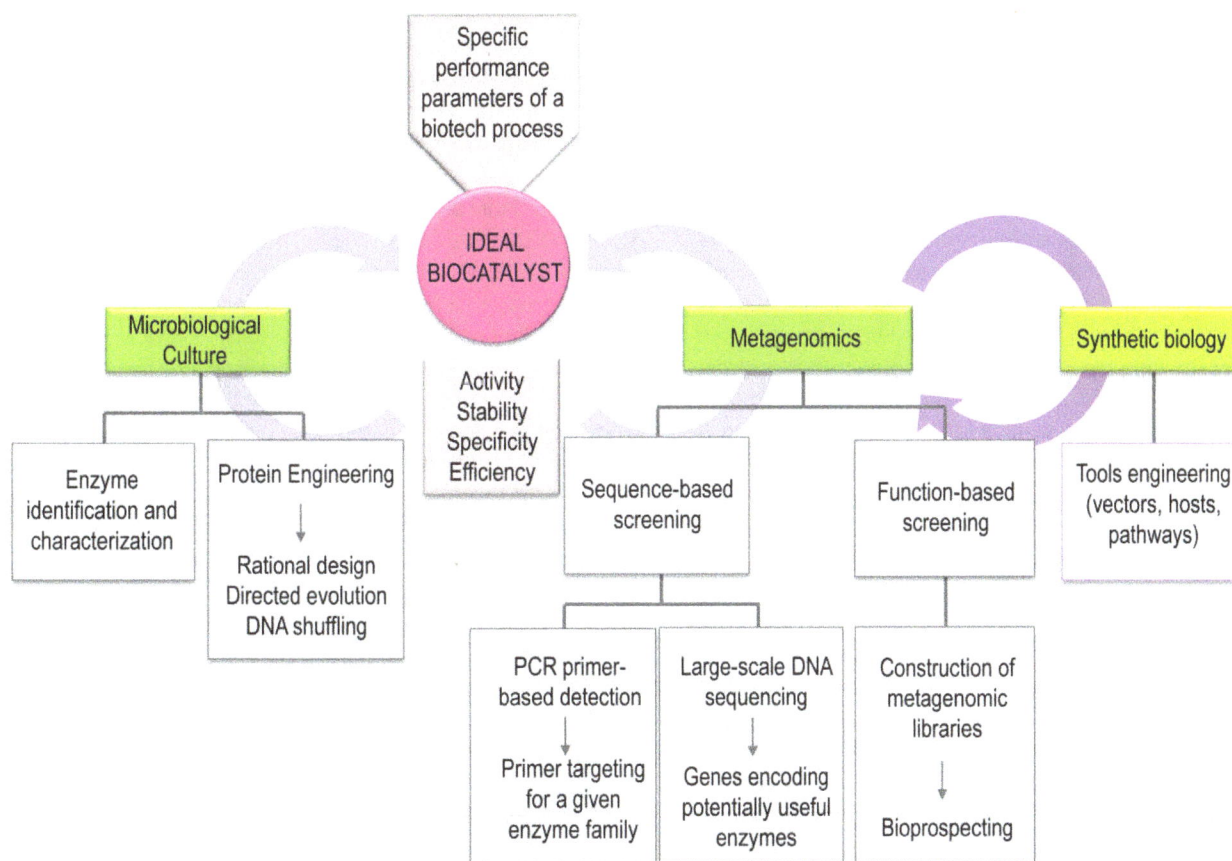

Fig. 1. Overall diagram for the identification strategy of an ideal biocatalyst. The identification of enzymes from cultured microorganisms and metagenomics are the two principal approaches currently employed for recovering of genes encoding the desired enzymatic activity for industrial processes. The genes encoding enzymes identified from cultured microorganisms may be cloned and expressed, and parameters such as activity, stability, specificity and efficiency improved using protein rational design and in vitro evolution techniques. Metagenomics strategies are based on either activity-based approaches, which involve the construction of expression libraries and its posterior activity screening, or sequence-based approaches. Sequence-based approaches involve either the design of DNA primers for conserved regions of known protein families or data mining of genes encoding potential biocatalysts identified in sequences from next generation sequencing projects. Synthetic biology can provide solutions to the current limitations in activity-based metagenomic approaches. Development of methods for the engineering of new bacterial hosts and molecular biology tools promise to increase the efficiency of discovery of biotechnologically relevant enzymes.

total number of microbial cells on Earth is estimated to be 10^{30} (Turnbaugh and Gordon, 2008), the huge natural wealth regarding protein diversity can be envisaged. Although prokaryotes represent the largest proportion of individual living organisms with an estimated 10^3–10^5 microbial species in 1 g of soil (Schloss and Handelsman, 2006), less than 1% can be cultured using existing methodologies (Sleator et al., 2008). If we assume that a single genome encodes 4000 proteins (as is the case for the typical bacteria Escherichia coli), then 4×10^8 potential proteins might be expected in just 1 g of soil. Supposing that 40% of these proteins display catalytic activity (Dinsdale et al., 2008), we might expect to find 1.6×10^8 biocatalysts, which highlights the vast inventory of biological functions available in nature. Metagenomics avoids the necessity of isolation and laboratory cultivation of individual microorganisms, and has become a powerful tool for accessing and exploring the biological and molecular biodiversity present in different natural environments. Over the past decades, many studies using metagenomic approaches have proven to be successful for the recovery of novel enzymes with potential use in industrial applications (Lorenz and Eck, 2005; Fernandez-Arrojo et al., 2010). Despite these successes, metagenomic strategies typically have low rates of target identification, and a number of issues need to be addressed in order to improve the screening efficiency of metagenomic libraries. The limits include: (i) bias imposed by host organism expression; (ii) vector performance in particular hosts; and (iii) suitable screening strategies in relation to the specific properties required in the target enzymes. As has already been demonstrated in several recent studies, all these issues may addressed using a synthetic biology approach (Williamson et al., 2005; Uchiyama and Watanabe, 2008; Uchiyama and Miyazaki, 2010). In addition to the application of existing synthetic biology approaches, the development of new methodologies is imperative for the next generation of metagenomic studies that aim to recover ideal biocatalysts for given industrial processes. This review focuses on how the interplay between synthetic biology and functional metagenomics can yield novel strategies to obtain ideal candidate enzymes with specific characteristics. We also discuss how innovative synthetic biology applications could help relieve current limitations on metagenomic screening.

Bioprospecting metagenomes to identify new biocatalysts

Recent advances and major limitations

Although most environmental bacteria are refractory to cultivation, the biotechnological potential of the uncultivated bacteria can be realized by directly cloning the DNA

retrieved from the microbial community (Guazzaroni et al., 2010a). The construction and subsequent screening of metagenomic libraries allows identification of the targeted genes encoding the desired catalytic activities (Fig. 1). Accordingly, a well-planned strategy should take into consideration the vector to be used, the host organism for transformation and the screening strategy in order to maximize the rate of identification of the target activities. For example, if single genes or small operons are of interest, the best option is to use a small-insert metagenomic library instead of a large-insert library (Guazzaroni et al., 2010a). Small-insert expression libraries, especially those using lambda phage vectors and plasmids, are especially suitable for activity-based screening. The small size of the cloned fragments (up to 8 kb) means that most genes that are present in the appropriate orientation will be under the influence of strong vector promoters, and thus have a good chance of being expressed and detected in activity screens (Ferrer et al., 2009). On the other hand, if the goal is biosynthetic pathway mining or functional expression of large multi-enzyme assemblies (for example, in the case of polyketide synthases clusters), the preferred option is library construction using cosmids or fosmid, which can harbour DNA inserts of up to 40 kb in size (Guazzaroni et al., 2010a). Since fosmids use both the F-plasmid origin of replication and the partitioning mechanisms present in E. coli (Kim et al., 1992), the use of these vectors are limited to this host organism. It is noteworthy that the bacterium E. coli is the most commonly used host strain for library screening, since many of the currently available genetic tools have been developed in this organism. One example is the use of modified phages that undergo the lytic cycle under controlled conditions, which allows the identification of proteins that would otherwise be toxic for the cell (Guazzaroni et al., 2010a). Furthermore, a number of protocols for obtaining high rates of efficiency of transformation in E. coli are well established. In addition, kits that facilitate library construction and efficient transformation with different vectors are commercially available (Guazzaroni et al., 2010a). However, it is important to appreciate that significant differences in expression modes exist between different taxonomic groups of prokaryotes, and that only 40% of enzymatic activities may be detected by random cloning in E. coli (Gabor et al., 2004). Therefore, it is likely that also performing metagenomic library screening in hosts other than E. coli will expand the range of detectable activities, although achieving this goal will require further optimization of the conditions for high transformation efficiency. Indeed, promising results of metagenomic library screening have been reported in Streptomyces spp. (Wang et al., 2000), Rhizobium leguminosarum (Wexler et al., 2005) and diverse Proteobacteria (Craig et al., 2010). For example, Wexler and collaborators constructed a library in the broad host-range cosmid pLAFR3 using metagenomic DNA obtained from the microbial community of an anaerobic digester in a wastewater treatment plant (Wexler *et al.*, 2005). After screening the metagenomic libraries in *R. leguminosarum*, a single cosmid that enabled *R. leguminosarum* to grow on ethanol as the sole carbon and energy source was recovered. Further analysis identified the presence of a gene encoding an atypical alcohol dehydrogenase that did not confer ethanol utilization ability to either *E. coli* or to *Pseudomonas aeruginosa*, even though the gene was transcribed in both these hosts (Wexler *et al.*, 2005). These results show that the use of broad host-range vectors enhances the flexibility of metagenomic library screening. Furthermore, a recent functional metagenomic study showed that recovery of genes conferring acid resistance to *E. coli*, and the subsequent transfer of some of these genes to *P. putida* and *B. subtilis*, expanded the capabilities of these two bacteria to survive harsh acid conditions (Guazzaroni *et al.*, 2013). However, in agreement with previous studies (Craig *et al.*, 2010), variable gene doses were present due to the use of different cloning plasmids in each host organism, and no quantitative comparison could be made regarding gene expression or activity levels between the hosts. Thus, the developing of robust broad-host-range vectors capable of replication in several different hosts is one of the major challenges in metagenomics, and one to which synthetic biology may make a significant contribution.

In addition to viable library construction, an additional bottleneck in metagenomic screening is related to the low frequency of positive clones that are typically recovered (Vieites *et al.*, 2009). Common screening methods are based on the degradation of specific enzyme substrates that result either in the appearance of halos surrounding the positive clones or alternatively on the use of chromogenic substrates (Guazzaroni *et al.*, 2010b). Depending on the type of substrate used and on the enzyme screened, detection of positive clones can be assayed directly in solid or in liquid media (Guazzaroni *et al.*, 2010b), and the choice of medium will have consequences on the throughput level of the screening method. Similarly, adequate selection of specific activity-driven substrates plays a key role in the success of recovering of the desired enzymatic activity and decreasing the frequency of false positives.

Although methods are available to the direct screening of many enzymes, there is an increasing need to expand the tools available for enzyme detection. An alternative approach to this problem using synthetic biology is the design and implementation of *in vivo* biosensors capable of generate a detectable output in response to the degradation or production of a particular metabolite (Galvao and de Lorenzo, 2006). In this context, substrate-induced gene expression and product-induced gene expression approaches have been developed and successfully applied to the detection of enzymatic activities associated with the metabolic modification of compounds of interest (Uchiyama *et al.*, 2005; Uchiyama and Miyazaki, 2010). In a broader approach, the concept of genetic traps has been used to guide the construction of synthetic circuits containing engineered regulators to control gene expression responses to metabolites generated by enzymatic activities present in cloned metagenomic fragments (Uchiyama and Watanabe, 2008). In general, these strategies are particularly useful for screening of libraries where processing of small molecules is targeted (Williamson *et al.*, 2005).

Activity-based versus sequence-based screening strategies

There are two distinct strategies in functional metagenomics for the recovery of sequences encoding the desired enzymatic activity (Fig. 1). First, activity-based approaches involve construction of small- to large-insert expression libraries that are suitable for direct activity screening, such as lambda phage, plasmid, cosmid or copy-controlled fosmid vectors (Lorenz and Eck, 2005). Once a library has been constructed, a critical step is the screening of a large number of clones, and in the case of activity-based screening, thousands of clones may be analysed in a single screen, with the advantage that sequence information is not required. Therefore, this is the strategy that has the potential to identify entirely novel classes of genes encoding known or novel functions (Handelsman, 2004; Daniel, 2005; Gloux *et al.*, 2010; Bhat *et al.*, 2013). Furthermore, activity-driven screening strategies can potentially provide a means to reveal undiscovered genes or gene families that cannot be detected by sequence-driven approaches.

The most widely used activity-driven screening approach is phenotype detection of the desired activity using chemical dyes and insoluble or chromophore-bearing derivatives of enzyme substrates. When incorporated into the growth medium, these compounds allow the detection of specific catalytic capabilities of individual clones (Ferrer *et al.*, 2009). In an elegant example using this approach, the diversity of rumen enzymes was characterized by screening for hydrolase activity in a metagenomic phage library from the rumen content of a dairy cow (Ferrer *et al.*, 2005b). In total, 22 clones with distinct hydrolytic activities were identified and characterized, among which four hydrolases exhibited no sequence similarity to enzymes deposited in the public databases, and the putative catalytic residues in the sequences from four other clones showing esterase and cellulase-like activities did not match any known enzymes. Similarly, activity-based screening of metagenomic fosmid libraries from cellulose-depleting microbial communities found in the fresh casts of two different earthworm species retrieved 10 carbohydrate-modifying enzymes (Beloqui et al., 2010). Interestingly, two of the GHs identified represented two novel families of β-galactosidases/α-arabinopyranosidases (Beloqui et al., 2010). In another example, metagenomic DNA extracted from activated sludge from industrial wastewater was cloned into fosmids and the resulting E. coli library was screened for extradiol dioxygenase activity using catechol as the substrate (Suenaga et al., 2007). Analysis of the sequences from a total of 38 clones identified various clusters encoding putative metabolic pathways that were dissimilar to previously reported pathways for degradation of the compound (Suenaga et al., 2009a,b). Additionally, several other recent examples have shown that the phenotype detection approach can be successfully applied to the identification of novel targeted enzymes such as dehalogenases (Sharma et al., 2013), meta-cleavage product hydrolases (Alcaide et al., 2013), GHs (Ferrer et al., 2012; Ko et al., 2013), xylanases (Gong et al., 2013b) carboxyl esterases and lipases (Martinez-Martinez et al., 2013; 2014).

An alternative activity-driven screening approach is based on heterologous complementation of host strains that require the targeted genes for growth under selective conditions. In this strategy, functional complementation permits growth of a clone transformed with a metagenomic DNA insert containing the necessary genes for survival (Wenzel and Muller, 2005). This technique allows the straightforward and rapid screening of complex metagenomic libraries comprising millions of clones, and since false positives are rare, this approach is highly selective for the targeted genes (Simon et al., 2009). However, this strategy is limited to the detection of enzymes that catalyse the synthesis of an essential product and for which an auxotrophic host is either available or can readily be constructed. Several examples in the literature have shown that this approach has been successful in the detection of different enzymes such as racemases (Chen et al., 2010), DNA polymerases (Simon et al., 2009), β-lactamases (Riesenfeld et al., 2004), alcohol dehydrogenases (Wexler et al., 2005) and enzymes involved in poly-3-hydroxybutyrate metabolism (Wang et al., 2006).

Sequence-based approaches have also led to the effective identification of genes encoding enzymes, such as dimethylsulfoniopropionate-degrading enzymes (Varaljay et al., 2010), glycerol dehydratases (Knietsch et al., 2003), dioxygenases (Morimoto and Fujii, 2009; Sul et al., 2009; Zaprasis et al., 2010), nitrite reductases (Bartossek et al., 2010), hydrazine oxidoreductases (Li et al., 2010), chitinases (Hjort et al., 2010), glycoside hydrolases (Lee et al., 2013), nitrilases (Gong et al.,

2013a), prephenate dehydrogenases (Jiang et al., 2013) and hemicellulases (Yan et al., 2013). The application of sequence-based approaches involves the design of polymerase chain reaction (PCR) primers for the target sequences that are derived from conserved regions of known protein families, and this dependence on prior knowledge limits the possibility for identifying new protein families (Ferrer et al., 2009). The large-scale sequencing of bulk DNA or metagenomic libraries through deep sequencing techniques provides the raw data for mining sequences encoding potentially useful enzymes. Since homology-based methods are effective only when the information regarding the reference sequences is accurate, a further disadvantage of this approach is its reliance both on existing genome annotations and on the quality and completeness of current databases (Hallin et al., 2008). It is worth a cautionary note, since a significant number of genomes in the current databases contain misannotations (Schnoes et al., 2009). Considering that the classification of protein families is based on amino acid similarity, novel enzyme families could not be detected by database searching with sequences from metagenomic sequencing or PCR-based detection methods, and might be annotated as hypothetical proteins. A review of prokaryotic protein diversity in different shotgun metagenome studies indicated that 30–60% of the proteins could not be assigned known functions using current public databases (Vieites et al., 2009). The advantage of using activity-based rather than sequence-based screening was highlighted in a recent report in which a novel cold-tolerant esterase with low sequence similarity (< 29% amino acid identity) was identified by functional screening of an Antarctic soil sample (Heath et al., 2009). This esterase had no identity to any GenBank nucleotide sequence, and it is unlikely that it would have been detected by low stringency PCR-based screening methods or by deep sequencing techniques.

Mining biocatalysts from extreme environments

The maximum yield of industrial processes is achieved by optimization of physicochemical parameters, and most currently available enzymes are incompatible with these conditions, and the use of enzymes requires that these processes be adapted, which can result in reduced production levels. Extreme environments, such as solfataric hot springs (Rhee et al., 2005), Urania hypersaline basins (Ferrer et al., 2005a), acid mine drainage biofilms (Guazzaroni et al., 2013), glacier soil (Yuhong et al., 2009), glacial ice (Simon et al., 2009) and Antarctic soil (Cieslinski et al., 2009; Heath et al., 2009) represent a rich and largely unexploited reservoirs of novel biocatalysts with biotechnologically valuable properties. Although the diversity of microbial communities present in many extreme habitats is likely to be low, samples from these environments are still a valuable source of novel enzymes that are active under extreme conditions (Steele *et al.*, 2009), and as might be expected, the properties of the enzymes retrieved are consistent with conditions of the source environments (Feng *et al.*, 2007; Heath *et al.*, 2009; Jiang *et al.*, 2009; Hu *et al.*, 2012).

Metagenomic libraries derived from extreme habitats have been constructed and enzymes displaying environment compatible properties have been recovered. A relevant consideration is the compatibility of the host organism growth with the optimum physicochemical characteristics used in the screening of the catalytic activity. *Escherichia coli* is highly sensitive to the conditions that are typically present in extreme environments, and experimental conditions should therefore be adjusted to perform library screening under the conditions of interest (for example, low pH) while still allowing growth of this organism. Alternatively, screening could be performed without subjecting the host to any selective pressure, for example in cases where the host organism is used to clone total metagenomic DNA and the screening is performed under regular growth conditions. In both cases, after the initial identification of the enzymes, characterization of the desired catalytic properties of the source clones can be made with crude culture extracts and selected enzymes may then be purified for a more detailed characterization. For instance, a novel alkaliphilic esterase active at 7°C was identified by screening a metagenomic library from Antarctic desert soil (Heath *et al.*, 2009). In a separate study, another cold-active esterase was also isolated from the same library using a similar approach (Hu *et al.*, 2012). Their low-temperature activity and alkaliphilic properties make these enzymes interesting candidates for industrial applications. Esterases are of particular industrial interest since they serve as useful biocatalysts in the detergent and food industries, and for the production of fine chemicals and in bioremediation processes (Aurilia *et al.*, 2008). Cold-adapted esterolytic enzymes could be of further value with regard to savings in energy, as compared with their mesophilic counterparts, due to their high catalytic activities at low temperatures (Hu *et al.*, 2012). At the opposite extreme, Rhee and colleagues (2005) constructed fosmid libraries using environmental samples from solfataric hot springs, and function-based screening identified a novel esterase that exhibited a high temperature optimum activity and high thermal stability.

Cellulases are enzymes with a wide range of applications such as in the textile industry for cotton softening and denim finishing; in the food industry for mashing; and in the cellulose pulp and paper industries for de-inking, fiber refining and fiber modification (Bhat, 2000). Recent interest has focussed on the use of cellulases in the hydrolysis of pretreated lignocellulosic materials into

sugars, which can then be fermented to produce second-generation ethanol and other bio-products. Metagenomic approaches have been widely employed to isolate cellulases from environmental samples in which ligno-cellulosic materials were under intensive degradation, including soil (Kim *et al.*, 2008; Jiang *et al.*, 2009), hindgut contents of higher termites (Warnecke *et al.*, 2007), compost (Pang *et al.*, 2009), contents of rabbit cecum (Feng *et al.*, 2007), fresh cast from earthworms (Beloqui *et al.*, 2010), Brazilian mangroves (Thompson *et al.*, 2013), calf rumen (Ferrer *et al.*, 2012) and cow rumen (Ferrer *et al.*, 2005b; Wang *et al.*, 2013). As anticipated, most cellulases retrieved from uncultured microbes from the rumen of herbivores were acidic and mesophilic, properties which are compatible with the conditions of yeast fermentation for the production of second-generation ethanol. Although a number of cellulase active clones have been isolated from metagenomic libraries by functional screening, the frequencies of true positive identifications were low due to low and biased expression of the majority of cellulase genes in *E. coli*. Moreover, most of the enzymes identified were endoglucanases (despite the use of specific substrates for screening other types of cellulases), which was probably a consequence of the expression of the correctly folded enzymes in *E. coli*. No cellobiohydrolases active against crystalline cellulose substrates have as yet been isolated from metagenomic libraries (Duan and Feng, 2010), which may be due to the fungal source of the majority of cellobiohydrolases and, to date, there are no reports using activity-based screening of cellulases from a cDNA metagenomic library in fungi. Moreover, since extreme environments are colonized mainly by microorganisms highly adapted to such harsh conditions, the majority of species are likely to be phylogenetically distant from *E. coli*. The recurring limitation of the metagenomic approach for certain enzyme families is therefore the lack of genetic tools for library screening in phylogenetically diverse hosts, and the development of such tools would significantly increase the probability of successful enzyme retrieval from a wide range of environmental samples.

Using synthetic biology to improve metagenomic screening strategies

The previous sections have highlighted a number of bottlenecks facing metagenomic screening that need to be resolved in order to improve the discovery rate of target enzymes. These limitations can be grouped in three main categories (as shown in Fig. 2). First, there is a need for improvement in host organism capabilities with the aim of improving the expression of the target enzymes. Second, the development of new genetic tools is necessary in order to improve the construction of metagenomic

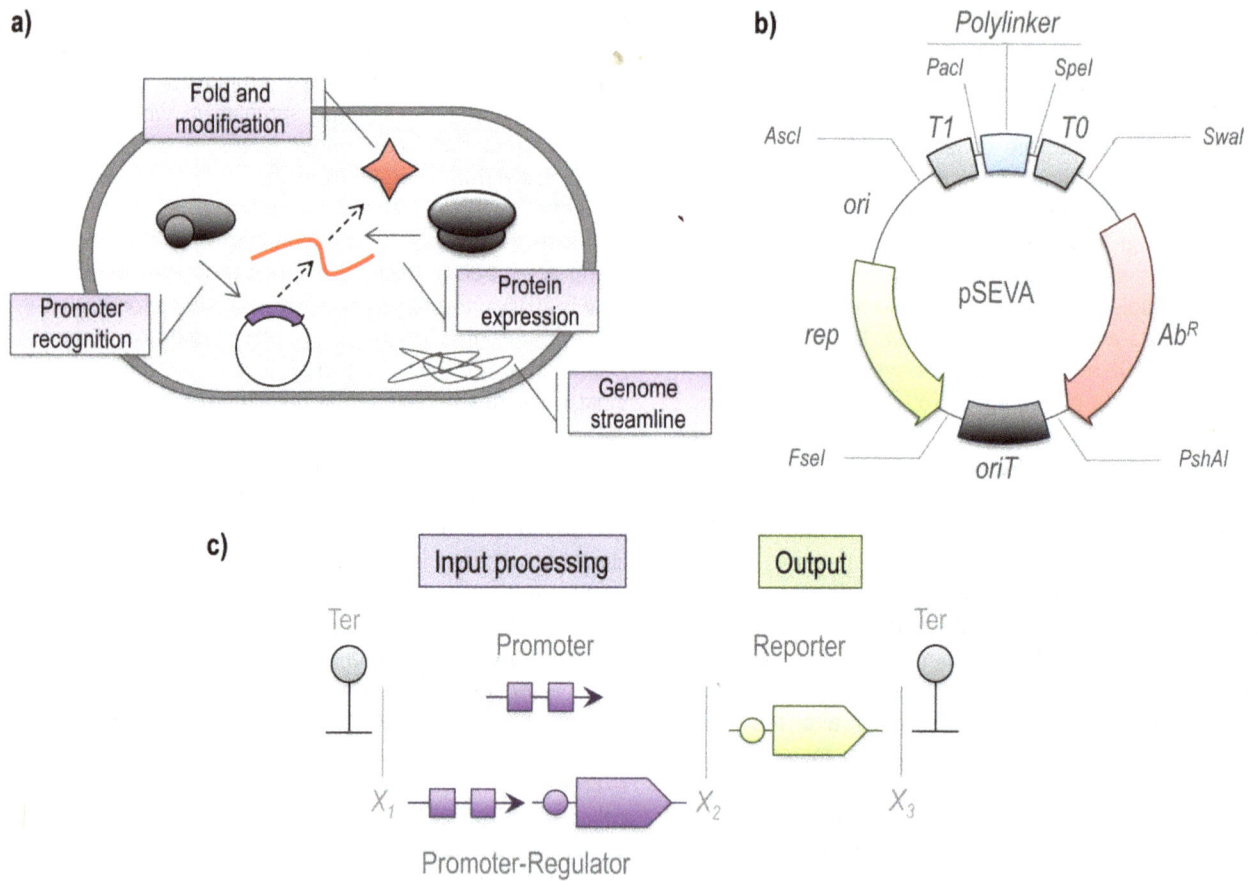

Fig. 2. Synthetic Biology may overcome limiting steps in activity-based metagenomic library screening. Current bottlenecks in functional metagenomics are related to (A) limitations in the host capabilities, (B) the performance of the genetic tools and (C) the availability of efficient screening methods.

A. In the case of the host, critical steps related to the recognition of transcriptional and translational signals, as well as the folding and modification of the expressed enzyme need to be enhanced. Host performance might be improved by reducing the metabolic burden related to the expression of unnecessary genes.

B. The use of semi-synthetic, high-efficiency genetic tools is essential for the construction of metagenomic libraries that can be maintained and screened in a wide number of microorganisms. The example shows the pSEVA bacterial vector, which is endowed with several functional features such as terminators, origin for transfer and an extensive polylinker optimized for use in several bacterial hosts.

C. Genetic circuits constructed by combining input modules (e.g. promoters and regulators) and output devices (such as reporter proteins) assembled with a standard format that uses the same sets of restriction enzymes (represented by X_1, X_2, etc.). Such circuits facilitate the screening of enzymatic activities in metagenomic libraries. The standardization of the assembly process facilitates the combination of several independent modules to construct sophisticated activity-triggered biosensors.

libraries suitable for screening in different hosts. Finally, continuation of ongoing research to elaborate novel screening strategies that enhance the discovery rate of the enzymes of interest is needed. The advances in synthetic biology over the past decade could provide the framework to address these constrains, and a particularly promising approach is the analysis of biological systems in an analogous way as electronic devices, whereby cells can be reprogrammed to perform new tasks with high efficiency (Purnick and Weiss, 2009; Weber and Fussenegger, 2010). Synthetic biology relies on a conceptual framework more closely related to engineering than biology, such as design, modelling, implementation and debugging (Canton *et al.*, 2008; Purnick and Weiss, 2009; Weber and Fussenegger, 2010). The design aspect

focuses on the planning and construction of new gene circuits for the desired application (Canton *et al.*, 2008). Modelling involves computational simulation of the proposed gene circuits in order to both evaluate performance and capabilities and to guide the selection of the suitable molecular components necessary for its construction (Koide *et al.*, 2009). The implementation step encompasses the physical assembly of the DNA elements encoding the appropriate components (such as promoters, regulators, terminators, enzymes, transporters, etc.), and follows a specific assembly standard (Arkin, 2008). Finally, the debugging step requires the testing and validation of the circuit *in vivo*, and includes the correction of undesirable traits that have their origin in the emergent properties of biology systems (Gardner *et al.*, 2000; Moon *et al.*, 2011; Siuti *et al.*, 2013). Several examples of new biological circuits that have been successfully designed and implemented are currently available (Gardner *et al.*, 2000; Cox *et al.*, 2007; 2010; Moon *et al.*, 2011; Silva-Rocha and de Lorenzo, 2011), and in recent years, the field has developed at a remarkable speed (Weber and Fussenegger, 2010; Zhan *et al.*, 2010; Regot *et al.*, 2011; Siuti *et al.*, 2013).

As mentioned in previous sections, failure of heterologous gene expression in host cells is among the main causes of the low recovery rates of enzymes of interest from metagenomic libraries (Handelsman, 2004; Lorenz and Eck, 2005). A combination of different strategies could be applied to optimize this critical event, (Fig. 2A). In the initial transcription step, a reduced affinity of the RNA polymerase (RNAP) for intrinsic promoters derived from metagenomic fragments represents an important limitation for heterologous protein expression in the host organism. Promoter recognition in prokaryotes is strongly biased among the different phylogenetic groups (Gabor *et al.*, 2004), and the ideal host for heterologous protein expression should be endowed with transcriptional machinery with broad promoter recognition capability. Such a goal could be attained, for example, by coexpressing heterologous sigma factors with different promoter specificities (Osterberg *et al.*, 2011; Rhodius *et al.*, 2013), thereby permitting protein expression from promoters derived from different bacterial sources. In a more direct approach, the expression of foreign genes in the host organism could be driven by a high-efficiency expression system such as the T7 RNAP (Terron-Gonzalez *et al.*, 2013), and has the advantage that genes controlled by complex signal transduction mechanisms could be easily expressed in response to a single inducer such as isopropyl-beta-D-thiogalactopyranoside (Tabor, 2001). The same line of reasoning may also be applied to mRNA translation, where poor recognition of the ribosome-binding site (RBS) can reduce protein expression levels (Zelcbuch *et al.*, 2013). In this case, the coexpression of additional proteins related to the mRNA recognition step could expand the host capability for foreign gene expression (Uchiyama and Miyazaki, 2009). While these strategies aim to improve the level of expression of target proteins, an additional factor is related to the activity levels of expressed enzymes. This is particularly true in the case of enzymes displaying either complex folding or requiring additional processing steps (such as cleavage, secretion, or peptide modification) (DeSantis and Jones, 1999; Bhat, 2000). For instance, in the case of protein folding, the coexpression of molecular chaperons (Dobson, 2003) has been reported to enhance the expression of heterologous proteins in *E. coli* (Ferrer *et al.*, 2004).

Among the additional approaches to improve host organism capabilities, genome edition (or streamlining)

(Cambray *et al.*, 2011) deserves special attention. Since many of the genes in bacteria such as *E. coli* are not essential for growth under standard laboratory conditions (Medini *et al.*, 2005), maintaining these genes and the consequent redundant expression of unnecessary proteins represents a significant energetic cost and a metabolic burden to the host cell (Posfai *et al.*, 2006). The reduction of genome size in bacteria by the removal of non-essential genes has been shown to endow bacteria with renewed metabolic vigor that enhances the production level of heterologous proteins (Posfai *et al.*, 2006; Martinez-Garcia *et al.*, 2014). Due to their superior expression capabilities, bacterial strains with minimized genomes are therefore promising host organisms for screening of metagenomic libraries.

In addition to genetic engineering of the more common hosts such as *E. coli*, several attempts have been focused on the use of alternative hosts to increase the rate of enzyme identification in metagenomic screening (Craig *et al.*, 2010; Guazzaroni *et al.*, 2013). Screening in alternative hosts requires either library construction in non-optimal, broad-host-range vectors (Craig *et al.*, 2010) or subcloning of target genes in appropriate vectors (Guazzaroni *et al.*, 2013). Universal and standardized tools are required in order to facilitate this type of multi-organism approach for the screening of metagenomic libraries in a larger number of hosts, and synthetic biology approaches are particularly suited to the development and use of such tools for the construction of adequate gene circuits in an increasing number of host cell platforms (Andrianantoandro *et al.*, 2006; Arkin, 2008; Shetty *et al.*, 2008). Among these new multihost tools, the pSEVA vectors are of particular interest as synthetic broad-host-range vectors that are expected to work in about 100 different bacterial species (Fig. 2B; Silva-Rocha *et al.*, 2013). Using this system as a starting point, new genetic tools could be developed for cloning and screening of environmental DNA from phylogenetically diverse bacteria.

Currently available methods that are used to screen for new catalytic activities in metagenomic libraries frequently rely either on the use of chromogenic enzyme substrates (Hu *et al.*, 2012; Ko *et al.*, 2013) or substrates that when degraded leave a clear halo around positive colonies (Wong *et al.*, 2009). More labor-intensive screening procedures such as colony PCR are also used for enzyme discovery (Hrvatin and Piel, 2007). The rapid expansion of synthetic biology has already resulted in the construction of regulatory circuits using well-characterized parts (Voigt, 2006), and there is a tremendous potential to use this accumulated knowledge for further advances in the design of biosensors to screen for enzymatic activities (Williamson *et al.*, 2005; Nasuno *et al.*, 2012). The use of biosensors is an *in vivo* strategy that has been used to identify specific enzymatic activities using engineered regulatory circuits coupled to a reporter gene, such as *lacZ*, green fluorescent protein or luciferase (Mohn *et al.*, 2006; de Las Heras *et al.*, 2010). This genetic trap approach (Uchiyama *et al.*, 2005; Uchiyama and Watanabe, 2008; Uchiyama and Miyazaki, 2010) eliminates the necessity for extensive manipulation during screening, yet allows the identification of positive clones in metagenomic libraries. Synthetic biology approaches have also been used to modulate enzyme levels in biosynthetic pathways by combinatorially pairing genes with a defined set of RBS (Zelcbuch *et al.*, 2013), resulting in the modulation of protein abundance by several orders of magnitude, which showed that engineering of metabolic pathways relies on precise control of enzyme levels (Zelcbuch *et al.*, 2013). These examples demonstrate how synthetic biology approaches can improve the ability to interconnect regulatory components (e.g. promoters and regulators) to generate new circuits with reliable performance characteristics (Fig. 2C; Schmidt and de Lorenzo, 2012). The combination of the available assembly platforms for circuit engineering (Arkin, 2008; Zhan *et al.*, 2010; Nikel and de Lorenzo, 2013) together with approaches for the redesign of regulatory systems to recognize new molecules (de Las Heras and de Lorenzo, 2012) is likely to lead to new concepts and allow the implementation of new genetic traps for the identification of enzymatic activities of interest (Fernandez-Arrojo *et al.*, 2010), and will have a significant impact in the detection rate of new biotechnologically relevant biocatalysts (Ferrer *et al.*, 2009).

Concluding remarks

Classic approaches for isolating industrial enzymes involve their identification from cultured microorganisms, and more recently, this strategy has been expanded to include *in vitro* evolution and rational design techniques for the improvement of their catalytic properties. However, the diversity of proteins identified using this strategy is restricted, the methods are time-consuming, costly and in the case of *in vitro* evolution, it is impossible to test all variants (Sommer *et al.*, 2010). The metagenomics approach encompasses the idea that the desired biocatalyst may already exist in nature, and is an alternative strategy that is used to explore the inherent diversity of natural environments. As a result, development of a broad repertoire of culture-independent techniques have been developed and advances towards the identification of novel and potent biocatalysts from metagenomic libraries have been made. However, the available metagenomic approaches require further refinement to achieve the goal of identifying industrially relevant biocatalysts. Existing limitations with respect to host expression, vector availability and specific screening restrictions cannot be solved by using a single approach, and requires the synergic implementation of multiple methodologies. A growing number of studies have shown that synthetic biology may significantly improve metagenomic library screening and allow exploitation of the rich biochemical potential present in natural environments.

Conflict of Interest

None declared.

References

Alcaide, M., Tornes, J., Stogios, P.J., Xu, X., Gertler, C., Di Leo, R., *et al.* (2013) Single residues dictate the co-evolution of dual esterases: MCP hydrolases from the alpha/beta hydrolase family. *Biochem J* **454:** 157–166.

Andrianantoandro, E., Basu, S., Karig, D.K., and Weiss, R. (2006) Synthetic biology: new engineering rules for an emerging discipline. *Mol Syst Biol* **2:** 2006. 0028.

Arkin, A. (2008) Setting the standard in synthetic biology. *Nat Biotechnol* **26:** 771–774.

Aurilia, V., Parracino, A., and D'Auria, S. (2008) Microbial carbohydrate esterases in cold adapted environments. *Gene* **410:** 234–240.

Bartossek, R., Nicol, G.W., Lanzen, A., Klenk, H.P., and Schleper, C. (2010) Homologues of nitrite reductases in ammonia-oxidizing archaea: diversity and genomic context. *Environ Microbiol* **12:** 1075–1088.

Beloqui, A., Nechitaylo, T.Y., Lopez-Cortes, N., Ghazi, A., Guazzaroni, M.E., Polaina, J., *et al.* (2010) Diversity of glycosyl hydrolases from cellulose-depleting communities enriched from casts of two earthworm species. *Appl Environ Microbiol* **76:** 5934–5946.

Bhat, A., Riyaz-Ul-Hassan, S., Ahmad, N., Srivastava, N., and Johri, S. (2013) Isolation of cold-active, acidic endocellulase from Ladakh soil by functional metagenomics. *Extremophiles* **17:** 229–239.

Bhat, M.K. (2000) Cellulases and related enzymes in biotechnology. *Biotechnol Adv* **18:** 355–383.

Cambray, G., Mutalik, V.K., and Arkin, A.P. (2011) Toward rational design of bacterial genomes. *Curr Opin Microbiol* **14:** 624–630.

Canton, B., Labno, A., and Endy, D. (2008) Refinement and standardization of synthetic biological parts and devices. *Nat Biotechnol* **26:** 787–793.

Chen, H.L., Chen, Y.C., Lu, M.Y., Chang, J.J., Wang, H.T., Ke, H.M., *et al.* (2012) A highly efficient beta-glucosidase from the buffalo rumen fungus *Neocallimastix patriciarum* W5. *Biotechnol Biofuels* **5:** 24.

Chen, I.C., Thiruvengadam, V., Lin, W.D., Chang, H.H., and Hsu, W.H. (2010) Lysine racemase: a novel non-antibiotic selectable marker for plant transformation. *Plant Mol Biol* **72:** 153–169.

Cieslinski, H., Bialkowskaa, A., Tkaczuk, K., Dlugolecka, A., Kur, J., and Turkiewicz, M. (2009) Identification and molecular modeling of a novel lipase from an Antarctic soil metagenomic library. *Pol J Microbiol* **58:** 199–204.

Cox, R.S., 3rd, Surette, M.G., and Elowitz, M.B. (2007) Programming gene expression with combinatorial promoters. *Mol Syst Biol* **3:** 145.

Cox, R.S., 3rd, Dunlop, M.J., and Elowitz, M.B. (2010) A synthetic three-color scaffold for monitoring genetic regulation and noise. *J Biol Eng* **4:** 10.

Craig, J.W., Chang, F.Y., Kim, J.H., Obiajulu, S.C., and Brady, S.F. (2010) Expanding small-molecule functional metagenomics through parallel screening of broad-host-range cosmid environmental DNA libraries in diverse proteobacteria. *Appl Environ Microbiol* **76:** 1633–1641.

Daniel, R. (2005) The metagenomics of soil. *Nat Rev Microbiol* **3:** 470–478.

DeSantis, G., and Jones, J.B. (1999) Chemical modification of enzymes for enhanced functionality. *Curr Opin Biotechnol* **10:** 324–330.

Dinsdale, E.A., Edwards, R.A., Hall, D., Angly, F., Breitbart, M., Brulc, J.M., *et al.* (2008) Functional metagenomic profiling of nine biomes. *Nature* **452:** 629–632.

Dobson, C.M. (2003) Protein folding and misfolding. *Nature* **426:** 884–890.

Duan, C.J., and Feng, J.X. (2010) Mining metagenomes for novel cellulase genes. *Biotechnol Lett* **32:** 1765–1775.

Feng, Y., Duan, C.J., Pang, H., Mo, X.C., Wu, C.F., Yu, Y., *et al.* (2007) Cloning and identification of novel cellulase genes from uncultured microorganisms in rabbit cecum and characterization of the expressed cellulases. *Appl Microbiol Biotechnol* **75:** 319–328.

Fernandez-Arrojo, L., Guazzaroni, M.E., Lopez-Cortes, N., Beloqui, A., and Ferrer, M. (2010) Metagenomic era for biocatalyst identification. *Curr Opin Biotechnol* **21:** 725–733.

Ferrer, M., Chernikova, T.N., Timmis, K.N., and Golyshin, P.N. (2004) Expression of a temperature-sensitive esterase in a novel chaperone-based *Escherichia coli* strain. *Appl Environ Microbiol* **70:** 4499–4504.

Ferrer, M., Golyshina, O.V., Chernikova, T.N., Khachane, A.N., Martins Dos Santos, V.A., Yakimov, M.M., *et al.* (2005a) Microbial enzymes mined from the Urania deep-sea hypersaline anoxic basin. *Chem Biol* **12:** 895–904.

Ferrer, M., Golyshina, O.V., Chernikova, T.N., Khachane, A.N., Reyes-Duarte, D., Santos, V.A., *et al.* (2005b) Novel hydrolase diversity retrieved from a metagenome library of bovine rumen microflora. *Environ Microbiol* **7:** 1996–2010.

Ferrer, M., Beloqui, A., Timmis, K.N., and Golyshin, P.N. (2009) Metagenomics for mining new genetic resources of microbial communities. *J Mol Microbiol Biotechnol* **16:** 109–123.

Ferrer, M., Ghazi, A., Beloqui, A., Vieites, J.M., Lopez-Cortes, N., Marin-Navarro, J., *et al.* (2012) Functional metagenomics unveils a multifunctional glycosyl hydrolase from the family 43 catalysing the breakdown of plant polymers in the calf rumen. *PLoS ONE* **7:** e38134.

Gabor, E.M., Alkema, W.B., and Janssen, D.B. (2004) Quantifying the accessibility of the metagenome by random expression cloning techniques. *Environ Microbiol* **6:** 879–886.

Galvao, T.C., and de Lorenzo, V. (2006) Transcriptional regulators a la carte: engineering new effector specificities in bacterial regulatory proteins. *Curr Opin Biotechnol* **17:** 34–42.

Gardner, T.S., Cantor, C.R., and Collins, J.J. (2000) Construction of a genetic toggle switch in Escherichia coli. *Nature* **403:** 339–342.

GIA (2012) *Global Industry Analysts. Industrial Enzymes – A Global Multi-Client Market Research Project*. California, USA: GIA.

Gloux, K., Berteau, O., El Oumami, H., Beguet, F., Leclerc, M., and Dore, J. (2010) A metagenomic beta-glucuronidase uncovers a core adaptive function of the human intestinal microbiome. *Proc Natl Acad Sci USA* **108** (Suppl. 1)**:** 4539–4546.

Gong, J.S., Lu, Z.M., Li, H., Zhou, Z.M., Shi, J.S., and Xu, Z.H. (2013a) Metagenomic technology and genome mining: emerging areas for exploring novel nitrilases. *Appl Microbiol Biotechnol* **97:** 6603–6611.

Gong, X., Gruniniger, R.J., Forster, R.J., Teather, R.M., and McAllister, T.A. (2013b) Biochemical analysis of a highly specific, pH stable xylanase gene identified from a bovine rumen-derived metagenomic library. *Appl Microbiol Biotechnol* **97:** 2423–2431.

Guazzaroni, M.-E., Golyshin, P.N., and Ferrer, M. (2010a) Analysis of complex microbial communities through metagenomic survey. In *Metagenomics: Theory, Methods and Applications*. Marco, D. (ed.). Norfolk, UK: Caister Academic, pp. 55–77.

Guazzaroni, M.E., Beloqui, A., Vieites, J.M., Al-ramahi, Y., Cortés, N.L., Ghazi, A., *et al.* (2010b) Metagenomic mining of enzyme diversity. In *Handbook of Hydrocarbon and Lipid Microbiology*. Timmis, K. (ed.). Heidelberg, Germany: Springer, pp. 2911–2927.

Guazzaroni, M.E., Morgante, V., Mirete, S., and Gonzalez-Pastor, J.E. (2013) Novel acid resistance genes from the metagenome of the Tinto River, an extremely acidic environment. *Environ Microbiol* **15:** 1088–1102.

Gupta, R., Beg, Q.K., and Lorenz, P. (2002) Bacterial alkaline proteases: molecular approaches and industrial applications. *Appl Microbiol Biotechnol* **59:** 15–32.

Hallin, P.F., Binnewies, T.T., and Ussery, D.W. (2008) The genome BLASTatlas-a GeneWiz extension for visualization of whole-genome homology. *Mol Biosyst* **4:** 363–371.

Handelsman, J. (2004) Metagenomics: application of genomics to uncultured microorganisms. *Microbiol Mol Biol Rev* **68:** 669–685.

Hasan, F., Shah, A.A., and Hameed, A. (2006) Industrial applications of microbial lipases. *Enzyme Microb Technol* **39:** 235–251.

Heath, C., Hu, X.P., Cary, S.C., and Cowan, D. (2009) Identification of a novel alkaliphilic esterase active at low temperatures by screening a metagenomic library from antarctic desert soil. *Appl Environ Microbiol* **75:** 4657–4659.

Herrera, S. (2004) Industrial biotechnology-a chance at redemption. *Nat Biotechnol* **22:** 671–675.

Hjort, K., Bergstrom, M., Adesina, M.F., Jansson, J.K., Smalla, K., and Sjoling, S. (2010) Chitinase genes revealed and compared in bacterial isolates, DNA extracts and a metagenomic library from a phytopathogen-suppressive soil. *FEMS Microbiol Ecol* **71:** 197–207.

Homann, M.J., Vail, R.B., Previte, E., Tamarez, M., Morgan, B., Dodds, D.R., and Zaks, A. (2004) Rapid identification of enantioselective ketone reductions using targeted microbial libraries. *Tetrahedron* **60:** 789–797.

Hrvatin, S., and Piel, J. (2007) Rapid isolation of rare clones from highly complex DNA libraries by PCR analysis of liquid gel pools. *J Microbiol Methods* **68:** 434–436.

Hu, X.P., Heath, C., Taylor, M.P., Tuffin, M., and Cowan, D. (2012) A novel, extremely alkaliphilic and cold-active esterase from Antarctic desert soil. *Extremophiles* **16:** 79–86.

Jiang, C., Ma, G., Li, S., Hu, T., Che, Z., Shen, P., *et al.* (2009) Characterization of a novel beta-glucosidase-like activity from a soil metagenome. *J Microbiol* **47:** 542–548.

Jiang, C., Yin, B., Tang, M., Zhao, G., He, J., Shen, P., and Wu, B. (2013) Identification of a metagenome-derived prephenate dehydrogenase gene from an alkaline-polluted soil microorganism. *Antonie Van Leeuwenhoek* **103:** 1209–1219.

Kim, S.J., Lee, C.M., Han, B.R., Kim, M.Y., Yeo, Y.S., Yoon, S.H., *et al.* (2008) Characterization of a gene encoding cellulase from uncultured soil bacteria. *FEMS Microbiol Lett* **282:** 44–51.

Kim, U.J., Shizuya, H., de Jong, P.J., Birren, B., and Simon, M.I. (1992) Stable propagation of cosmid sized human DNA inserts in an F factor based vector. *Nucleic Acids Res* **20:** 1083–1085.

Kirk, O., Borchert, T.V., and Fuglsang, C.C. (2002) Industrial enzyme applications. *Curr Opin Biotechnol* **13:** 345–351.

Knietsch, A., Waschkowitz, T., Bowien, S., Henne, A., and Daniel, R. (2003) Metagenomes of complex microbial consortia derived from different soils as sources for novel genes conferring formation of carbonyls from short-chain polyols on *Escherichia coli*. *J Mol Microbiol Biotechnol* **5:** 46–56.

Ko, K.C., Han, Y., Cheong, D.E., Choi, J.H., and Song, J.J. (2013) Strategy for screening metagenomic resources for exocellulase activity using a robotic, high-throughput screening system. *J Microbiol Methods* **94:** 311–316.

Koide, T., Pang, W.L., and Baliga, N.S. (2009) The role of predictive modelling in rationally re-engineering biological systems. *Nat Rev Microbiol* **7:** 297–305.

Krogh, K.B., Harris, P.V., Olsen, C.L., Johansen, K.S., Hojer-Pedersen, J., Borjesson, J., and Olsson, L. (2010) Characterization and kinetic analysis of a thermostable GH3 beta-glucosidase from *Penicillium brasilianum*. *Appl Microbiol Biotechnol* **86:** 143–154.

de Las Heras, A., and de Lorenzo, V. (2012) Engineering whole-cell biosensors with no antibiotic markers for monitoring aromatic compounds in the environment. *Methods Mol Biol* **834:** 261–281.

de Las Heras, A., Carreno, C.A., Martinez-Garcia, E., and de Lorenzo, V. (2010) Engineering input/output nodes in prokaryotic regulatory circuits. *FEMS Microbiol Rev* **34:** 842–865.

Lee, S., Cantarel, B., Henrissat, B., Gevers, D., Birren, B.W., Huttenhower, C., and Ko, G. (2013) Gene-targeted metagenomic analysis of glucan-branching enzyme gene profiles among human and animal fecal microbiota. *ISME J* **8:** 493–503.

Li, M., Hong, Y., Klotz, M.G., and Gu, J.D. (2010) A comparison of primer sets for detecting 16S rRNA and hydrazine oxidoreductase genes of anaerobic ammonium-oxidizing

bacteria in marine sediments. *Appl Microbiol Biotechnol* **86:** 781–790.

Lorenz, P., and Eck, J. (2005) Metagenomics and industrial applications. *Nat Rev Microbiol* **3:** 510–516.

Lorenz, P., Liebeton, K., Niehaus, F., and Eck, J. (2002) Screening for novel enzymes for biocatalytic processes: accessing the metagenome as a resource of novel functional sequence space. *Curr Opin Biotechnol* **13:** 572–577.

Lynd, L.R., Weimer, P.J., van Zyl, W.H., and Pretorius, I.S. (2002) Microbial cellulose utilization: fundamentals and biotechnology. *Microbiol Mol Biol Rev* **66:** 506–577. table of contents.

Martinez-Garcia, E., Nikel, P.I., Chavarria, M., and de Lorenzo, V. (2014) The metabolic cost of flagellar motion in *Pseudomonas putida* KT2440. *Environ Microbiol* **16:** 291–303.

Martinez-Martinez, M., Alcaide, M., Tchigvintsev, A., Reva, O., Polaina, J., Bargiela, R., *et al.* (2013) Biochemical diversity of carboxyl esterases and lipases from Lake Arreo (Spain): a metagenomic approach. *Appl Environ Microbiol* **79:** 3553–3562.

Martinez-Martinez, M., Lores, I., Pena-Garcia, C., Bargiela, R., Reyes-Duarte, D., Guazzaroni, M.E., *et al.* (2014) Biochemical studies on a versatile esterase that is most catalytically active with polyaromatic esters. *Microb Biotechnol* **7:** 184–191.

Maurer, K.H. (2004) Detergent proteases. *Curr Opin Biotechnol* **15:** 330–334.

Medini, D., Donati, C., Tettelin, H., Masignani, V., and Rappuoli, R. (2005) The microbial pan-genome. *Curr Opin Genet Dev* **15:** 589–594.

Mohn, W.W., Garmendia, J., Galvao, T.C., and de Lorenzo, V. (2006) Surveying biotransformations with a la carte genetic traps: translating dehydrochlorination of lindane (gamma-hexachlorocyclohexane) into lacZ-based phenotypes. *Environ Microbiol* **8:** 546–555.

Moon, T.S., Clarke, E.J., Groban, E.S., Tamsir, A., Clark, R.M., Eames, M., *et al.* (2011) Construction of a genetic multiplexer to toggle between chemosensory pathways in *Escherichia coli*. *J Mol Biol* **406:** 215–227.

Morimoto, S., and Fujii, T. (2009) A new approach to retrieve full lengths of functional genes from soil by PCR-DGGE and metagenome walking. *Appl Microbiol Biotechnol* **83:** 389–396.

Nasuno, E., Kimura, N., Fujita, M.J., Nakatsu, C.H., Kamagata, Y., and Hanada, S. (2012) Phylogenetically novel LuxI/LuxR-type quorum sensing systems isolated using a metagenomic approach. *Appl Environ Microbiol* **78:** 8067–8074.

Nikel, P.I., and de Lorenzo, V. (2013) Implantation of unmarked regulatory and metabolic modules in Gram-negative bacteria with specialised mini-transposon delivery vectors. *J Biotechnol* **163:** 143–154.

Osterberg, S., del Peso-Santos, T., and Shingler, V. (2011) Regulation of alternative sigma factor use. *Annu Rev Microbiol* **65:** 37–55.

Pang, H., Zhang, P., Duan, C.J., Mo, X.C., Tang, J.L., and Feng, J.X. (2009) Identification of cellulase genes from the metagenomes of compost soils and functional characterization of one novel endoglucanase. *Curr Microbiol* **58:** 404–408.

Percival Zhang, Y.H., Himmel, M.E., and Mielenz, J.R. (2006) Outlook for cellulase improvement: screening and selection strategies. *Biotechnol Adv* **24:** 452–481.

Posfai, G., Plunkett, G., 3rd, Feher, T., Frisch, D., Keil, G.M., Umenhoffer, K., *et al.* (2006) Emergent properties of reduced-genome *Escherichia coli. Science* **312:** 1044–1046.

Purnick, P.E., and Weiss, R. (2009) The second wave of synthetic biology: from modules to systems. *Nat Rev Mol Cell Biol* **10:** 410–422.

Regot, S., Macia, J., Conde, N., Furukawa, K., Kjellen, J., Peeters, T., *et al.* (2011) Distributed biological computation with multicellular engineered networks. *Nature* **469:** 207–211.

Rhee, J.K., Ahn, D.G., Kim, Y.G., and Oh, J.W. (2005) New thermophilic and thermostable esterase with sequence similarity to the hormone-sensitive lipase family, cloned from a metagenomic library. *Appl Environ Microbiol* **71:** 817–825.

Rhodius, V.A., Segall-Shapiro, T.H., Sharon, B.D., Ghodasara, A., Orlova, E., Tabakh, H., *et al.* (2013) Design of orthogonal genetic switches based on a crosstalk map of sigmas, anti-sigmas, and promoters. *Mol Syst Biol* **9:** 702.

Riesenfeld, C.S., Goodman, R.M., and Handelsman, J. (2004) Uncultured soil bacteria are a reservoir of new antibiotic resistance genes. *Environ Microbiol* **6:** 981–989.

Schloss, P.D., and Handelsman, J. (2003) Biotechnological prospects from metagenomics. *Curr Opin Biotechnol* **14:** 303–310.

Schloss, P.D., and Handelsman, J. (2006) Toward a census of bacteria in soil. *PLoS Comput Biol* **2:** e92.

Schmidt, M., and de Lorenzo, V. (2012) Synthetic constructs in/for the environment: managing the interplay between natural and engineered Biology. *FEBS Lett* **586:** 2199–2206.

Schnoes, A.M., Brown, S.D., Dodevski, I., and Babbitt, P.C. (2009) Annotation error in public databases: misannotation of molecular function in enzyme superfamilies. *PLoS Comput Biol* **5:** e1000605.

Schoemaker, H.E., Mink, D., and Wubbolts, M.G. (2003) Dispelling the myths–biocatalysis in industrial synthesis. *Science* **299:** 1694–1697.

Sharma, P., Jindal, S., Bala, K., Kumari, K., Niharika, N., Kaur, J., *et al.* (2013) Functional screening of enzymes and bacteria for the dechlorination of hexachlorocyclohexane by a high-throughput colorimetric assay. *Biodegradation* **25:** 179–187.

Shetty, R.P., Endy, D., and Knight, T.F., Jr (2008) Engineering BioBrick vectors from BioBrick parts. *J Biol Eng* **2:** 5.

Silva-Rocha, R., and de Lorenzo, V. (2011) Implementing an OR-NOT (ORN) logic gate with components of the SOS regulatory network of Escherichia coli. *Mol Biosyst* **7:** 2389–2396.

Silva-Rocha, R., Martinez-Garcia, E., Calles, B., Chavarria, M., Arce-Rodriguez, A., de Las Heras, A., *et al.* (2013) The Standard European Vector Architecture (SEVA): a coherent platform for the analysis and deployment of complex prokaryotic phenotypes. *Nucleic Acids Res* **41:** D666–D675.

Simon, C., Herath, J., Rockstroh, S., and Daniel, R. (2009) Rapid identification of genes encoding DNA polymerases by function-based screening of metagenomic libraries derived from glacial ice. *Appl Environ Microbiol* **75:** 2964–2968.

Singhania, R.R., Patel, A.K., Sukumaran, R.K., Larroche, C., and Pandey, A. (2013) Role and significance of beta-glucosidases in the hydrolysis of cellulose for bioethanol production. *Bioresour Technol* **127:** 500–507.

Siuti, P., Yazbek, J., and Lu, T.K. (2013) Synthetic circuits integrating logic and memory in living cells. *Nat Biotechnol* **31:** 448–452.

Sleator, R.D., Shortall, C., and Hill, C. (2008) Metagenomics. *Lett Appl Microbiol* **47:** 361–366.

Sommer, M.O., Church, G.M., and Dantas, G. (2010) A functional metagenomic approach for expanding the synthetic biology toolbox for biomass conversion. *Mol Syst Biol* **6:** 360.

Steele, H.L., Jaeger, K.E., Daniel, R., and Streit, W.R. (2009) Advances in recovery of novel biocatalysts from metagenomes. *J Mol Microbiol Biotechnol* **16:** 25–37.

Suenaga, H., Ohnuki, T., and Miyazaki, K. (2007) Functional screening of a metagenomic library for genes involved in microbial degradation of aromatic compounds. *Environ Microbiol* **9:** 2289–2297.

Suenaga, H., Koyama, Y., Miyakoshi, M., Miyazaki, R., Yano, H., Sota, M., *et al.* (2009a) Novel organization of aromatic degradation pathway genes in a microbial community as revealed by metagenomic analysis. *ISME J* **3:** 1335–1348.

Suenaga, H., Mizuta, S., and Miyazaki, K. (2009b) The molecular basis for adaptive evolution in novel extradiol dioxygenases retrieved from the metagenome. *FEMS Microbiol Ecol* **69:** 472–480.

Sul, W.J., Park, J., Quensen, J.F., 3rd, Rodrigues, J.L., Seliger, L., Tsoi, T.V., *et al.* (2009) DNA-stable isotope probing integrated with metagenomics for retrieval of biphenyl dioxygenase genes from polychlorinated biphenyl-contaminated river sediment. *Appl Environ Microbiol* **75:** 5501–5506.

Tabor, S. (2001) Expression using the T7 RNA polymerase/promoter system. *Curr Protoc Mol Biol* 16.2.1–16.2.11.

Terron-Gonzalez, L., Medina, C., Limon-Mortes, M.C., and Santero, E. (2013) Heterologous viral expression systems in fosmid vectors increase the functional analysis potential of metagenomic libraries. *Sci Rep* **3:** 1107.

Thompson, C.E., Beys-da-Silva, W.O., Santi, L., Berger, M., Vainstein, M.H., Guimaraes, J.A., and Vasconcelos, A.T. (2013) A potential source for cellulolytic enzyme discovery and environmental aspects revealed through metagenomics of Brazilian mangroves. *AMB Express* **3:** 65.

Turnbaugh, P.J., and Gordon, J.I. (2008) An invitation to the marriage of metagenomics and metabolomics. *Cell* **134:** 708–713.

Uchiyama, T., and Miyazaki, K. (2009) Functional metagenomics for enzyme discovery: challenges to efficient screening. *Curr Opin Biotechnol* **20:** 616–622.

Uchiyama, T., and Miyazaki, K. (2010) Product-induced gene expression, a product-responsive reporter assay used to screen metagenomic libraries for enzyme-encoding genes. *Appl Environ Microbiol* **76:** 7029–7035.

Uchiyama, T., and Watanabe, K. (2008) Substrate-induced gene expression (SIGEX) screening of metagenome libraries. *Nat Protoc* **3:** 1202–1212.

Uchiyama, T., Abe, T., Ikemura, T., and Watanabe, K. (2005) Substrate-induced gene-expression screening of environmental metagenome libraries for isolation of catabolic genes. *Nat Biotechnol* **23:** 88–93.

Varaljay, V.A., Howard, E.C., Sun, S., and Moran, M.A. (2010) Deep sequencing of a dimethylsulfoniopropionate-degrading gene (dmdA) by using PCR primer pairs designed on the basis of marine metagenomic data. *Appl Environ Microbiol* **76:** 609–617.

Vieites, J.M., Guazzaroni, M.E., Beloqui, A., Golyshin, P.N., and Ferrer, M. (2009) Metagenomics approaches in systems microbiology. *FEMS Microbiol Rev* **33:** 236–255.

Voigt, C.A. (2006) Genetic parts to program bacteria. *Curr Opin Biotechnol* **17:** 548–557.

Wang, C., Meek, D.J., Panchal, P., Boruvka, N., Archibald, F.S., Driscoll, B.T., and Charles, T.C. (2006) Isolation of poly-3-hydroxybutyrate metabolism genes from complex microbial communities by phenotypic complementation of bacterial mutants. *Appl Environ Microbiol* **72:** 384–391.

Wang, G.Y., Graziani, E., Waters, B., Pan, W., Li, X., McDermott, J., *et al.* (2000) Novel natural products from soil DNA libraries in a streptomycete host. *Org Lett* **2:** 2401–2404.

Wang, L., Hatem, A., Catalyurek, U.V., Morrison, M., and Yu, Z. (2013) Metagenomic insights into the carbohydrate-active enzymes carried by the microorganisms adhering to solid digesta in the rumen of cows. *PLoS ONE* **8:** e78507.

Ward, R. (2011) Cellulase engineering for biomass saccharification. In *Routes to Cellulosic Ethanol*. Buckeridge, M.S., and Goldman, G.H. (eds). New York, USA: Springer, pp. 135–151.

Warnecke, F., and Hess, M. (2009) A perspective: metatranscriptomics as a tool for the discovery of novel biocatalysts. *J Biotechnol* **142:** 91–95.

Warnecke, F., Luginbuhl, P., Ivanova, N., Ghassemian, M., Richardson, T.H., Stege, J.T., *et al.* (2007) Metagenomic and functional analysis of hindgut microbiota of a wood-feeding higher termite. *Nature* **450:** 560–565.

Weber, W., and Fussenegger, M. (2010) Synthetic gene networks in mammalian cells. *Curr Opin Biotechnol* **21:** 690–696.

Wenzel, S.C., and Muller, R. (2005) Recent developments towards the heterologous expression of complex bacterial natural product biosynthetic pathways. *Curr Opin Biotechnol* **16:** 594–606.

Wexler, M., Bond, P.L., Richardson, D.J., and Johnston, A.W. (2005) A wide host-range metagenomic library from a waste water treatment plant yields a novel alcohol/aldehyde dehydrogenase. *Environ Microbiol* **7:** 1917–1926.

Williamson, L.L., Borlee, B.R., Schloss, P.D., Guan, C., Allen, H.K., and Handelsman, J. (2005) Intracellular screen to identify metagenomic clones that induce or inhibit a quorum-sensing biosensor. *Appl Environ Microbiol* **71:** 6335–6344.

Wong, D.W., Chan, V.J., and McCormack, A.A. (2009) Functional cloning and expression of a novel Endo-alpha-1,5-L-arabinanase from a metagenomic library. *Protein Pept Lett* **16:** 1435–1441.

Yan, X., Geng, A., Zhang, J., Wei, Y., Zhang, L., Qian, C., *et al.* (2013) Discovery of (hemi-) cellulase genes in a metagenomic library from a biogas digester using 454 pyrosequencing. *Appl Microbiol Biotechnol* **97:** 8173–8182.

Yuhong, Z., Shi, P., Liu, W., Meng, K., Bai, Y., Wang, G., *et al.* (2009) Lipase diversity in glacier soil based on analysis of metagenomic DNA fragments and cell culture. *J Microbiol Biotechnol* **19:** 888–897.

Zaprasis, A., Liu, Y.J., Liu, S.J., Drake, H.L., and Horn, M.A. (2010) Abundance of novel and diverse tfdA-like genes, encoding putative phenoxyalkanoic acid herbicide-degrading dioxygenases, in soil. *Appl Environ Microbiol* **76:** 119–128.

Zelcbuch, L., Antonovsky, N., Bar-Even, A., Levin-Karp, A., Barenholz, U., Dayagi, M., *et al.* (2013) Spanning high-dimensional expression space using ribosome-binding site combinatorics. *Nucleic Acids Res* **41:** e98.

Zhan, J., Ding, B., Ma, R., Ma, X., Su, X., Zhao, Y., *et al.* (2010) Develop reusable and combinable designs for transcriptional logic gates. *Mol Syst Biol* **6:** 388.

9

Microbial response to single-cell protein production and brewery wastewater treatment

Jackson Z. Lee,[1][†] Andrew Logan,[2] Seth Terry[2] and John R. Spear[1]*

[1]*Department of Civil and Environmental Engineering, Colorado School of Mines, Golden, CO, USA.*
[2]*Nutrinsic, Corp., Aurora, CO, USA.*

Summary

As global fisheries decline, microbial single-cell protein (SCP) produced from brewery process water has been highlighted as a potential source of protein for sustainable animal feed. However, biotechnological investigation of SCP is difficult because of the natural variation and complexity of microbial ecology in wastewater bioreactors. In this study, we investigate microbial response across a full-scale brewery wastewater treatment plant and a parallel pilot bioreactor modified to produce an SCP product. A pyrosequencing survey of the brewery treatment plant showed that each unit process selected for a unique microbial community. Notably, flow equalization basins were dominated by *Prevotella*, methanogenesis effluent had the highest levels of diversity, and clarifier wet-well samples were sources of sequences for the candidate bacterial phyla of TM7 and BD1-5. Next, the microbial response of a pilot bioreactor producing SCP was tracked over 1 year, showing that two different production trials produced two different communities originating from the same starting influent. However, SCP production resulted generally in enrichment of several clades of rhizospheric diazotrophs of *Alphaproteobacteria* and *Betaproteobacteria* in the bioreactor and even more so in the final product. These diazotrophs are potentially useful as the basis of a SCP product for commercial feed production.

*For correspondence. E-mail jspear@mines.edu

Funding Information This work is supported by the National Science Foundation grant to J. R. S. (NSF-MUSES 0628282) and a Dorthy Bertine Internship award to J. Z. L. from the Sussman Fund.

Introduction

Already half of all global fish stocks have been deemed fully exploited (Cressey, 2009), which has led to the collapse of several fisheries and the potential collapse of others over the next several decades (Worm *et al.*, 2006). Concomitantly, aquaculture (the farm rearing of fish) has grown at an annual rate of 14% since 1970 (FAO Fisheries Department, 2003). Because aquaculture feed production relies on significant amounts of non-sustainable fish meal protein harvested from ocean fisheries, further aquaculture growth will result in more fish meal shortages and further depletion of ocean fisheries. Therefore, there has been renewed interest in the development of less expensive and more sustainable fish meal replacements.

In the brewing industry, solid byproducts of various forms (spent grains, hops, yeasts, etc.), once a costly landfill waste, have become a livestock feed source. Even after this removal of solids, a large amount of dissolved carbon still remains in the typical brewery wastewater (Hough, 1985). This brewery waste can be aerobically and microbiologically treated in a process-wastewater treatment facility and the carbon-degrading microbiota harvested as dried microbial biomass, called single-cell protein (SCP). Researchers have recognized SCP's potential as an animal feed for decades, although SCP has never fully replaced fish meal at production scale (El-Sayed, 1999). A major concern has been the negative performance and connotations associated with wastewater and, in particular, reuse of raw sewage. Initial studies in SCP production from domestic wastewater biomass sources were often plagued by heavy metal contamination and faecal pathogens as part of their process stream (Tacon, 1979; Lovell, 1989). However, food-processing wastes have minor or no exposure to domestic sewage (Vriens *et al.*, 1989) and are known to have a much higher chemical oxygen demand (COD) and lower total nitrogen profile than domestic sewage (Gray, 2004). Specifically, brewery process water possesses distinct characteristics that make the technology more feasible than most food-processing process water types, such as a continuous global year-round production of dissolved carbon process water (Huige, 2006), and an amino acid profile rich in lysine and methionine (two essential amino acids absent from many plant and fungal sources) that is comparable with fish meal (Vriens *et al.*, 1989; Tacon *et al.*, 2009). Several major challenges to bring SCP to market are to reliably maintain a high crude protein and essential amino acid content, and to continue to produce a treated process-wastewater that meets local water regulations. An understanding of the underlying microbial community responsible for SCP product formation is needed to help achieve these goals.

Fortuitously, knowledge of microbial communities is enabled by the rapid increase in DNA sequencing throughput with decreased cost, which has greatly expanded the detection coverage of microbial diversity in environmental samples and has the potential to identify the functioning and turnover of species in environmental engineering applications such as wastewater technology. Several recent studies have established that high-throughput sequencing can track how wastewater community consortia fluctuate in real-world systems (Werner *et al.*, 2011) and how wastewater treatment diversity can have notable biogeographical trends between plants (Werner *et al.*, 2011; Zhang *et al.*, 2012). The production of novel bioproducts from wastewater such as fuel and chemicals requires a closer look at how these microbial systems function in order to identify relevant communities responsible for mixed community biotechnologies [e.g.

microbial fuel cells (Lee *et al.*, 2010), and lignin-cellulose degradation (Hollister *et al.*, 2011)]. In this study, the microbial turnover and diversity of an entire brewery wastewater treatment works was analysed by pyrotag sequencing technology in order to identify metabolisms of biotechnological interest and to develop possible strategies to improve quality or economic competitiveness. Specifically, this study attempted to identify what common microbial community responses were seen in an SCP production pilot bioreactor running under a non-conventional aerobic treatment regime and the relationship of this community compared with the microbial consortia of the larger treatment plant facility.

Results

Physical characteristics and operational conditions of the wastewater treatment facility

Table 1 shows the operational conditions of unit processes in this study. For the wastewater treatment facility, operational parameters were tracked for each stage by plant operators. This data indicated high COD wastewaters and a sizable suspended solids fraction in both the influent and in the acidogenic basins. Organic acid and volatile fatty acids (VFA) monitoring of the basin indicated

Table 1. Operating parameters of each unit process.

Sample location	Operating conditions	Average (SD)	
Plant influent sample	Flow rate	402 (174)	$m^3 d^{-1}$
	pH	10.85 (1.3)	
	COD total	10 147 (3848)	$mg l^{-1}$
	COD soluble	7859 (2573)	$mg l^{-1}$
	TSS	1989 (1011)	$mg l^{-1}$
	Temperature	35	°C
Acidogenic basin mixed liquor	Volume	600	m^3
	pH	5.68 (0.43)	
	VFA	1654 (284)	$mg l^{-1}$ acetic acid
	Total organic acids	4489 (1439)	$mg l^{-1}$
	COD total	10 554 (1854)	$mg l^{-1}$
	COD soluble	8005 (1714)	$mg l^{-1}$
	TSS	1275 (352)	$mg l^{-1}$
	Total N	164 (18)	$mg l^{-1}$
	P-PO$_4$	152 (21)	$mg l^{-1}$
	NH$_3$-N	8.76 (9.48)	$mg l^{-1}$
Methanogenesis UASB outfall	Volume	800	m^3
	VFA	4418 (2567)	$mg l^{-1}$ acetic acid
	TSS	4705 (2785)	$mg l^{-1}$
Aerobic basin mixed liquor	Volume	3820	m^3
	SRT	16	d
Clarifier outfall	COD total	131 (22)	$mg l^{-1}$
	COD soluble	123 (22)	$mg l^{-1}$
	TSS	12 (5)	$mg l^{-1}$
Pilot bioreactor sample port	Volume	38	m^3
	Flow rate	4.75	$m^3 d^{-1}$
	pH	7.15 (0.36)	
	MLSS	2600 (461)	$mg l^{-1}$
	Total organic acids	3666 (1301)	$mg l^{-1}$
	DO	0.22 (0.05)	$mg l^{-1}$
	Excess N, P added as urea and phosphoric acid		

SD, standard deviation; SRT, solids retention time.

production of organic acids and (in conjunction with free ammonia measurements) also indicated a relatively low ratio of free nitrogen to carbon available for microbes. In the methanogenic upflow anaerobic sludge blanket (UASB) basin, further acid production was noted, and the majority of this was consumed in the UASB with a notable sludge blanket observable as total suspended solids (TSS). These constituents were then aerobically treated to discharge standards in the aerobic basin. In contrast, the pilot bioreactor was marked by a higher mixed liquor suspended solids (MLSS) content than most conventional domestic wastewater treatment regimes, as well as much lower dissolved oxygen (DO) levels and higher VFA levels due to influent from the acidogenic basin.

Pyrotag microbial diversity across the wastewater treatment plant

Table 2 shows alpha diversity information and sequencing depth for each sample. A total of 808 near full-length Sanger and a total of ~ 54 000 pyrotag 16S amplicon sequences were completed. Microbial diversity coverage estimates based on alpha diversity were bracketed on the low end by the Chao1 estimator and on the high end by the CatchAll estimate. Chao1 and abundance-based coverage estimator (ACE) metrics of samples rarified to the same sequencing depth showed that the influent and acidogenic basin had the lowest levels of diversity, while the majority of operational taxonomic units (OTUs) in this study were found from both the UASB bioreactor and the aerobic basin of the brewery wastewater treatment plant (WWTP). Figure 2 describes phylum-level diversity from all samples arranged by unweighted pair group method with arithmetic mean (UPGMA) cladogram based on the unweighted UniFrac distance, which shows that samples clustered primarily because of unit process type (jackknife sensitivity analysis of sequencing depth in Supporting Information Fig. S1). Samples tended to primarily cluster by treatment regime, except the pilot reactor mixed liquor samples clustered closely to the corresponding time point of the final dried SCP product. The four main dominant phyla found from pyrotags in this study were the *Bacteroidetes*, *Firmicutes*, *Actinobacteria* and *Proteobacteria* (from the alpha, beta and gamma classes), and were consistent with previous research on wastewater aerobic treatment (Seviour and Nielsen, 2010) and process wastewater specifically (Manz *et al.*, 1994). The candidate divisions of SR1 and RF3, seen previously in methanogenesis bioreactor surveys (Chouari *et al.*, 2005; Riviere *et al.*, 2009), comprised 2.8% and 2.3% of sequences respectively from the methanogenesis UASB process. TM7 (Hugenholtz *et al.*, 2001) and BD1-5 (Li *et al.*, 1999) accounted for 5.2% and 5.1% of sequences detected in the treatment plant

wet well. Krona hierarchical pie charts (Supplemental Information, http://inside.mines.edu/~jspear/resources .html) showed the relative abundance of OTUs across lower taxonomic levels. The influent was dominated by several groups of *Firmicutes* from the *Lactobacillus* and *Enterococcus* families. Acidogenic basin pyrotag sequences were dominated by the genus *Prevotella*, a saccharolytic fermenter. The remaining stages showed no highly dominant clades across time, although several clades of closely related enriched OTUs (from *Bacteroidetes*, *Firmicutes* and *Proteobacteria*) appear to dominate in the pilot bioreactor and SCP product. In addition, approximately 14.3% of all OTUs (representing 4% of total sequences) remained unclassified beyond the domain level (5.9% of pilot plant OTUs, 5.2% of methanogenesis OTUs and 4.0% of aerobic basin OTUs).

Tracking microbial response in the pilot bioreactor by principal component analysis

When tracked over time, each trial run within the pilot plant produced the same pattern of shifts with the three beta diversity metrics studied (Fig. 3A–C). In trial 1, the pilot reactor community was initially similar to the influent environment and then shifted to a new community composition. Before trial 2, the reactor was drained and refilled but not re-inoculated or sterilized. In trial 2, a second microbial community developed that was different from the influent and trial 1. For principal component analyses (PCoAs) of the pilot bioreactor and influent, results did not change with jackknifing to 800 sequences per sample (Supporting Information Fig. S2). No significant time–decay relationship (representing steady succession or turnover) was seen in the study, and most comparisons between time points showed the same level of dissimilarity distance (Fig. 3D). Bi-plots of the most common family-level classifications of OTUs found in this study were graphed in conjunction with PCoA data (Fig. 3A–C) and showed a correlation of *Rhodospirillales* with trial 1, several family types from *Proteobacteria* with trial 2 and *Prevotellaceae* with the influent.

Co-abundance, metastats and microbial lifestyle analysis of pilot bioreactor and SCP product OTUs

To address the question of which OTUs might be biotechnologically relevant, the variations of OTU abundances in the pilot reactor and SCP product samples over time were used to generate Bray–Curtis distances (a generalized distance metric) between repeatedly occurring OTUs. These data were clustered using UPGMA (Supporting Information Fig. S3) and identified that a small number of recurrent OTUs with high abundance (6.2% of OTUs and 67.6% of sequences) contributed to a deep

Table 2. Sampling schedule, sequencing depth and alpha diversity of each unit process.

Sample location	Date, day number	Number of Sanger reads	Number of pyrotags (V1-V2)	Observed OTUs (97%)	% coverage	Alpha diversity estimator Chao1 Average (SD)[a]	ACE Average (SD)[a]
Plant influent sample port	1/30/08, 0	–	459	52	46–76	68.2	76.0
	3/19/08, 49	–	816	38	54–66	48.4 (11.8)	54.0 (12.2)
	4/30/09, 456	–	674	8	56–84	8.8 (2.0)	10.2 (2.6)
Acidogenic basin mixed liquor	1/30/08, 0	–	1127	38	70–73	39.4 (11.0)	43.4 (10.8)
	3/19/08, 49	92	3162	85	45–58	66.1 (16.5)	71.2 (15.2)
	3/19/09, 414	52	3450	66	67–93	51.4 (12.1)	52.5 (7.3)
	4/30/09, 456	43	884	41	58–76	48.4 (12.9)	50.1 (10.5)
Methanogenesis UASB outfall	1/30/08, 0	–	975	128	51–60	159.8 (23.2)	172.4 (21.4)
	3/19/08, 49	14	2373	220	59–71	193.5 (31.5)	191.8 (27.0)
	4/30/09, 456	72	805	131	32–58	192.1 (25.1)	192.2 (18.3)
Aerobic basin mixed liquor	4/30/09, 456	34	1097	194	45–48	255.4 (38.7)	267.5 (30.2)
	4/30/09, 456	–	722	155	43–65	222.8 (29.9)	239.7 (22.4)
	4/30/09, 456	–	1226	211	46–63	266.2 (42.4)	275.8 (37.4)
Clarifier outfall	1/30/08, 0	–	1176	138	40–59	174.9 (46.7)	193.8 (32.3)
	3/19/08, 49	52	2439	315	53–61	276.2 (41.7)	305.6 (45.0)
	4/30/09, 456	38	960	176	45–66	220.3 (25.9)	223.8 (21.8)
Pilot bioreactor sample port	03/19/08, 49	48	4090	88	52–71	68.6 (30.7)	71.0 (16.7)
	04/08/08, 69	–	1217	127	39–63	167.8 (34.2)	177.3 (27.7)
	09/10/08, 224	24	4874	197	59–71	127.8 (31.1)	136.2 (26.6)
	01/06/09, 342	29	2840	193	50–70	163.2 (25.4)	164.1 (21.9)
	02/16/09, 383	56	3005	163	53–75	130.9 (32.1)	133.7 (25.9)
	03/19/09, 414	51	1711	151	63–70	159.2 (26.2)	165.2 (19.1)
	04/30/09, 456	25	1231	115	33–59	137.5 (28.1)	143.9 (25.8)
Dried SCP product	05/07/08, 98	46	3266	92	63–79	72.4 (13.4)	78.9 (12.8)
	09/10/08, 224	19	3580	123	55–70	90.3 (19.1)	102.3 (19.8)
	01/06/09, 342	41	2122	171	64–74	160.5 (25.9)	171.1 (24.3)
	02/16/09, 383	–	1309	103	52–71	119.2 (22.4)	128.3 (23.8)
	03/19/09, 414	37	2262	147	56–68	142.2 (32.0)	152.9 (28.7)
	04/30/09, 456	35	1104	94	46–60	118.8 (22.9)	123.9 (18.6)

[a]Jackknifed to 500 sequences sample^{-1} for all samples except first.
SD, standard deviation.

clade within the clustergram. For these OTUs to be responsible for SCP formation, they should have a distribution that was enriched within the pilot bioreactor and SCP product when compared with the community of the acidogenic basin (which served as the influent community to the pilot bioreactor). When the MetaStats statistic was used to compare these OTUs with the pilot bioreactor influent distribution of OTUs, about 44.7% of the sequences (from the combined set of SCP and pilot bioreactor OTUs) were significantly enriched over the influent, and 29.6% of the sequences from the combined set were from OTUs that were significantly depleted compared with the influent (Supporting Information Fig. S4). When a phylogenetic tree was constructed from representative pyrotags of these OTUs, a phylogenetically coherent trend was observed. As shown in Fig. 4, when sequences from enriched or depleted OTUs were cross-referenced with the known microbial lifestyle of the taxonomic classification of the sequence, the depleted OTUs tended to be more from saccharolytic fermenters, primarily *Prevotellaceae*. Enriched OTUs tended to be from genera consisting of rhizospheric diazotrophs such as *Azospirillum*, *Azonexus* and *Telmatospirillum*, and a smaller fraction of saccharolytic fermenters. OTUs identified as having a rhizospheric diazotroph lifestyle were absent from both the pilot bioreactor influent and from depleted OTUs, but accounted for 39.4% of sequences from enriched OTUs in the pilot bioreactor and 73.0% of sequences from enriched OTUs in the final product.

Full-length 16S sequences of rhizospheric diazotrophs from the pilot bioreactor

Phylogenetic trees of full-length Sanger 16S sequences from rhizospheric diazotrophs in *Alphaproteobacteria* and *Betaproteobacteria* are shown in Fig. 5 (full phylogenetic tree in Supporting Information Fig. S5). In the *Betaproteobacteria* (Fig. 5A), near full-length sequences were related to known isolates of *Azospira* sp., *Azovibrio restrictus* and *Azonexus caeni* (but not genera

Azoarcus). In the *Alphaproteobacteria* (Fig. 5B and C), near full-length clone sequences were not related to any of the known rhizospheric strains for *Magnetospirillum* and were related to one strain of *Azospirillum* [*Azospirillum fermentarium* CC-LY743 isolated from a fermentation tank (Lin *et al.*, 2013)]. Closest basic local assignment search tool (BLAST) matches were commonly from strains isolated in large part from either wastewater and microbial fuel cell sources (Kaksonen *et al.*, 2004; Quan *et al.*, 2006; Borole *et al.*, 2009; de Cárcer *et al.*, 2011; Croese *et al.*, 2011; Sun *et al.*, 2011), as well as rhizosphere studies (Ashida *et al.*, 2010; Knief *et al.*, 2012) for which rhizospheric diazotrophs are primarily associated.

Discussion

Insights into microbial diversity in WWTPs

Prior to this work, it was unclear if large segments of the microbial community would pass through the entire treatment plant or if communities would shift entirely from one treatment regime to another. This study reveals that the influent waste beer from brewery process water was limited in bacterial diversity and had little impact on the colonization of subsequent unit processes in a brewery wastewater treatment works, and that the microbial community of each unit process remained largely self-selective. In concurrence with previous work on wastewater microbial variation (Fernandez *et al.*, 2000; Wells *et al.*, 2011; Werner *et al.*, 2011), constant turnover of even the most common species was a normal occurrence in samples from this study. In terms of novel diversity discovery, unclassifiable OTUs were found in all environments downstream of the acidogenic basin. Particularly surprising was the number of unclassified OTUs found in the methanogenesis UASB and the clarifier wetwell outfall (purple and orange squares, Fig. 1). The design of the UASB may be one contributor to the diversity, as the sludge granules are known to be highly

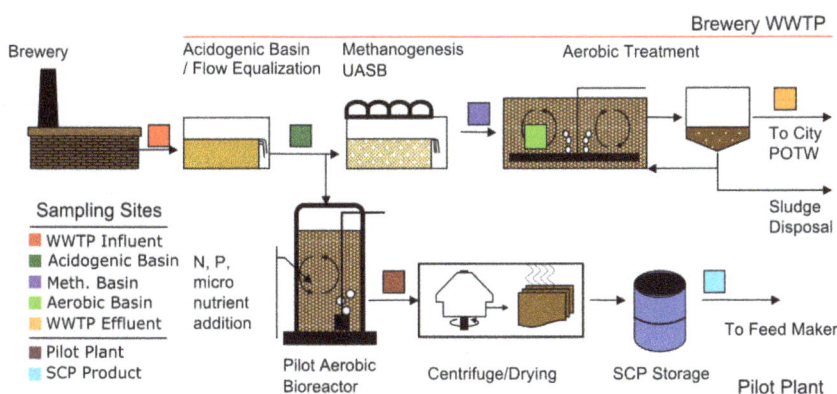

Fig. 1. Overview of the study site showing the brewery treatment works (top) and the pilot-scale reactor (bottom). Coloured squares indicate sampling location and are used to illustrate figures throughout this paper.

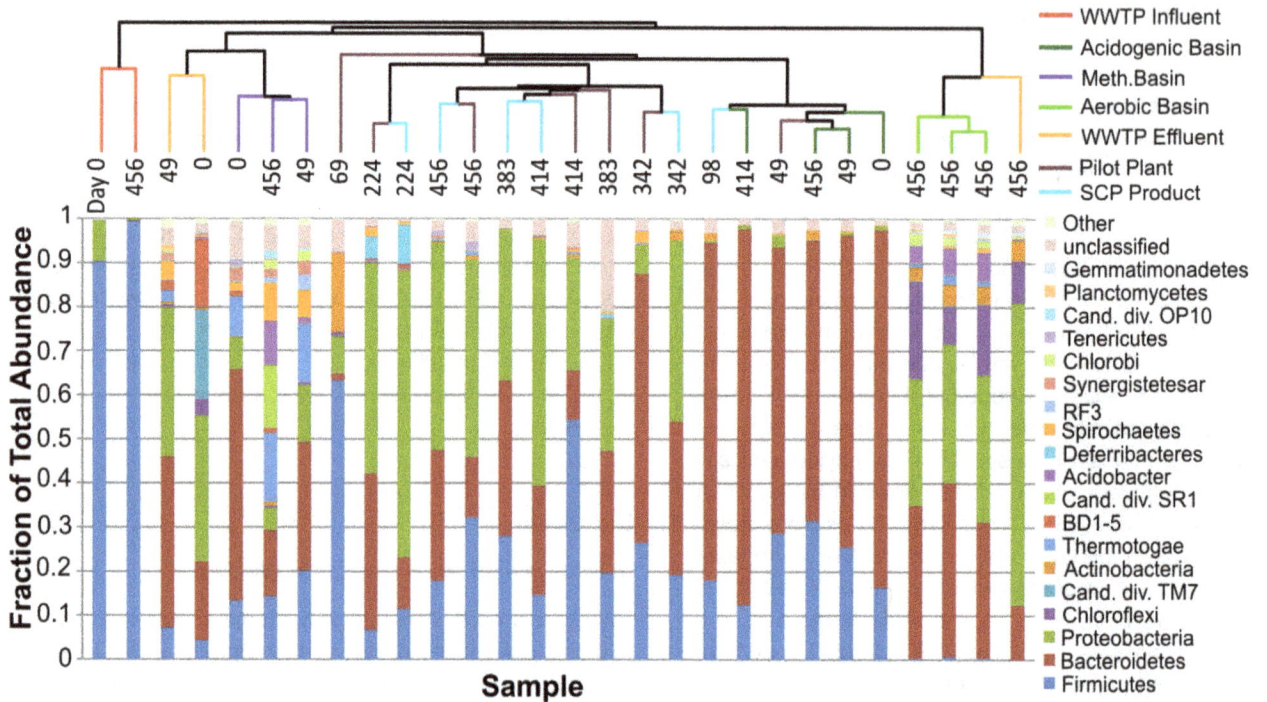

Fig. 2. Phylum-level distribution shown sorted by UPGMA clustering of UniFrac unweighted distance. Clades common to a unit process colored based on Fig. 1. Stacked bar graphs of phyla distributions are sorted by total dataset rank abundance starting with the most common from the bottom. Leaves are labeled with the day of sampling from initial date.

complex physical structures with differing chemical and microbial composition based on depth from the surface of the granule (Sekiguchi et al., 1999; Liu et al., 2002; Diaz et al., 2006). The clarifier wet well receives low settling

COD effluent from clarifier operations prior to discharge from the plant and may be an overlooked environment for sampling, likely harbouring organisms related to wastewater treatment washout.

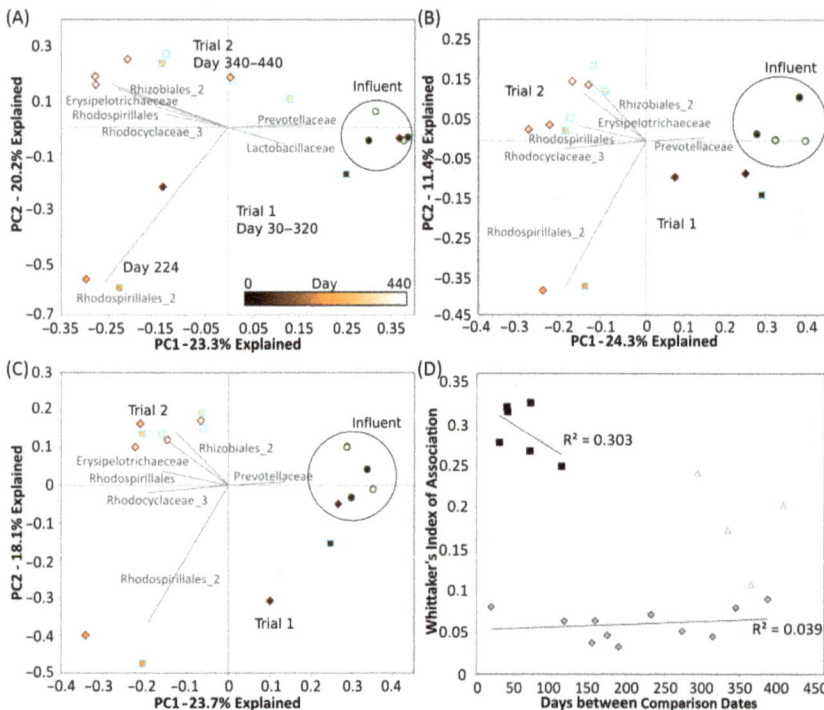

Fig. 3. Microbial turnover across a bioreactor over two production trials. PCoA plots of (A) Whittaker's index of association, (B) unweighted UniFrac, (C) Sørensen index and (D) distance–decay of Whittaker's index of association. (A, B, C) PCoA plots include influent samples as a baseline (green circle) and pilot plant mixed liquor (brown diamond) and SCP (blue square) time courses (fill color indicates time progression). PCoA bi-plots of the level 4 SILVA 104 non-redundant database taxonomy of the most abundant OTUs are denoted. (D) Distance–decay pairwise comparisons are graphed according to the difference in days of time points. Distance–decay trends are seen only for sample comparisons for trial 2 (square) but not trial 1 (triangle), and all other comparisons (diamond).

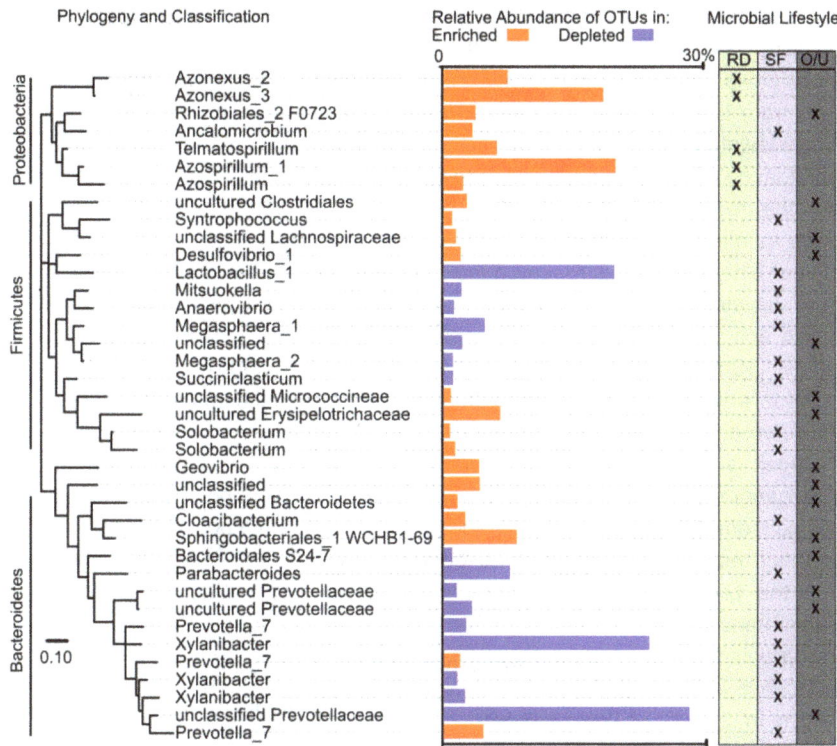

Fig. 4. Phylogenetic tree and sequence abundance of depleted and enriched OTUs. Enriched OTUs (orange) and depleted OTUs (blue) show clustering by phylogeny and have differential metabolic roles when classified by rhizospheric diazotroph (RD, pea green), saccharolytic fermenter (SF, light purple) and other/unclassified (O/U, grey). References for metabolic assignment and phylogenetic tree bootstrap values in Supporting Information Fig. S6.

Organisms responsible for SCP cell growth

The significantly depleted OTUs were found to be primarily saccharolytic fermentative anaerobes such as *Prevotella* (Fig. 4), whereas several separate types of diazotrophs (as well as some fermentative anaerobes) comprised much of the enriched sequences in the pilot bioreactor and SCP product. The presence of

saccharolytic bacteria in aerobic conditions was unusual as *Prevotella* are strict anaerobes. In the final SCP product, even higher abundances of rhizospheric diazotroph sequences were seen than in the pilot bioreactor. We interpret the reduction of *Prevotella* and an increase in rhizospheric diazotrophs in SCP as indicating that the primary treatment effect responsible for SCP production was from rhizospheric diazotrophs and that the

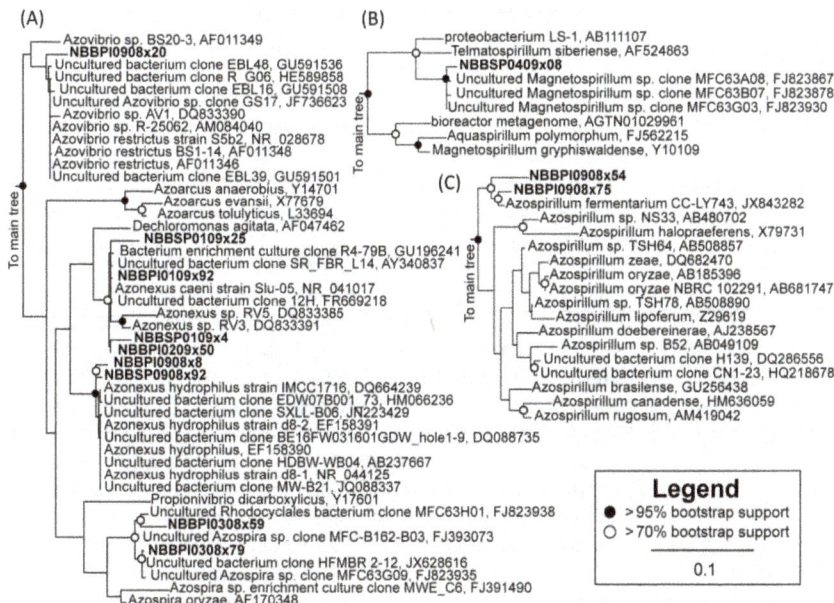

Fig. 5. Phylogenetic diversity of full-length Sanger sequences related to rhizospheric diazotrophs detected from the pilot bioreactor and final product for sequences from (A) *Azoarcus*, *Azonexus* and *Azospira* clades, (B) *Magnetospirillum* and *Telmatospirillum* clades, (C) and *Azospirillum* clades. Symbols denoting bootstrap support values are for maximum likelihood analyses. Representative sequences for OTUs identified from this study are in bold.

saccharolytic fermenter sequences, although common, may be from inactive cells washed in from the acidogenic basin. While diazotrophy exists across numerous phyla in nature, in this pilot bioreactor only diazotrophs from the *Proteobacteria* commonly associated with wastewater consortia were detected. The large relative abundance of diazotrophs seen in this study has a parallel in the treatment of certain high-strength wastewaters such as olive oil wastes and paper and pulp mill wastes where *Proteobacteria* rhizospheric diazotrophs were isolated and *nifH* genes and transcripts and nitrogenase activity detected (Papadelli *et al.*, 1996; Clark *et al.*, 1997; Gauthier *et al.*, 2000; Bowers *et al.*, 2008). These environments are often replete with simple carbon sources, yet are relatively limited in nitrogen and oxygen and require a diverse suite of microorganisms to generate fixed nitrogen to support the broader microbial consortia. In the pilot reactor of our studied system, aeration estimates produced DO levels 0.1–0.5 mg l^{-1} in the pilot reactor. There may be similarities with this environment and the rhizospheric habitat where such organisms autochthonously exist, particularly the abundance of simple organic acids from acidogenesis, which in the rhizosphere are released from plant roots to diazotrophs (Christiansen-Weniger *et al.*, 1992).

However, we must caution that without further investigation, nitrogen fixation cannot conclusively be attributed as the primary biological advantage, especially since unlike conventional brewery wastewater treatment, nitrogen (as urea) was added to excess, and nitrogen fixation would be rapidly suppressed by the presence of fixed nitrogen. A major challenge in this study was monitoring the entire mixed liquor of such large bioreactors; therefore, it is possible that urea was not optimally dissolved, mixed or sorbed to microbial flocs. Another possible association is that the saccharolytic fermenting anaerobic bacteria in the acidogenic upstream treatment produced easily accessible organic acids that become the influent of the pilot bioreactor, where microbes adapted to have affinity for organic acids, and other pilot bioreactor conditions are then enriched. Under both scenarios, the wide presence of wastewater-specific diazotrophs found in this study indicates a potential target for future metabolic biotechnology exploitation. For example, because of the variable flow nature of batch fermentation in beer production, breweries have large flow equalization basins to regulate the supply of wastewater to continuous flow treatment stages. The conversion of such basins to an intentional acidogenic pretreatment stage upstream of the treatment bioreactors is one possibility to produce a constant supply of organic acids for uptake and conversion to SCP by aerobic or microaerophilic organic acid consuming microbes. Another potential avenue of research centres on the use of methanogenic UASBs as

an inoculant source for these types of diazotrophs to seed their growth.

In summary, this study highlights the overall distinctiveness of each treatment stage within a single process WWTP, especially the occurrence of *Prevotella* and related saccharolytic fermenters in flow equalization basins and the role of wastewater-associated rhizospheric diazotrophs in high-strength wastewaters. This research indicated that rather than growing a single culture of these organisms, endogenous enrichment of rhizospheric diazotrophic bacteria in high-strength wastewater treatment systems can be used to produce SCP and should be studied as a way to deliberately manipulate microbial ecology for the production of a high protein content ingredient for aquaculture feed. However, before large-scale implementation of this technology can occur, several challenges remain to bring such a product to market such as regulatory and feed safety approval, scale up of production and large-scale animal feeding trials.

Experimental procedures

Project site description and sample handling

Figure 1 shows the brewery WWTP and pilot reactor research site used in this study (New Belgium Brewing, Fort Collins, CO, USA). The brewery treatment works itself consists of an unaerated influent flow equalization basin with acidogenic conditions, followed in series by a methanogenesis basin consisting of an UASB bioreactor. Next follows an aerobic treatment basin and clarifier for activated sludge growth and settling prior to discharge to the city publicly owned treatment works.

The pilot reactor received wastewater from the acidogenic basin to aerobically treat the wastewater at low mean cell residence time (< 8 days) with nutrient addition (N as urea, P as phosphoric acid and a customized micronutrient cocktail) to increase the protein concentration via growth of cell mass. The pilot bioreactor was initially fed and seeded from the methanogenesis UASB for about 1 month prior to the start of trial 1 before being switched to the acidogenic basin.

Two separate production trials using the pilot bioreactor were completed over a 450 day study period (trial 1: day 30–320, trial 2: day 340–440), and enough SCP was consistently produced at > 55% crude protein content for use in commercial feeding trials (Logan *et al.*, 2011). Sampling locations and reactor conditions are shown in Fig. 1 (colored squares) and Table 1. Treatment plant influent samples (Fig. 1, red square) were taken from the waste influent receiving line upstream of the acidogenic basin. Acidogenic basin (dark green square) samples were taken from basin mixed liquor. Methanogenesis UASB (purple square) samples were taken from the basin outfall stream. Aerobic basin (light green square) samples were taken from basins directly, and clarifier wet well (orange square) was taken from the wet-well outfall channel. Pilot bioreactor samples (brown square) were taken from an effluent sampling port leading to the drying and centrifuge assembly. Finally, dried SCP product (blue square) was sampled directly from storage containers containing the most recent batch of manufacture. Liquid grab samples from basins and outfalls were collected over the study period in autoclaved 1-L serum bottles (rinsed with sample first) and returned to the lab and spun down at 10 000 relative centrifugal force (RCF) for 5 min. Grab samples and dry SCP product samples were stored at –20°C until DNA extraction with a Mo-Bio PowerSoil Kit (Carlsbad, CA, USA) per the manufacture's protocol with the exception of a 1 min bead beating step in a Biospec (Bartlesville, OK, USA) MiniBeadbeater-8 instead of vortex bead beating.

Wastewater chemical analysis

WWTP characterization (N, P, COD and VFA) analysis was conducted using commercial Hach TNT kits (Loveland, CO, USA). TSS was measured using filter paper and drying at 105°C (APHA, 1985) collected daily from January to March 2009. Pilot bioreactor MLSS (Royce 711, Australia) measurements and pH were collected daily from October 2008 to May 2009. Grab samples collected at the same time as nucleic acid samples were returned to the lab for organic acid analysis. An ion exclusion organic acid column Animex HPX-87H (Bio-Rad, Hercules, CA, USA) with a pre-column filter (Bio-Rad) was used with an Agilent 1100 HPLC. Samples were spun (10 000 RCF, 5 min) and filtered (0.45 μm) before use. Standards for formic, acetic, lactic, propionic, succinic, butyric, isobutyric, valeric, isovaleric, 2-methylbutyric and citric acid were run at three intervals in replicates of three with an injection volume of 50 um. Running buffer consisted of 0.04 N phosphoric acid, 0.60 ml min^{-1}, 40°C, 35 min. Diode array detector signal frequency was 210 nm, reference 360 nm. Concentrations of all detected organic acids were summed to determine total organic acid content in samples.

Sampling and Sanger sequencing

Samples were collected over a period of 1 year from throughout the brewery treatment works as well as from the pilot plant and final dried product to track changes in community composition across treatment stages and in time. Sample DNA extraction, Sanger sequencing and full-length 16S small subunit (SSU) ribosomal RNA (rRNA) gene bioinformatic methods were adapted from Sahl and colleagues (2010) for 8F and 1492R bacterial primers and with details listed in the Supplemental Methods. Briefly, DNA was extracted from samples and amplified by polymerase chain reaction (PCR) followed by vector cloning of amplicons and Sanger sequencing of the T3/T7/515F reads, and finally contig formation to determine full-length 16S SSU rRNA gene reads. Reads were clustered at the 97% OTU level and unique representative sequences extracted. Representative sequences derived from Sanger sequencing OTUs of full-length 16S sequence clones from *Proteobacteria* were used with 64 full-length 16S sequences of BLAST closest matches and 37 SILVA reference sequences for phylogenetic reconstruction. These sequences were trimmed to a contiguous aligned [Nearest Alignment Space Termination (NAST)-based aligner of mothur (Schloss, 2009)] 1142 bp length and used with the pos_var_bac '0' filter in PHYML (Guindon *et al.*, 2010) with default parameters and 100 bootstraps for detailed phylogenetic comparison.

Pyrosequencing and bioinformatics pipeline

Pyrosequencing PCR of the 16S SSU rRNA gene using the bacterial 27F and 338R primers with sequence barcoding (Hamady *et al.*, 2008) adapted from Sahl and colleagues (2010) and processed using QIIME (Caporaso *et al.*, 2010), with details listed in the Supplemental Methods. Briefly, sequences were denoised filtered, and clustered using QIIME, then aligned and classified using the NAST-based aligner and classifier of mothur (Schloss, 2009) trained on the customized SILVA 104 non-redundant database (80% confidence level cut-off). Note that taxonomic assignments in figures and text in this paper with underscores or number assignments at the end (e.g. firmicutes_bacilli or azonexus_2) indicate paraphyletic classification in the SILVA reference tree system.

Alpha and beta diversity analysis

QIIME was used to compute alpha diversity estimates of the Chao1 and ACE metrics (Colwell, 2009). The CatchAll statistic (Bunge, 2011) was also used with default parameters, and the best model total species observed values were used. All alpha diversity estimators were computed with samples rarified up to the same sequencing depth with 50 replicates each. Coverage estimates were based on the percentage of observed OTUs to the Chao1 estimator as a low estimate and to the CatchAll as the high estimate using all sequences per sample for computation. Several metrics of beta diversity originally derived from classical macroecology [particularly plant communities (Sørensen, 1948; Whittaker, 1972)], such as the Sørensen (a qualitative sharing metric) and Whittaker's Index of Association (quantitative relative abundance sharing metric), were chosen to study the pilot plant system. These have been applied for the study of variations in bacterioplankton communities across distance in the North Pacific (Hewson *et al.*, 2006) and could indicate temporal-spatial distance–decay relationships. Additionally, based on recent work on wastewater variation (Zhang *et al.*, 2012), the unweighted UniFrac metric (Lozupone *et al.*, 2011) was used to compare wastewater samples. QIIME was used to calculate the beta diversity metrics of unweighted UniFrac (using a whole tree of the entire dataset), the complement of Whittaker's index of association (Legendre and Legendre, 1998) (using a custom python script rather than the qualitative Whittaker's found in QIIME), and the qualitative Sørensen index of similarity. UPGMA clusters and principal coordinate analysis were determined using QIIME scripts. Distance matrices from each metric were also jackknife subsampled to the smallest sequence count to examine the sensitivity of sequencing depth on clustering and principal coordinate analysis. Bi-plots of only the top 10 most commonly found taxonomic families were displayed.

Determining biotechnologically relevant OTUs

While it is useful to compare environments over time in a holistic manner with beta diversity metric analysis, we also wanted to identify the distribution of OTUs responsible for the community shift, especially the fraction that indicates SCP formation. To determine organisms that might be of biotechnological interest, two tests were used. The first was to test if OTUs are co-abundant (i.e. co-incident in time and relative abundance). This can highlight patterns of variation of consortia in response to natural wastewater source variations. Additionally, a significance test was used to determine OTUs that were differentially enriched in the mixed liquor as compared with the reactor influent.

To understand the co-abundance of OTUs in both the pilot plant and final SCP product, OTUs that were present in these two environments were used. OTUs were included if an OTU was present in these two environments (pilot bioreactor and final product) more than once and had more than two sequences associated with it. 319 out of 1604 OTUs passed this criteria. The Bray–Curtis pairwise distance metric (Legendre and Legendre, 1998) was computed for each OTU based on the abundances of each OTU across SCP and pilot bioreactor samples ($n = 13$).

Each OTU was then grouped into a functional category based on the metabolic profile of the nearest taxonomic representative found in Bergey's Index (Brenner et al., 2005) (characteristic of the entire known genus or family level) or related primary literature (of the genus level only). A conservative approach was taken where clades with significant known metabolic variation or poor classification (at the genus level) were labeled as 'unknown or unclassified'.

OTUs were additionally compared by online METASTATS (http://metastats.cbcb.umd.edu/detection.html) (White et al., 2009), a program developed to examine differential abundance of elements associated between patients from a control and treatment category for medical microbiome studies. Abundance information from the series of acidogenic basin samples (representing the influent to the pilot plant) and the pilot plant mixed liquor samples were used as the pre- and post-treatment cases with a threshold p-value < 0.05 and a difference in average relative abundance of $\pm 1\%$ absolute magnitude as being significantly 'enriched' or 'depleted' respectively.

A major concern in this work was the role of PCR and extraction biases in calculating co-abundance dissimilarity distances. This work did not seek and does not serve to capture the true compositional nature of the sample, but rather relative abundance data is used to understand patterns in OTU distributions undergoing the same DNA extraction, amplification and analysis conditions. Another concern was misidentification or the inclusion of spurious sequences leading to poor taxonomic identification. Therefore, no conclusions are made on any single OTU, and full-length 16S SSU Sanger-based gene sequencing was used to verify the findings of short sequencing reads.

Nucleotide accession numbers and online resources

A total of 809 Sanger sequences were deposited in GenBank under accession numbers JQ072092–JQ072899. Sanger sequences were named according to the location and date in which they were sampled [i.e. NBBEQMMYY_44; where NBB = New Belgium Brewing, EQ = flow equalization/ acidogenic basin, ME = methanogenic UASB, PI = pilot plant, SP = SCP product, AB = aerobic basin, OT = wet well outfall, MMYY = (month/year) 44 = sequence number]. 454 pyrotags taken from quality score processing were submitted in FASTQ format to MG-RAST (ID 4477674.3). Custom python scripts and interactive Krona pie charts (Ondov et al., 2011) used in this study can be obtained from http://inside .mines.edu/~jspear/resources.html.

Acknowledgements

Authors wish to thank Brandon Weaver and New Belgium Brewing, Fort Collins, CO, for facilities, space and process wastewater treatment support (http://www.newbelgium .com). Thanks to the Pace Lab at the University of Colorado, Boulder for assistance with Sanger sequencing. The authors would like to acknowledge J. M. R., R. C. E. and members of the GEM Lab for useful comments on the manuscript.

Conflict of interest

Andrew Logan and Seth Terry are both Vice-Presidents of Research of Nutrinsic, Corp.

References

APHA (1985) *Standard Methods for the Examination of Water and Wastewater*, 16th edn. Washington, DC, USA: American Public Health Association.

Ashida, N., Ishii, S., Hayano, S., Tago, K., Tsuji, T., Yoshimura, Y., et al. (2010) Isolation of functional single cells from environments using a micromanipulator: application to study denitrifying bacteria. *Appl Microbiol Biotechnol* **85:** 1211–1217.

Borole, A.P., Mielenz, J.R., Vishnivetskaya, T.A., and Hamilton, C.Y. (2009) Controlling accumulation of fermentation inhibitors in biorefinery recycle water using microbial fuel cells. *Biotechnol Biofuels* **2:** 7.

Bowers, T.H., Reid, N.M., and Lloyd-Jones, G. (2008) Composition of nifH in a wastewater treatment system reliant on N2 fixation. *Appl Microbiol Biotechnol* **79:** 811–818.

Brenner, D.J., Krieg, N.R., Garrity, G.M., Staley, J.T., Boone, D.R., Vos, P., et al. (2005) *Bergey's Manual of Systematic Bacteriology*. New York, USA: Springer.

Bunge, J. (2011) Estimating the number of species with catchall. *Pac Symp Biocomput* **2011:** 121–130.

de Cárcer, D.A., Ha, P.T., Jang, J.K., and Chang, I.S. (2011) Microbial community differences between propionate-fed microbial fuel cell systems under open and closed circuit conditions. *Appl Microbiol Biotechnol* **89:** 605–612.

Caporaso, J.G., Kuczynski, J., Stombaugh, J., Bittinger, K., Bushman, F.D., Costello, E.K., et al. (2010) QIIME allows analysis of high-throughput community sequencing data. *Nat Methods* **7:** 335–336.

Chouari, R., Le Paslier, D., Dauga, C., Daegelen, P., Weissenbach, J., and Sghir, A. (2005) Novel major bacterial candidate division within a municipal anaerobic sludge digester. *Appl Environ Microbiol* **71:** 2145–2153.

Christiansen-Weniger, C., Groneman, A.F., and Veen, J.A. (1992) Associative N2 fixation and root exudation of organic acids from wheat cultivars of different aluminium tolerance. *Plant Soil* **139:** 167–174.

Clark, T.A., Dare, P.H., and Bruce, M.E. (1997) Nitrogen fixation in an aerated stabilization basin treating bleached kraft mill wastewater. *Water Environ Res* **69:** 1039–1046.

Colwell, R.K. (2009) EstimateS: statistical estimation of species richness and shared species from samples. Version 8.2 User's Guide.

Cressey, D. (2009) Aquaculture: future fish. *Nature* **458:** 398–400.

Croese, E., Pereira, M.A., Euverink, G.-J.W., Stams, A.J.M., and Geelhoed, J.S. (2011) Analysis of the microbial community of the biocathode of a hydrogen-producing microbial electrolysis cell. *Appl Microbiol Biotechnol* **92:** 1083–1093.

Diaz, E.E., Stams, A.J.M., Amils, R., and Sanz, J.L. (2006) Phenotypic properties and microbial diversity of methanogenic granules from a full-scale upflow anaerobic sludge bed reactor treating brewery wastewater. *Appl Environ Microbiol* **72:** 4942–4949.

El-Sayed, A.-F.M. (1999) Alternative dietary protein sources for farmed tilapia, *Oreochromis* spp. *Aquaculture* **179:** 149–168.

FAO Fisheries Department (2003) Review of the state of world aquaculture. FAO Fisheries Circular No. 886. Food and Agriculture Organization of the United Nations, Rome.

Fernandez, A.S., Hashsham, S.A., Dollhopf, S.L., Raskin, L., Glagoleva, O., Dazzo, F.B., *et al.* (2000) Flexible community structure correlates with stable community function in methanogenic bioreactor communities perturbed by glucose. *Appl Environ Microbiol* **66:** 4058–4067.

Gauthier, F., Neufeld, J.D., Driscoll, B.T., and Archibald, F.S. (2000) Coliform bacteria and nitrogen fixation in pulp and paper mill effluent treatment systems. *Appl Environ Microbiol* **66:** 5155 5160.

Gray, N. (2004) *Biology of Wastewater Treatment*, 2nd edn. London, UK: Imperial College Press.

Guindon, S., Dufayard, J.-F., Lefort, V., Anisimova, M., Hordijk, W., and Gascuel, O. (2010) New algorithms and methods to estimate maximum-likelihood phylogenies: assessing the performance of PhyML 3.0. *Syst Biol* **59:** 307–321.

Hamady, M., Walker, J.J., Harris, J.K., Gold, N.J., and Knight, R. (2008) Error-correcting barcoded primers for pyrosequencing hundreds of samples in multiplex. *Nat Methods* **5:** 235–237.

Hewson, I., Steele, J.A., Capone, D.G., and Fuhrman, J.A. (2006) Remarkable heterogeneity in meso- and bathypelagic bacterioplankton assemblage composition. *Limnol Oceanogr* **51:** 1274–1283.

Hollister, E.B., Hammett, A.M., Holtzapple, M.T., Gentry, T.J., and Wilkinson, H.H. (2011) Microbial community composition and dynamics in a semi-industrial-scale facility operating under the MixAlco™ bioconversion platform. *J Appl Microbiol* **110:** 587–596.

Hough, J. (1985) *The Biotechnology of Malting and Brewing*. New York, USA: Cambridge University Press.

Hugenholtz, P., Tyson, G.W., Webb, R.I., Wagner, A.M., and Blackall, L.L. (2001) Investigation of candidate division TM7, a recently recognized major lineage of the domain bacteria with no known pure-culture representatives. *Appl Environ Microbiol* **67:** 411–419.

Huige, N. (2006) Brewery by-products and effluents. In *Handbook of Brewing*. Priest, F. and Stewart, G. (eds). Boca Raton, USA: CRC/Taylor & Francis, pp. 655–714.

Kaksonen, A.H., Plumb, J.J., Franzmann, P.D., and Puhakka, J.A. (2004) Simple organic electron donors support diverse sulfate-reducing communities in fluidized-bed reactors treating acidic metal- and sulfate-containing wastewater. *FEMS Microbiol Ecol* **47:** 279–289.

Knief, C., Delmotte, N., Chaffron, S., Stark, M., Innerebner, G., Wassmann, R., *et al.* (2012) Metaproteogenomic analysis of microbial communities in the phyllosphere and rhizosphere of rice. *ISME J* **6:** 1378–1390.

Lee, T.K., Van Doan, T., Yoo, K., Choi, S., Kim, C., and Park, J. (2010) Discovery of commonly existing anode biofilm microbes in two different wastewater treatment MFCs using FLX titanium pyrosequencing. *Appl Microbiol Biotechnol* **87:** 2335–2343.

Legendre, P., and Legendre, L. (1998) *Numerical Ecology 2nd English Ed.* New York, USA: Elsevier.

Li, L., Kato, C., and Horikoshi, K. (1999) Bacterial diversity in deep-sea sediments from different depths. *Biodivers Conserv* **8:** 659–677.

Lin, S.-Y., Liu, Y.-C., Hameed, A., Hsu, Y.-H., Lai, W.-A., Shen, F.-T., and Young, C.-C. (2013) *Azospirillum fermentarium* sp. nov., a nitrogen-fixing species isolated from a fermenter. *Int J Syst Evol Microbiol* **63:** 3762–3768.

Liu, W.-T., Chan, O.-C., and Fang, H.H.P. (2002) Characterization of microbial community in granular sludge treating brewery wastewater. *Water Res* **36:** 1767–1775.

Logan, A., Lawrence, A., Dominy, W., and Tacon, A. (2011) Bacterial single cell protein produced from brewery wastes as a protein source in white shrimp diets. *Aquafeed Spring* **2011:** 7–12.

Lovell, T. (1989) *Nutrition and Feeding of Fish*. New York, USA: Chapman & Hall.

Lozupone, C., Lladser, M.E., Knights, D., Stombaugh, J., and Knight, R. (2011) UniFrac: an effective distance metric for microbial community comparison. *ISME J* **5:** 169–172.

Manz, W., Wagner, M., Amann, R., and Schleifer, K.-H. (1994) In situ characterization of the microbial consortia active in two wastewater treatment plants. *Water Res* **28:** 1715–1723.

Ondov, B.D., Bergman, N.H., and Phillippy, A.M. (2011) Interactive metagenomic visualization in a Web browser. *BMC Bioinformatics* **12:** 385.

Papadelli, M., Roussis, A., Papadopoulou, K., Venieraki, A., Chatzipavlidis, I., Katinakis, P., and Ballis, K. (1996) Biochemical and molecular characterization of an Azotobacter vinelandii strain with respect to its ability to grow and fix nitrogen in olive mill wastewater. *Int Biodeterior Biodegradation* **38:** 179–181.

Quan, Z.-X., Im, W.-T., and Lee, S.-T. (2006) *Azonexus caeni* sp. nov., a denitrifying bacterium isolated from sludge of a wastewater treatment plant. *Int J Syst Evol Microbiol* **56:** 1043–1046.

Riviere, D., Desvignes, V., Pelletier, E., Chaussonnerie, S., Guermazi, S., Weissenbach, J., *et al.* (2009) Towards the definition of a core of microorganisms involved in anaerobic digestion of sludge. *ISME J* **3:** 700–714.

Sahl, J.W., Fairfield, N., Harris, J.K., Wettergreen, D., Stone, W.C., and Spear, J.R. (2010) Novel microbial diversity retrieved by autonomous robotic exploration of the world's deepest vertical phreatic sinkhole. *Astrobiology* **10**: 201–213.

Schloss, P.D. (2009) A high-throughput DNA sequence aligner for microbial ecology studies. *PLoS ONE* **4**: e8230.

Sekiguchi, Y., Kamagata, Y., Nakamura, K., Ohashi, A., and Harada, H. (1999) Fluorescence in situ hybridization using 16S rRNA-targeted oligonucleotides reveals localization of methanogens and selected uncultured bacteria in mesophilic and thermophilic sludge granules. *Appl Environ Microbiol* **65**: 1280–1288.

Seviour, R., and Nielsen, P.H. (2010) *Microbial Ecology of Activated Sludge [New ed.]*. London, UK: IWA Publishing.

Sørensen, T. (1948) A method of establishing groups of equal amplitude in plant sociology based on similarity of species and its application to analyses of the vegetation on Danish commons. *Biol Skr* **5**: 1–34.

Sun, Y., Wei, J., Liang, P., and Huang, X. (2011) Electricity generation and microbial community changes in microbial fuel cells packed with different anodic materials. *Bioresour Technol* **102**: 10886–10891.

Tacon, A., Metian, M., and Hasan, M. (2009) Feed ingredients and fertilizers for farmed aquatic animals: sources and composition. UN FAO, Technical Paper 540.

Tacon, A.G.J. (1979) Activated sewage sludge, a potential animal foodstuff II. Nutritional characteristics. *Agric Environ* **4**: 271–279.

Vriens, L., Nihoul, R., and Verachtert, H. (1989) Activated sludges as animal feed: a review. *Biol Waste* **27**: 161–207.

Wells, G.F., Park, H.-D., Eggleston, B., Francis, C.A., and Criddle, C.S. (2011) Fine-scale bacterial community dynamics and the taxa–time relationship within a full-scale activated sludge bioreactor. *Water Res* **45**: 5476–5488.

Werner, J.J., Knights, D., Garcia, M.L., Scalfone, N.B., Smith, S., Yarasheski, K., *et al.* (2011) Bacterial community structures are unique and resilient in full-scale bioenergy systems. *PNAS* **108**: 4158–4163.

White, J.R., Nagarajan, N., and Pop, M. (2009) Statistical methods for detecting differentially abundant features in clinical metagenomic samples. *PLoS Comput Biol* **5**: e1000352.

Whittaker, R.H. (1972) Evolution and measurement of species diversity. *Taxon* **21**: 213–251.

Worm, B., Barbier, E.B., Beaumont, N., Duffy, J.E., Folke, C., Halpern, B.S., *et al.* (2006) Impacts of biodiversity loss on ocean ecosystem services. *Science* **314**: 787–790.

Zhang, T., Shao, M.-F., and Ye, L. (2012) 454 pyrosequencing reveals bacterial diversity of activated sludge from 14 sewage treatment plants. *ISME J* **6**: 1137–1147.

Supporting information

Additional Supporting Information may be found in the online version of this article at the publisher's web-site:

Fig. S1. Jackknife UPGMA node sensitivity analysis for Figure 2.

Fig. S2. Jackknife PCOA ellipsoid plots, subsampled to 800 sequences per sample with 100 jackknife replicates.

Fig. S3. OTUs in a pilot reactor and SCP product clustered over time.

Fig. S4. Breakdown of sequences in co-abundance and lifestyle analysis showing the distribution of enriched and depleted sequences across pilot bioreactor mixed liquor and final SCP product.

Fig. S5. Phylogenetic diversity of full-length Sanger sequences related to rhizospheric diazotrophs detected from the pilot bioreactor and final product.

Fig. S6. Phylogenetic tree and sequence abundance of depleted and enriched OTUs.

10

Evaluation of industrial *Saccharomyces cerevisiae* strains as the chassis cell for second-generation bioethanol production

Hongxing Li,[1] Meiling Wu,[1] Lili Xu,[1] Jin Hou,[1] Ting Guo,[1,2] Xiaoming Bao[1] and Yu Shen[1]*

[1]*State Key Laboratory of Microbial Technology, Shandong University, Jinan 250100, China.*

[2]*Guangzhou Sugarcane Industry Research Institute, Guangzhou 510316, China.*

Summary

To develop a suitable *Saccharomyces cerevisiae* industrial strain as a chassis cell for ethanol production using lignocellulosic materials, 32 wild-type strains were evaluated for their glucose fermenting ability, their tolerance to the stresses they might encounter in lignocellulosic hydrolysate fermentation and their genetic background for pentose metabolism. The strain BSIF, isolated from tropical fruit in Thailand, was selected out of the distinctly different strains studied for its promising characteristics. The maximal specific growth rate of BSIF was as high as 0.65 h^{-1} in yeast extract peptone dextrose medium, and the ethanol yield was 0.45 g g^{-1} consumed glucose. Furthermore, compared with other strains, this strain exhibited superior tolerance to high temperature, hyperosmotic stress and oxidative stress; better growth performance in lignocellulosic hydrolysate; and better xylose utilization capacity when an initial xylose metabolic pathway was introduced. All of these results indicate that this strain is an excellent chassis strain for lignocellulosic ethanol production.

Introduction

Biofuels (including ethanol, butanol and biodiesel), chemicals and other commodities produced from renewable and abundant lignocellulosic feedstocks have become increasingly important because of the depletion of fossil fuel energy sources and growing public concerns about the environment and food security (Zhou *et al.*, 2012; Jönsson *et al.*, 2013). One of the most practical solutions is to produce bioethanol from lignocellulosic feedstocks with *Saccharomyces cerevisiae* (Palmqvist and Hahn-Hägerdal, 2000a). However, this natural ethanol producer faces several new challenges when the substrate is lignocellulose instead of starch. Not only a high glucose metabolism capacity and ethanol yield, but also the capacity to tackle the challenges associated with lignocellulose fermentation are necessary properties for a lignocellulosic ethanol-producing strain.

The lignocellulosic ethanol conversion process generally includes raw material pretreatment, cellulose hydrolysis, sugar fermentation by microorganisms and distillation. Most of the hemicellulose fraction is hydrolysed to monosaccharides in the pretreatment step, and the xylose obtained during this process is the second most abundant sugar in lignocellulosic materials (Kim *et al.*, 2013). Efficient use of xylose would greatly increase the economic benefits of lignocellulosic bioethanol production (Sun and Cheng, 2002; Hahn-Hägerdal *et al.*, 2007). However, xylose is not naturally fermented by *S. cerevisiae*. The inability of *S. cerevisiae* to use xylose is not only due to its lack of relevant enzymes, but also related to the low efficiency of other necessary metabolic pathways, such as the pentose phosphate pathway. In addition, inhibitors that are formed during the pretreatment and hydrolysis processes with the release of sugars are toxic to microorganisms. Therefore, the yeast strain used in lignocellulosic bioethanol production requires not only high ethanol yields from both glucose and xylose, but also robustness in its harsh working environment.

The inhibitors to ethanol production are usually divided into three major groups: weak acids, furan derivatives and phenolic compounds (Palmqvist and Hahn-Hägerdal, 2000b; Almeida *et al.*, 2007; Alriksson *et al.*, 2010). The inhibitory effects of these compounds on *S. cerevisiae* strains are quite different. Undissociated weak acids are liposoluble and can diffuse across the plasma membrane, causing intracellular anion accumulation and inhibiting cell growth. Furan derivatives, such as furfural and HMF (5-hydroxymethylfurfural), have been shown to reduce the

*For correspondence. E-mail shenyu@sdu.edu.cn

Funding Information This work was supported by the National Key Basic Research Program (2011CB707405), the National High-Technology Research and Development Program of China (2012AA022106, 2014AA021903), Project of National Energy Administration of China (NY20130402) and the State Key Laboratory of Motor Vehicle Biofuel Technology (No.2013004).

Table 1. *Saccharomyces cerevisiae* strain sources and numbers used in this work.

Strain sources	Strain number	Strains evaluated[a]
Obtained from ATCC, CICC, CGMCC	6	CICC31034
Used in starch-based ethanol production	6	NAN-27, 6508
Isolated from commercially available active dry yeast (China, Sweden, America, France, Japan)	5	
Isolated from grape or wine production regions	4	RC212
Isolated from liquor production regions	3	
Isolated from tropical fruit in Thailand	3	BSIF
Isolated from waste materials in sugar cane industry	5	

a. The five strains selected for all evaluations and their corresponding sources.
ATCC, American Type Culture Collection; CGMCC, China General Microbiological Culture Collection Center; CICC, China Center of Industrial Culture Collection.

specific growth rate, the cell mass yield, and the volumetric and specific ethanol productivities. Phenolic compounds destroy the cell membrane integrity, thereby impeding the membrane's function (Palmqvist and Hahn-Hägerdal, 2000b; Almeida *et al.*, 2007). The interaction effects of these inhibitor compounds often lead to more serious inhibition (Palmqvist *et al.*, 1999). In addition, some hydrolysate components that have not been extensively analysed also cause undetermined stresses on the cells (Palmqvist and Hahn-Hägerdal, 2000b). In addition to the osmotic stress caused by the ions and sugars in the hydrolysate, the ethanol product also has negative effects on yeast (Sonderegger *et al.*, 2004; Jönsson *et al.*, 2013).

Metabolic and evolutionary engineering strategies have been extensively performed in constructing and enhancing the xylose fermentation capacity of both laboratory and industrial *S. cerevisiae* strains (Eliasson *et al.*, 2000; Kuyper *et al.*, 2003; Peng *et al.*, 2012; Zhou *et al.*, 2012). The fermentation properties and inhibitor resistance depend on the strains' individual genetic backgrounds. The diploid or polyploid industrial *S. cerevisiae* strains are usually more robust than the haploid strains, and the whole-genome duplication in yeast was proposed to lead to an efficient fermentation system (Piškur *et al.*, 2006). The choice of the chassis strain has a significant effect on the performance of lignocellulosic hydrolysate-based ethanol production (Brandberg *et al.*, 2004), and it is important to possess a self-owned, suitable chassis strain because of intellectual property concerns.

Therefore, in this paper, 32 wild-type *S. cerevisiae* strains were evaluated. Among them, candidate strains with an ethanol yield of more than 0.41 g g^{-1} consumed glucose were selected in the preliminary screening. Then, the tolerance to inhibitors, the growth performance in hydrolysate and the background xylose utilization capacity of the candidate strains were further compared. In addition, the ploidy of the strains was determined. Based on these evaluations, an *S. cerevisiae* strain was chosen as the chassis cell for constructing a lignocellulosic ethanol-producing strain.

Results and discussion

Collection of S. cerevisiae *strains*

In this work, 32 wild-type *S. cerevisiae* strains were collected or isolated from several strain preservation institutions, ethanol production companies, commercial active dry yeasts and different habitats. The detailed strain sources and numbers are listed in Table 1. The strains isolated from specific environments were identified by sequence analysis of the 26S rDNA D1/D2 domain.

Fermentation performance of S. cerevisiae *strains on glucose*

Efficient ethanol production from glucose, a main component in lignocellulosic hydrolysate, is the most important characteristic for the candidate strains for bioethanol production. Therefore, the fermentation performances of the *S. cerevisiae* strains were first evaluated with glucose as the sole carbon source. The fermentation was performed in shake flasks with an initial cell density of OD$_{600}$ 0.1 (~ 0.02 g l^{-1} dry cell biomass). Rubber stoppers with syringe needles were used for oxygen-limited cultivations. As expected, these strains exhibited significant differences in the ethanol yield, ranging from 0.33 to 0.43 g g^{-1} consumed glucose after 10 h. Five strains, NAN-27, BSIF, RC212, CICC31034 and 6508 (Table 1), with ethanol yields of more than 0.41 g g^{-1} consumed glucose, were selected and evaluated in further detail. The results in Fig. 1 and Table 2 show that strains BSIF and RC212 had the best performance in terms of the maximum specific growth rate (μ_{max}) and corresponding glucose volumetric consumption rate and that the ethanol yield of both strains was higher than 0.44 g g^{-1} consumed glucose.

Evaluation of tolerance of S. cerevisiae *strains to individual stress factors*

Considering the stress conditions that the fermentation strain may encounter in the second-generation fuel ethanol production process, the growth performances of

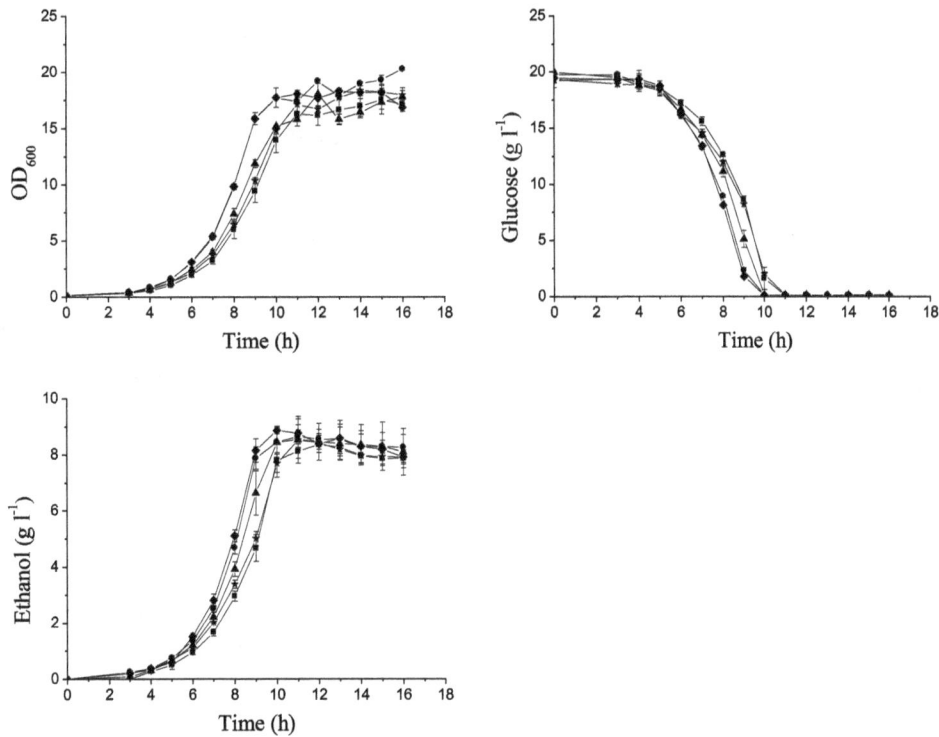

Fig. 1. Oxygen-limited fermentation on glucose in shake flasks. The cells were cultured at 30°C in shake flasks with initial OD_{600} of 0.1 (~ 0.02 g l^{-1} dry cell biomass). Oxygen-limited conditions were obtained with a rubber stopper and a syringe needle, and the cells were agitated at 200 r.p.m. The experiments were performed in triplicate. Symbol: *S. cerevisiae* strain NAN-27, ■; BSIF, ◆; RC212, ▲; CICC31034, ●; 6508, ★.

the five selected strains were tested under aerobic conditions with several stress factors (Fig. 2). The performances of the five *S. cerevisiae* strains were greatly different. Strains BSIF and CICC31034 clearly grew better at a higher temperature, 42°C (Fig. 2A). The hypertonic stress tolerance of BSIF and RC212 was slightly higher than the tolerance of other strains on plates with 1.0 mol^{-1} KCl (Fig. 2B). Strain CICC31034 was more sensitive to oxidative stress by 30% H_2O_2 than the other four strains (Fig. 2C). The tested strains showed distinct tolerance to

the inhibitors in the hydrolysate, but no single strain showed the highest tolerance to all of the different inhibitors. Strain NAN-27 showed more tolerance to acetic acid (Fig. 2E) but less to furfural (Fig. 2D) and vanillin (Fig. 2F), strains RC212 and CICC31034 grew better on vanillin plates (Fig. 2F). All of the tested strains exhibited strong ethanol tolerance, all growing well in 10% ethanol (Fig. 2G). The results suggested the relative robustness of strains; however, under anaerobic fermentation conditions, the specific tolerance of strains to stresses might be lower because of a lower energy supply.

Evaluation of tolerance of S. cerevisiae *strains to corn stover hydrolysate*

Lignocellulosic hydrolysate is the strains' working medium for second-generation bioethanol production. In this hydrolysate, weak acids, furan derivatives and phenolic compounds are generally recognized as major toxicants, and their mixture usually increases their inhibitory impact (Almeida *et al.*, 2007). Furthermore, there are other toxic substances not readily identified in hydrolysate (Palmqvist and Hahn-Hägerdal, 2000b). The fermentation characteristics of the five industrial strains and the laboratory strain CEN.PK 113-5D were therefore tested in hydrolysed corn stover. The hydrolysate contained 3.7 g l^{-1} acetic acid,

Table 2. Metabolic characteristics of *S. cerevisiae* strains on glucose in shake flasks.

Strains	μ_{max} (h^{-1})	Ethanol yield (g g^{-1} consumed glucose)[a]	Glucose volume consumption rate (g l^{-1} h^{-1})[b]
NAN27	0.590 ± 0.005	0.421 ± 0.003	1.183 ± 0.106
BSIF	0.652 ± 0.004	0.451 ± 0.016	1.966 ± 0.055
RC212	0.632 ± 0.005	0.441 ± 0.001	1.939 ± 0.021
CICC31034	0.609 ± 0.006	0.423 ± 0.007	1.653 ± 0.071
6508	0.556 ± 0.022	0.425 ± 0.001	1.238 ± 0.048

a. Ethanol yield was calculated at the time of maximal ethanol concentration determined by high-performance liquid chromatography (HPLC).
b. Glucose volume consumption rate was calculated at the fermentation time of 9 h.

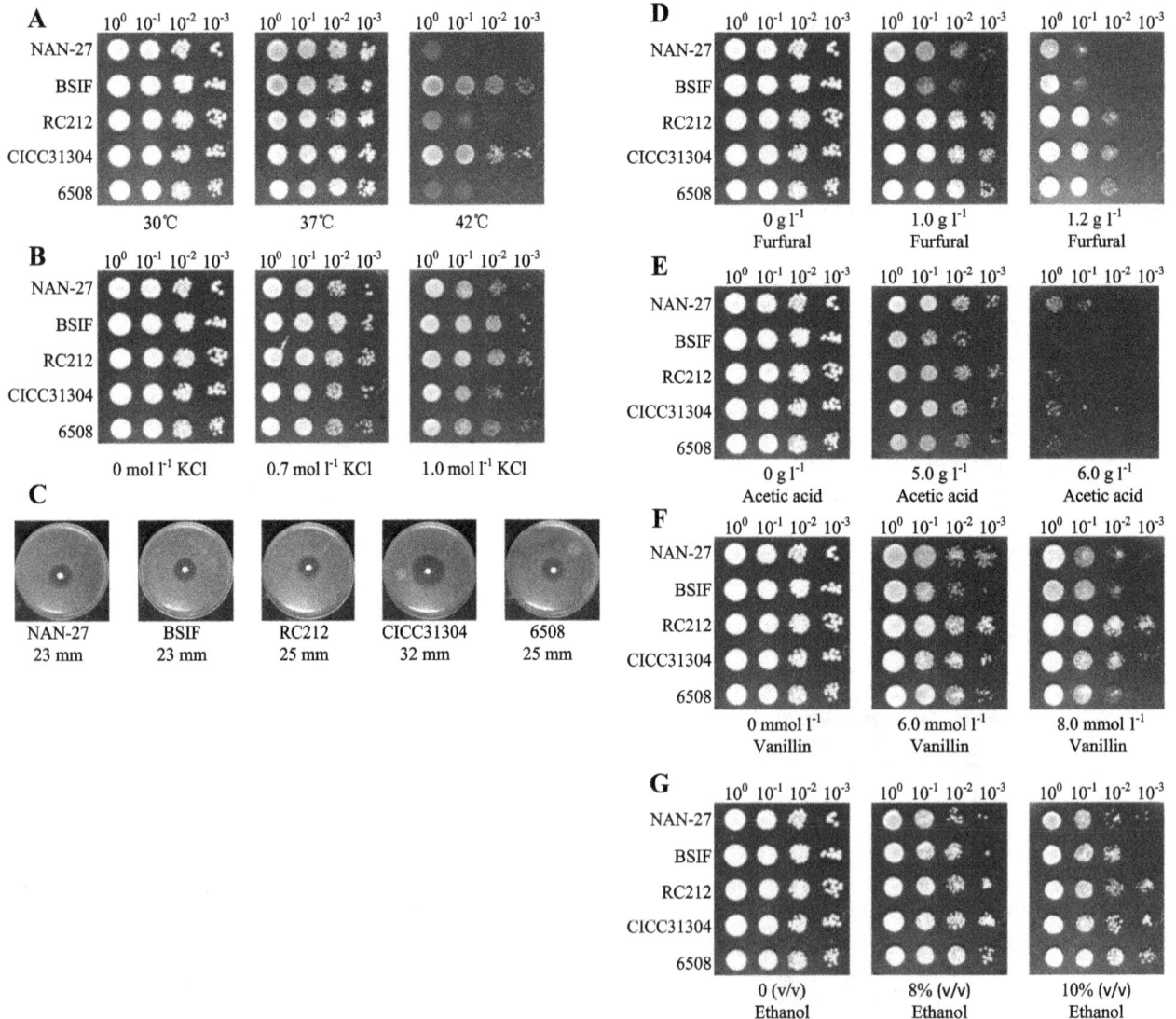

Fig. 2. Growth of *S. cerevisiae* strains on the plate at stress conditions. (A) Temperatures; (B) osmotic stress with high concentration of KCl; (C) oxidative stress with H_2O_2; (D) furfural; (E) acetic acid; (F) vanillin; (G) ethanol. Except for measuring the oxidative stress, 4 µl of each suspension from 10-fold serial dilution with initial OD_{600} of 1.0 was spotted onto the plates. For oxidative stress (C), approximately 100 µl OD_{600} 2.0 cells were mixed with 20 ml agar, and 6 µl 30% H_2O_2 was dropped in the centre of the 0.5 mm diameter sterile filter paper. All of the cell populations were incubated at 30°C for 2 days, but two more temperatures, 37°C and 42°C, were considered for thermotolerance evaluation.

0.39 g l^{-1} furfural and 0.29 g l^{-1} HMF in addition to sugars. The total phenolics and solubilized lignin in the hydrolysate were 15.5 mmol l^{-1} and 4 g l^{-1} respectively (Table 3). The *S. cerevisiae* strains were inoculated into this hydrolysate (the pH was adjusted to 6.0 with 5 mol l^{-1} NaOH) with an initial cell density of OD_{600} 2.5 (~ 0.5 g l^{-1} dry cell biomass). Shake flasks with rubber stoppers, and syringe needles were used for oxygen-limited cultivations. As expected, the growth and fermentation of the auxotrophic haploid laboratory strain CEN.PK 113-5D was strongly inhibited. All five of the wild-type strains exhibited good fermentation characteristics in the hydrolysate, with strains BSIF and NAN-27 showing the best characteristics (Fig. 3). These strains consumed almost all of the glucose

Table 3. Main components in corn stover hydrolysate.

Components		Concentration
Monosaccharides (g l^{-1})	Glucose	77.06 ± 0.63
	Xylose	29.32 ± 0.24
	Galactose	3.34 ± 0.05
	Arabinose	5.29 ± 0.02
	Mannose	0.79 ± 0.04
Weak acids (g l^{-1})	Formic acid	ND
	Acetic acid	3.72 ± 0.02
	Levulinic acid	ND
Furan aldehydes (g l^{-1})	Furfural	0.39 ± 0.01
	5-HMF	0.29 ± 0.01
Total phenolics (mmol l^{-1})		15.51 ± 0.86
Solubilized lignin (g l^{-1})		3.99 ± 0.07

ND, not detected.

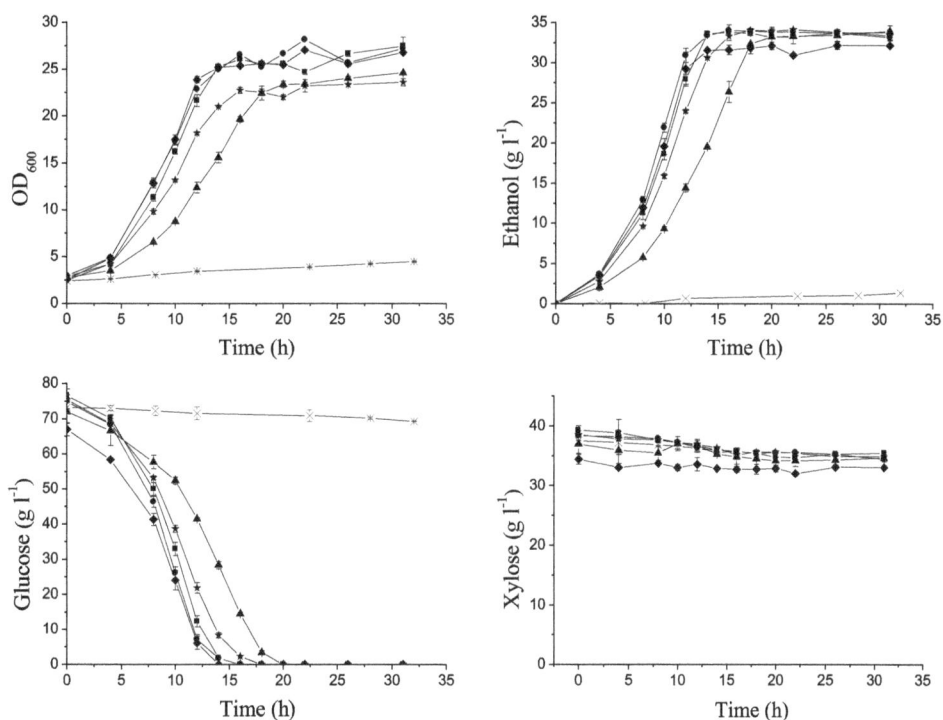

Fig. 3. Oxygen-limited fermentation in corn stover hydrolysate. The cells were cultured at 30°C in shake flasks with initial OD_{600} of 2.5 (~ 0.5 g l^{-1} dry cell biomass), and the pH was adjusted to 6.0 with 5 mol l^{-1} NaOH. Oxygen-limited conditions were obtained with a rubber stopper and a syringe needle, and the cells were agitated at 200 r.p.m. The experiments were performed in triplicate. Symbol: *S. cerevisiae* strain NAN-27, ■; BSIF, ◆; RC212, ▲; CICC31034, ●; 6508, ★; CEN.PK 113-5D, ×.

but no xylose because of the absence of enzymes converting the xylose to xylulose (Peng *et al.*, 2012).

Evaluation of inherent capacity for xylose utilization

It is generally considered that *S. cerevisiae* cannot metabolize xylose. However, some strains have weak background xylose utilization (Toivari *et al.*, 2004) or show different xylulose fermentation characteristics (Matsushika *et al.*, 2009). A stronger xylose or xylulose metabolic background is an advantage to constructing a glucose and xylose co-utilizing strain. Our strains could not utilize xylose, as mentioned above, and xylulose is expensive and difficult to prepare. Therefore, for selecting a strain with a better capacity for using xylose, a basic initial xylose metabolic pathway was introduced into the chromosomes of the candidate strains using the plasmid pYMIK-xy127 (Wang *et al.*, 2004), which contains the *Scheffersomyces (Pichia) stipitis* genes *XYL1* (encoding xylose reductase), *XYL2* (encoding xylitol dehydrogenase) and the *S. cerevisiae* gene *XKS1* (encoding xylulokinase). Then, their inherent capacity for xylulose utilization was evaluated using YEPX (yeast extract-peptone xylose) medium (10 g l^{-1} yeast extract, 20 g l^{-1} peptone and 20 g l^{-1} xylose). Among the recombinant strains, strain BSIF with the addition of the *XYL1-XYL2-XKS1* showed the highest maximum specific

growth rate (μ_{max}) (0.060 h^{-1}) on xylose and highest xylose volumetric consumption rate (0.201 g l^{-1} h^{-1}) (Fig. 4 and Table 4), suggesting a better inherent capacity for xylose utilization. Because no more modifications had been made on the recombinant strains, low ethanol yields were obtained because of the accumulation of xylitol. However, less xylitol was accumulated in BSIF-127 than in the other strains except 6508-127, which used only a small quantity of xylose. This result suggested that the BSIF strain has a more flexible cofactor pool than other strains.

Ploidy confirmation of *S. cerevisiae* strains

Keeping the characteristics obtained by molecular modification stable is one of the most important issues for industrial use. Moreover, the molecular modifications of wild-type industrial strains are generally performed on their chromosomal DNA. Therefore, it is necessary to determine the ploidy of the industrial strain. The strain ploidy was determined by the G0/G1 peak value on FL2-A versus that of the haploid strain CEN.PK 113-5D, which was set as 200. The results showed that all of the strains we evaluated were diploid or triploid. The five *S. cerevisiae* strains we selected were all diploid as shown by their G0/G1 peak values on FL2-A of nearly 400, which was about twice that of the control haploid strain (Fig. 5).

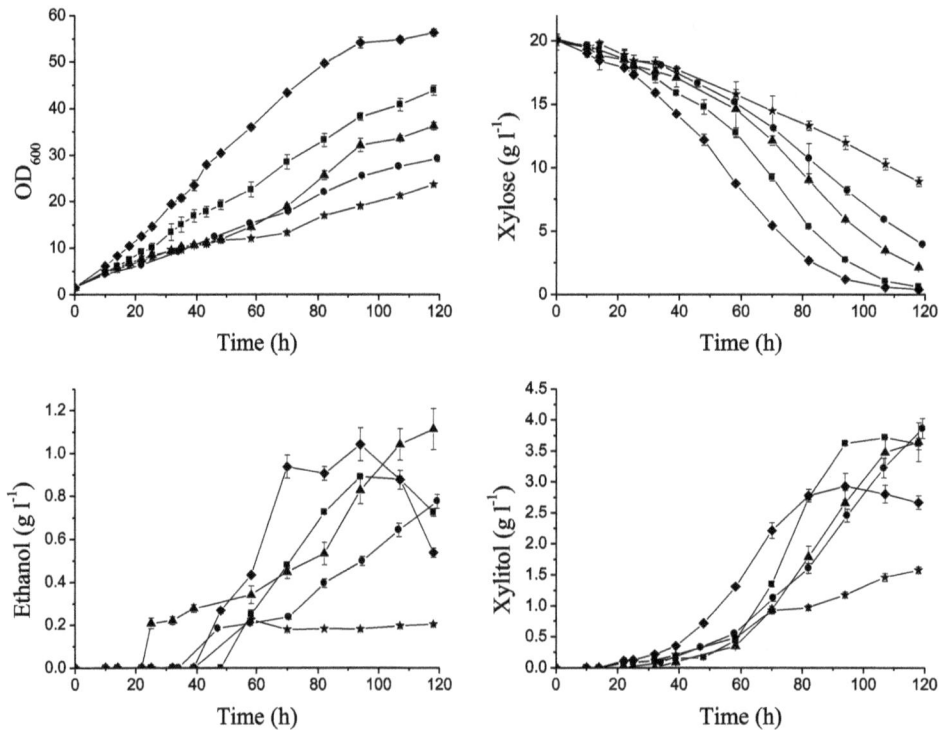

Fig. 4. Oxygen-limited fermentation on xylose in shake flasks. The cells were cultured at 30°C in shake flasks with initial OD_{600} of 1.5($\sim$ 0.3 g l^{-1} dry cell biomass). The oxygen-limited conditions were obtained with a rubber stopper and a syringe needle, and the cells were agitated at 200 r.p.m. The experiments were performed in triplicate. The plasmid pYMIK-xy127, containing genes of reductase, xylitol dehydrogenase and xylulokinase (Wang *et al.*, 2004), was integrated into the chromosome. Symbol: *S. cerevisiae* strain NAN-127, ■; BSIF-127, ◆; RC212-127, ▲; CICC31034-127, ●; 6508-127, ★.

The performances of the five strains are summarized in Table 5. Strain BSIF was isolated from tropical fruit in Thailand, RC 212 was selected from a wine production region, and NAN-27 (Wang *et al.*, 2004; Zhang *et al.*, 2010) and 6508 are currently being used in first-generation bioethanol production in China. Among the five strains, BSIF showed superior properties for most of the evaluated characteristics and was better than the first-generation bioethanol strains NAN-27 and 6508.

Second-generation bioethanol production faces more complex challenges than producing ethanol from starch.

Table 4. Metabolic characteristics of recombinant *S. cerevisiae* strains on xylose in shake flasks.

Strains	μ_{max} (h^{-1})	Ethanol yield (g g^{-1} consumed xylose)[a]	Xylose volume consumption rate (g l^{-1} h^{-1})[b]
NAN-127	0.055 ± 0.001	0.052 ± 0.004	0.183 ± 0.023
BSIF-127	0.060 ± 0.003	0.064 ± 0.009	0.201 ± 0.031
RC212-127	0.037 ± 0.002	0.063 ± 0.003	0.151 ± 0.018
CICC-127	0.030 ± 0.000	0.048 ± 0.008	0.124 ± 0.026
6508-127	0.033 ± 0.002	0.051 ± 0.010	0.087 ± 0.038

a. Ethanol yield was calculated at the time of maximal ethanol concentration determined by high-performance liquid chromatography (HPLC).
b. Xylose volume consumption rate was calculated at the fermentation time of 94 h.

To produce lignocellulose bioethanol in a cost-effective manner, the *S. cerevisiae* strain needs to utilize multiple types of monosaccharides, especially glucose and xylose, under various stresses arising from the inhibitors in the lignocellulosic substrate. However, the regulation mechanism of stress tolerance is very complex and not very well understood. Choosing a strain with naturally suitable properties as a chassis is more convenient than conferring these properties on it by engineering. In the present study, we give an example of a systematic evaluation to select a suitable chassis strain for lignocellulosic ethanol production. Because of its high ethanol yield from sugar, good growth capacity in hydrolysate, better potential xylose utilization capacity and superior tolerance to most of the stress factors, strain BSIF was selected, and more genetic modifications and evolutionary work will be implemented on it to produce cost-effective lignocellulosic bioethanol.

Experimental procedures

The preculture of S. cerevisiae strains

Thirty-two wild-type *S. cerevisiae* strains from different sources were used in the present study (Table 1). A single colony from an agar plate was cultured in YEPD medium (10 g l^{-1} yeast extract, 20 g l^{-1} peptone and 20 g l^{-1} glucose)

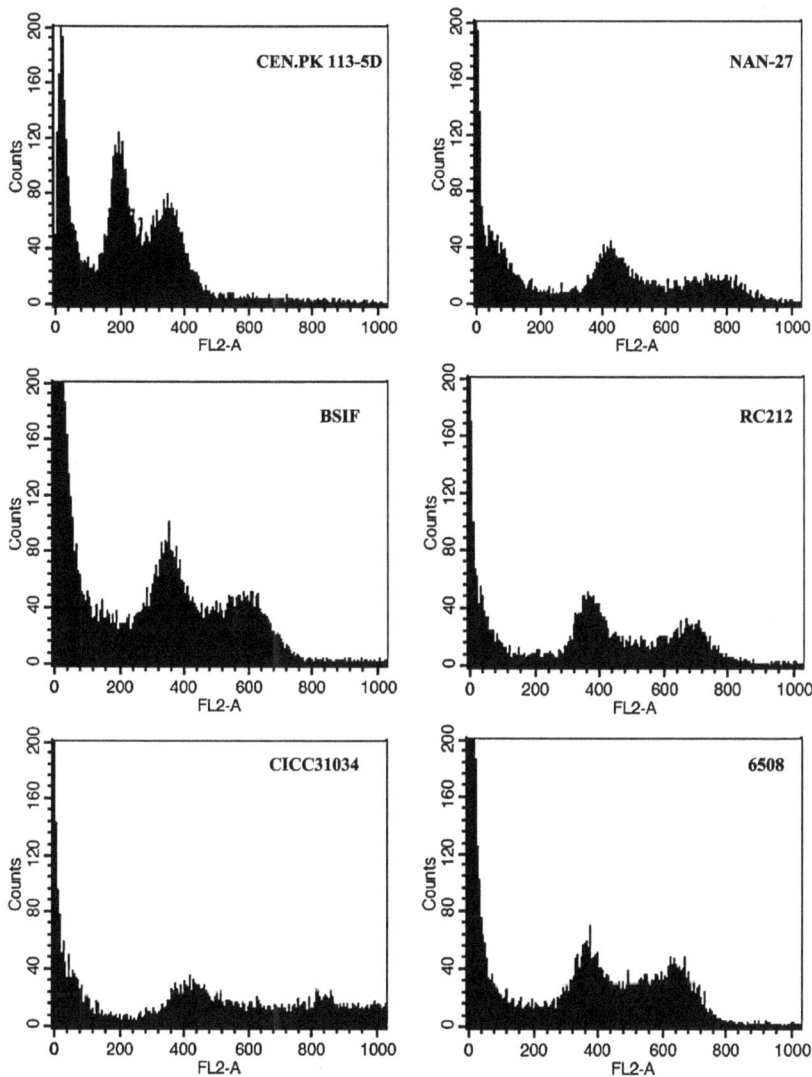

Fig. 5. Determination of DNA content by flow cytometry of *S. cerevisiae* strains. Strains were grown to exponential phase then fixed with ethanol, and the DNA was stained with propidium iodide. The x-coordinate FL2-A represents total cell fluorescence, and it is proportional to DNA content. Two Gaussian curves in density plot represent the G0/G1 and G2/M phase of cells respectively. The strain ploidy was then determined by the G0/G1 peak value on FL2-A versus that of the haploid strain, which was set as 200.

overnight at 30°C. The culture was then transferred into 250 ml shake flasks containing 100 ml YEPD medium at an initial OD_{600} of 0.2 and incubated for 12 h at 30°C. Then, the precultured cells were harvested and used as inoculants.

Identification of yeast strains

The genomic DNA of the yeast strains was extracted as described previously (Hong *et al.*, 2007a). The sequences of

Table 5. The summary of growth or fermentation performances and ploidy of the *S. cerevisiae* strains.[a]

Line		NAN-27	BSIF	RC212	CICC31304	6508
1	Oxygen limited fermentation on glucose	+	+++	+++	++	+
2	High temperature	+	+++	+	+++	+
3	Hyper-osmotic stress	+	+++	+++	++	++
4	Oxidative stress	+++	+++	++	+	++
5	High concentration of ethanol	++	++	+++	+++	+++
6	Furfural stress	+	+	++	++	++
7	Acetic acid stress	++	+	++	++	++
8	Vanillin stress	++	++	+++	++	+
9	Growth performance in hydrolysate	+++	+++	+	+++	++
10	Potential xylose transport and downstream metabolism capacity	++	+++	+	+	+
11	Ploidy	Diploid	Diploid	Diploid	Diploid	Diploid

a. The greater the number of '+' symbols, the better the growth or fermentation performance.

the rDNA D1/D2 domain were amplified and sequenced (Hu et al., 2012). The wild-type yeast strains were identified using the sequence analysis of the 26S rDNA D1/D2 domain.

Stress tolerance assays

The growth performance of selected *S. cerevisiae* strains under the typical stresses of furfural, acetic acid, vanillin and ethanol, as well as hyper-osmotic stress (Chen, 1989; Liu et al., 2006), oxidative stress (Stephen et al., 1995; Liu et al., 2006) and high temperature (Hong et al., 2007b; Abdel-Banat et al., 2010) were characterized through a spot dilution growth assay. Cells were harvested from an overnight culture in YEPD liquid medium and washed twice with sterile water. The density of re-suspended cells was normalized to an OD_{600} of 1.0. A 10-fold serial dilution of this suspension (10^0, 10^{-1}, 10^{-2} and 10^{-3}) was prepared, and 4 μl of each dilute suspension was spotted onto the appropriate solid medium (Liu et al., 2009).

To determine the oxidative stress resistance of each strain, *S. cerevisiae* cells were washed with 100 μl sterile water and re-suspended. The density was adjusted to OD_{600} 2.0, and the cells were mixed with 20 ml YEPD agar (cooled to approximately 50°C) and immediately poured into a plate. Then, a sterile filter paper (0.5 mm diameter) with 6 μl 30% H_2O_2 was placed in the centre of each plate. The oxidative stress resistance of each strain was demonstrated by the diameter of the zone of growth inhibition (mm) after cultivation for 2 days at 30°C. A smaller inhibition zone was interpreted as a higher resistance to oxidative stress (Liu et al., 2006).

The construction of xylose-fermenting recombinant strains

An integrated plasmid, pYMIK-xy127 (Wang et al., 2004), containing the *Scheffersomyces (Pichia) stipitis* genes *XYL1* (encoding xylose reductase) and *XYL2* (encoding xylitol dehydrogenase) and the *S. cerevisiae* gene *XKS1* (encoding xylulokinase) as described previously was linearized with *Hpa*I and then transformed into the *S. cerevisiae* strains using the lithium acetate method. Colonies of recombinant cells were selected on YEPD plates containing 0.4 g l⁻¹ of the antibiotic G418 (Wang et al., 2004).

Batch fermentation

All of the batch fermentations were performed in 50 ml of medium in 100 ml flasks with rubber stoppers (with a syringe needle to release CO_2 during fermentation) to maintain the oxygen-limited conditions, and the flasks were shaken at 200 r.p.m. To determine the fermentation performance using glucose, the cells were cultured in YEPD medium with an initial cell density of OD_{600} 0.1 (~ 0.02 g l⁻¹ dry cell biomass). To evaluate tolerance to hydrolysate, the cells were cultured in hydrolysate with 5 g l⁻¹ ammonium sulfate added, and the initial cell density was adjusted to OD_{600} 2.5 (~ 0.5 g l⁻¹ dry cell biomass). To evaluate the potential xylose utilization capacity, the cells were cultured in YEPX medium (10 g l⁻¹

yeast extract, 20 g l⁻¹ peptone and 20 g l⁻¹ xylose) with an initial OD_{600} 1.5 (~ 0.3 g l⁻¹ dry cell biomass).

Analytical methods

Biomass was measured by a turbidity determination at 600 nm. For the analysis of metabolites, culture samples taken at different time points were first centrifuged at 13 000 r.p.m. for 15 min, and then the supernatants were filtered through a 0.45 μm membrane. The concentrations of glucose, xylose, xylitol, glycerol, acetate and ethanol were determined using HPLC (Shimadzu, Japan) with a BIO-RAD Aminex HPX-87H ion exclusion column (300 × 7.8 mm) and a refractive index detector (RID-10A). The mobile phase was 5 mmol l⁻¹ H_2SO_4 with a flow rate of 0.6 ml min⁻¹. The temperature of the column oven was 45°C (Zhang et al., 2010). To analyse the monosaccharides in the hydrolysate, a BIO-RAD Aminex HPX-87P ion exclusion column (300 × 7.8 mm) was used at 80°C with a mobile phase of water at a flow rate of 0.6 ml min⁻¹ (Wang et al., 2013). The weak acids, furfural and HMF in the hydrolysate were measured using the HPX-87H column as mentioned above. Total phenolics were determined using the Folin phenol method, and vanillin was used to prepare the standard curve (Singleton et al., 1999). Solubilized lignin was determined by measuring the absorbance at 320 nm (Tan et al., 2013).

Ploidy determination of S. cerevisiae *strains*

The ploidy of the *S. cerevisiae* strains was determined by fluorescently labelling the cell nuclei and then analysing the fluorescence of each cell in the population by flow cytometry. The haploid strain CEN.PK 113-5D was used as a reference strain. Samples of *S. cerevisiae* cells (2×10^6 cells ml⁻¹) were fixed in 70% (v/v) ethanol (−20°C) and kept at −20°C for 2 h. Then, the cells were washed twice with PBS (137 mmol l⁻¹ NaCl, 2.7 mmol l⁻¹ KCl, 10 mmol l⁻¹ phosphate buffer, pH = 7.4) and re-suspended in 1 ml PBS containing 1 mg ml⁻¹ RNAase A (Sigma). After 2 h of incubation at 37°C, cells were washed with PBS and re-suspended in 200 μl of PBS containing 50 μg ml⁻¹ propidium iodide. Data were collected on a linear scale. Under these conditions, the fluorescence is proportional to the DNA content (Santos et al., 2013).

Conflict of interest

None declared.

References

Abdel-Banat, B., Hoshida, H., Ano, A., Nonklang, S., and Akada, R. (2010) High-temperature fermentation: how can processes for ethanol production at high temperatures become superior to the traditional process using mesophilic yeast? *Appl Microbiol Biotechnol* **85:** 861–867.

Almeida, J.R.M., Modig, T., Petersson, A., Hahn-Hägerdal, B., Lidén, G., and Gorwa-Grauslund, M.F. (2007) Increased tolerance and conversion of inhibitors in lignocellulosic hydrolysates by *Saccharomyces cerevisiae*. *J Chem Technol Biotechnol* **82:** 340–349.

Alriksson, B., Horváth, I.S., and Jönsson, L.J. (2010) Overexpression of *Saccharomyces cerevisiae* transcription factor and multidrug resistance genes conveys enhanced resistance to lignocellulose-derived fermentation inhibitors. *Process Biochem* **45**: 264–271.

Brandberg, T., FranzéN, C.J., and Gustafsson, L. (2004) The fermentation performance of nine strains of *Saccharomyces cerevisiae* in batch and fed-batch cultures in dilute-acid wood hydrolysate. *J Biosci Bioeng* **98**: 122–125.

Chen, C.S. (1989) Water activity – concentration models for solutions of sugars, salts and acids. *J Food Sci* **54**: 1318–1321.

Eliasson, A., Christensson, C., Wahlbom, C.F., and Hahn-Hägerdal, B. (2000) Anaerobic xylose fermentation by recombinant *Saccharomyces cerevisiae* carrying *XYL1*, *XYL2*, and *XKS1* in mineral medium chemostat cultures. *Appl Environ Microbiol* **66**: 3381–3386.

Hahn-Hägerdal, B., Karhumaa, K., Fonseca, C., Spencer-Martins, I., and Gorwa-Grauslund, M. (2007) Towards industrial pentose-fermenting yeast strains. *Appl Microbiol Biotechnol* **74**: 937–953.

Hong, J., Tamaki, H., and Kumagai, H. (2007a) Cloning and functional expression of thermostable β-glucosidase gene from *Thermoascus aurantiacus*. *Appl Microbiol Biotechnol* **73**: 1331–1339.

Hong, J., Wang, Y., Kumagai, H., and Tamaki, H. (2007b) Construction of thermotolerant yeast expressing thermostable cellulase genes. *J Biotechnol* **130**: 114–123.

Hu, N., Yuan, B., Sun, J., Wang, S.-A., and Li, F.-L. (2012) Thermotolerant *Kluyveromyces marxianus* and *Saccharomyces cerevisiae* strains representing potentials for bioethanol production from Jerusalem artichoke by consolidated bioprocessing. *Appl Microbiol Biotechnol* **95**: 1359–1368.

Jönsson, L., Alriksson, B., and Nilvebrant, N.-O. (2013) Bioconversion of lignocellulose: Inhibitors and detoxification. *Biotechnol Biofuels* **6**: 16.

Kim, S.R., Park, Y.-C., Jin, Y.-S., and Seo, J.-H. (2013) Strain engineering of *Saccharomyces cerevisiae* for enhanced xylose metabolism. *Biotechnol Adv* **31**: 851–861.

Kuyper, M., Harhangi, H.R., Stave, A.K., Winkler, A.A., Jetten, M.S.M., de Laat, W.T.A.M., *et al.* (2003) High-level functional expression of a fungal xylose isomerase: the key to efficient ethanolic fermentation of xylose by *Saccharomyces cerevisiae*? *FEMS Yeast Res* **4**: 69–78.

Liu, X., Yang, H., Zhang, X., Liu, L., Lei, M., Zhang, Z., and Bao, X. (2009) Bdf1p deletion affects mitochondrial function and causes apoptotic cell death under salt stress. *FEMS Yeast Res* **9**: 240–246.

Liu, X., Zhang, X., and Bao, X. (2006) Study on the stress resistance of *Saccharomyces cerevisiae* industrial strains. *China Brewing* **1**: 8–11.

Matsushika, A., Inoue, H., Watanabe, S., Kodaki, T., Makino, K., and Sawayama, S. (2009) Efficient bioethanol production by a recombinant flocculent *Saccharomyces cerevisiae* strain with a genome-integrated NADP⁺-dependent xylitol dehydrogenase gene. *Appl Environ Microbiol* **75**: 3818–3822.

Palmqvist, E., and Hahn-Hägerdal, B. (2000a) Fermentation of lignocellulosic hydrolysates. I: inhibition and detoxification. *Bioresour Technol* **74**: 17–24.

Palmqvist, E., and Hahn-Hägerdal, B. (2000b) Fermentation of lignocellulosic hydrolysates. II: inhibitors and mechanisms of inhibition. *Bioresour Technol* **74**: 25–33.

Palmqvist, E., Grage, H., Meinander, N.Q., and Hahn-Hägerdal, B. (1999) Main and interaction effects of acetic acid, furfural, and p-hydroxybenzoic acid on growth and ethanol productivity of yeasts. *Biotechnol Bioeng* **63**: 46–55.

Peng, B., Shen, Y., Li, X., Chen, X., Hou, J., and Bao, X. (2012) Improvement of xylose fermentation in respiratory-deficient xylose-fermenting *Saccharomyces cerevisiae*. *Metab Eng* **14**: 9–18.

Piškur, J., Rozpędowska, E., Polakova, S., Merico, A., and Compagno, C. (2006) How did *Saccharomyces* evolve to become a good brewer? *Trends Genet* **22**: 183–186.

Santos, A., Alonso, A., Belda, I., and Marquina, D. (2013) Cell cycle arrest and apoptosis, two alternative mechanisms for PMKT2 killer activity. *Fungal Genet Biol* **50**: 44–54.

Singleton, V.L., Orthofer, R., and Lamuela-Raventós, R.M. (1999) Analysis of total phenols and other oxidation substrates and antioxidants by means of folin-ciocalteu reagent. In *Methods in Enzymology* 299 (Oxidants and Antioxidants, Part A). Lester, P. (ed). San Diego, CA: Academic Press, pp. 152–178.

Sonderegger, M., Jeppsson, M., Larsson, C., Gorwa-Grauslund, M.-F., Boles, E., Olsson, L., *et al.* (2004) Fermentation performance of engineered and evolved xylose-fermenting *Saccharomyces cerevisiae* strains. *Biotechnol Bioeng* **87**: 90–98.

Stephen, D.W.S., Rivers, S.L., and Jamieson, D.J. (1995) The role of the *YAP1* and *YAP2* genes in the regulation of the adaptive oxidative stress responses of *Saccharomyces cerevisiae*. *Mol Microbiol* **16**: 415–423.

Sun, Y., and Cheng, J. (2002) Hydrolysis of lignocellulosic materials for ethanol production: a review. *Bioresour Technol* **83**: 1–11.

Tan, L., Yu, Y., Li, X., Zhao, J., Qu, Y., Choo, Y.M., and Loh, S.K. (2013) Pretreatment of empty fruit bunch from oil palm for fuel ethanol production and proposed biorefinery process. *Bioresour Technol* **135**: 275–282.

Toivari, M., Salusjarvi, L., Ruohonen, L., and Penttila, M. (2004) Endogenous xylose pathway in *Saccharomyces cerevisiae*. *Appl Environ Microbiol* **70**: 3681–3686.

Wang, C., Shen, Y., Zhang, Y., Suo, F., Hou, J., and Bao, X. (2013) Improvement of L-arabinose fermentation by modifying the metabolic pathway and transport in *Saccharomyces cerevisiae*. *Biomed Res Int* **2013**: 461204.

Wang, Y., Shi, W.L., Liu, X.Y., Shen, Y., Bao, X.M., Bai, F.W., and Qu, Y.B. (2004) Establishment of a xylose metabolic pathway in an industrial strain of *Saccharomyces cerevisiae*. *Biotechnol Lett* **26**: 885–890.

Zhang, X., Shen, Y., Shi, W., and Bao, X. (2010) Ethanolic cofermentation with glucose and xylose by the recombinant industrial strain *Saccharomyces cerevisiae* NAN-127 and the effect of furfural on xylitol production. *Bioresour Technol* **101**: 7104–7110.

Zhou, H., Cheng, J.-S., Wang, B.L., Fink, G.R., and Stephanopoulos, G. (2012) Xylose isomerase overexpression along with engineering of the pentose phosphate pathway and evolutionary engineering enable rapid xylose utilization and ethanol production by *Saccharomyces cerevisiae*. *Metab Eng* **14**: 611–622.

11

The application of nitric oxide to control biofouling of membrane bioreactors

Jinxue Luo,[1,2,3] Jinsong Zhang,[4] Robert J. Barnes,[3]
Xiaohui Tan,[3] Diane McDougald,[2,5,6] Anthony G.
Fane,[4] Guoqiang Zhuang,[1] Staffan Kjelleberg,[2,5,6]
Yehuda Cohen[2,6] and Scott A. Rice[2,5,6]*

[1]Research Center for Eco-Environmental Sciences,
Chinese Academy of Sciences, Beijing, China
[2]School of Biological Sciences, 60 Nanyang Drive,
SBS-01N-27 Singapore, 637551,
[3]Advanced Environmental Biotechnology Centre,
[4]Singapore Membrane Technology Centre, Nanyang
Environment and Water Research Institute and
[6]Singapore Centre on Environmental Life Sciences
Engineering, Nanyang Technological University,
Singapore
[5]Centre for Marine Bio-Innovation, School of
Biotechnology and Biomolecular Sciences, The
University of New South Wales, Sydney, NSW, Australia

Summary

A novel strategy to control membrane bioreactor
(MBR) biofouling using the nitric oxide (NO) donor
compound PROLI NONOate was examined. When
the biofilm was pre-established on membranes
at transmembrane pressure (TMP) of 88–90 kPa,
backwashing of the membrane module with 80 μM
PROLI NONOate for 45 min once daily for 37 days
reduced the fouling resistance (R_f) by 56%. Similarly,
a daily, 1 h exposure of the membrane to 80 μM PROLI
NONOate from the commencement of MBR operation
for 85 days resulted in reduction of the TMP and R_f by
32.3% and 28.2%. The microbial community in the
control MBR was observed to change from days 71 to
85, which correlates with the rapid TMP increase.
Interestingly, NO-treated biofilms at 85 days had a
higher similarity with the control biofilms at 71 days
relative to the control biofilms at 85 days, indicating
that the NO treatment delayed the development of
biofilm bacterial community. Despite this difference,
sequence analysis indicated that NO treatment did
not result in a significant shift in the dominant fouling
species. Confocal microscopy revealed that the
biomass of biopolymers and microorganisms in
biofilms were all reduced on the PROLI NONOate-
treated membranes, where there were reductions of
37.7% for proteins and 66.7% for microbial cells,
which correlates with the reduction in TMP. These
results suggest that NO treatment could be a promising strategy to control biofouling in MBRs.

*For correspondence. E-mail RSCOTT@ntu.edu

Funding Information This research was supported by a research
grant (MEWRC651/06/177) from the Environment and Water Industry
Programme Office of Singapore. The Singapore Membrane Technol-
ogy Centre, Nanyang Environment and Water Research Institute,
Nanyang Technological University is supported by the Economic
Development Board of Singapore. This work was also funded by NRF
and Ministry of Education Singapore under its Research Centre of
Excellence Programme.

Introduction

The membrane bioreactor (MBR) is a combined technology for the treatment of wastewater, which integrates the
biological degradation of organics by activated sludge and
separation of clean water from mixed liquor sludge suspension by membrane filtration into one system (Williams
and Pirbazari, 2007). The difference between MBR-based
water remediation and the traditional wastewater treatment process (WWTP) is in the separation of purified
water from the sludge biomass, which is accomplished by
using microfiltration or ultrafiltration membranes (MBR)
instead of a settling tank (traditional WWTP). The MBR is
advantageous because it reduces the treatment space,
the production of sludge and the process time of water
release, thus saving on capital expenditure and increasing the effluent quality (Van Nieuwenhuijzen et al., 2008).

However, biofouling remains a major limitation for the
MBR technology (Wang et al., 2008). This biological phenomenon results from the formation of microbial biofilms
on the membrane surface, which consists of cells bound
by a self-produced matrix of extracellular polymeric substances (EPS), leading to plugging of the membrane
pores (Meng et al., 2009). This process leads to a reduction of permeation flux when the MBR is operated under
constant pressure conditions (Lee et al., 2008), or a rise
of transmembrane pressure (TMP) when the system is
operated under constant flux (Le Clech et al., 2003).
Under both operational conditions, the fouling ultimately
results in reduced productivity, increased treatment costs
as well as a shorter lifespan of the membranes, limiting
the general adoption of MBRs as a preferred wastewater
treatment option (Meng et al., 2009).

A number of strategies have been tested to reduce biofouling in MBRs, such as modification of membrane surface properties (Tan and Obendorf, 2007), ultrasonic vibrations (Veerasamy et al., 2009) and chemical cleaning with acid or alkaline (Madaeni et al., 2009). Chemical cleaning is the most popular method in biofouling control, which can kill microbes and remove biofouling materials from the membranes (Lee et al., 2013). However, overuse of chemicals may damage the polymeric membrane structure (Puspitasari et al., 2010) and also pose an environmental and health risk (Estrela et al., 2002). Recently, a number of biological strategies have been regarded as promising methods to control the biofilm formation on membranes. For example, the quorum sensing (QS) inhibition agents, such as signal receptor antagonists and QS quenching enzymes, were reported to control biofilm formation in MBRs (Yeon et al., 2009a; Kim et al., 2012). Nitric oxide (NO) is an intracellular signalling molecule that has recently been shown to be involved in biofilm dispersal, inducing the transition from the biofilm mode of growth to the free swimming planktonic state (Barraud et al., 2006; 2014). NO induces biofilm dispersal by stimulating phosphodiesterase activity, resulting in the degradation of cyclic di-guanylate monophosphate (c-di-GMP), culminating in changes in gene expression that favour the planktonic mode of growth (Barraud et al., 2009b). The exogenous addition of NO, through the employment of low, sublethal concentrations of the NO donor compounds, such as 25–500 nM sodium nitroprusside and 5–80 µM PROLI NONOate, were shown to induce the dispersal of Pseudomonas aeruginosa biofilm and a range of other bacterial species and mixed species biofilms (Barraud et al., 2009b; Barnes et al., 2013). The difference in concentrations used in those studies primarily reflects the differences in NO donor chemistry and does not directly indicate actual NO concentrations achieved.

The aim of this research was to investigate the application of NO as a novel strategy to control membrane biofouling by complex microbial communities in an MBR system. The NO donor compound, PROLI NONOate, was applied to the membrane module at different biofouling stages (low and high TMP stages) to investigate its effect on the reduction of TMP and fouling resistance on the membrane.

Results

The effect of NO on established biofilms at high TMP stage

Data from previous experimental work showed that, during the biofouling process, the TMP increased in two distinct phases, where the TMP steadily increased from 3 to 15 kPa and a phase of rapid increase from 15 to 90 kPa. Membrane plugging caused by biomass from the activated sludge attaching to the hollow fibre (HF) membranes to form a biofilm has been proposed to be the primary reason for the rapid increase in the TMP (Chen et al., 2013). Since NO has been shown to reduce biofilms by inducing their dispersal in a non-toxic fashion, the effect of NO on the established MBR biofilms was studied by backwashing the NO donor compound PROLI NONOate into the membrane module in vitro once the MBR had reached the high TMP stage, 88–90 kPa. Two MBRs, operated in parallel, were used, where one was treated with the NO donor (MBR-1) and the second (MBR-2) served as the untreated control. During operation, the Mixed Liquor Suspended Solids (MLSS) and hydraulic retention time were maintained at approximately 2–4 g l^{-1} and 10 h respectively. The removal efficiency of the total organic carbon (TOC) was maintained at 95–98%.

The MBR system, operated under a constant flux of 15 l m^{-2} h^{-1}, reached the maximum TMP (88–90 kPa) after 116 days with a fouling resistance (R$_f$) of 3.37×10^{10} m^{-1} for MBR-1 and 3.24×10^{10} m^{-1} for MBR-2 (Fig. 1) at which time, backwashing with NO was tested as a means to remove the existing biofilm on the HF membranes. After 2 days of backwashing with distilled water (dH$_2$O), the R$_f$ was reduced to 2.55×10^{10} m^{-1} for membrane MBR-1 and 1.7×10^{10} m^{-1} for membrane MBR-2 (Fig. 1). R$_f$ is an indicator of the hydraulic resistance at the membrane and increases as a consequence of biofouling of the membrane (Martin et al., 2014). MBR-1, which had a slightly higher R$_f$, was selected for the NO treatment and MBR-2 was used as the control module. During the operation, the TOC removal efficiency

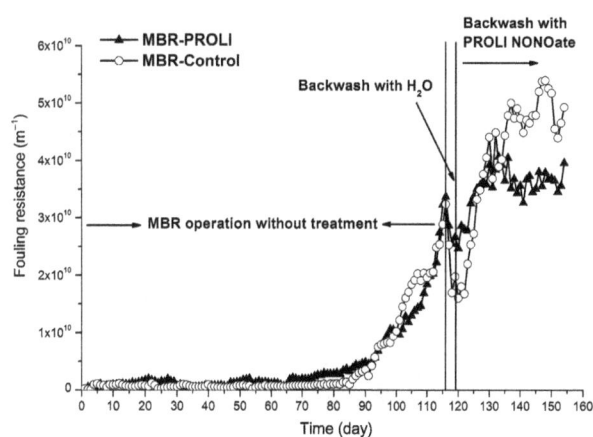

Fig. 1. The effect of PROLI-NONOate backwashing treatment on pre-established biofilm at high TMP. Fouling was quantified by determining the resistance of the membrane to the passage of water. The vertical lines separate the experiment into three phases: normal MBR operation, backwashing with dH$_2$O (distilled water) and backwashing with PROLI NONOate. The R$_f$ values were determined after daily backwashing.

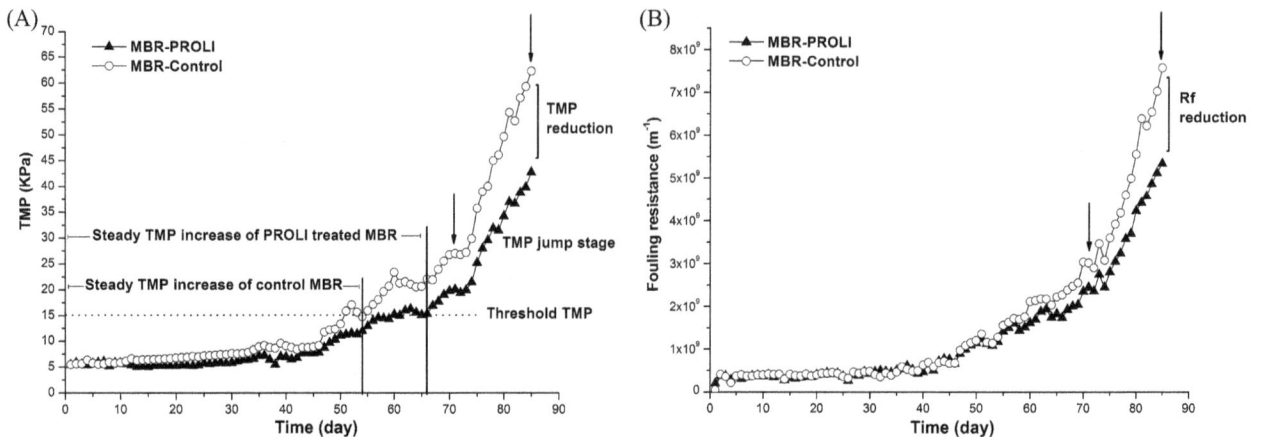

Fig. 2. The effect of PROLI-NONOate treatment on the formation of biofilms from the commencement of MBR operation. The horizontal dashed line indicates the threshold TMP between the low pressure TMP stage and the TMP jump stage. The vertical lines show when PROLI-treated MBR and control MBR reached the TMP jump stage. The downward pointing arrows indicate the times at which membrane and sludge samples were collected and analysed to determine TMP (A) and R_f (B). The R_f values were determined after the daily PROLI NONOate treatment.

in the MBRs was within the expected range of 95–98%. Since the two modules had reached the maximum TMP achievable for the system, the effect of NO addition was monitored as a change in the resistance.

After 9 days of treatment (PROLI NONOate backwashing 45 min per day), the resistance increased for both the PROLI NONOate (NO)-treated membrane module and control membrane module, with values of 3.58×10^{10} m^{-1} and 3.49×10^{10} m^{-1} respectively (Fig. 1). However, the R_f of the NO-treated module increased at a lower rate than the control module, 1.4-fold increase versus 2.1-fold. After 10 days of treatment, the R_f for the control module was higher than the R_f for the NO-treated module. When the backwashing experiments were terminated at 155 days (37 days of treatment), the R_f of the NO-treated membrane module had increased to 3.97×10^{10} m^{-1}, while the Rf for the control membrane module had increased to 4.93×10^{10} m^{-1} (Fig. 1). During the PROLI treatment, the TMP was not reduced and remained at the maximum level, 88–90 kPa (Fig. S1). However, compared with the first day of NO treatment, the R_f had increased by 1.42×10^{10} m^{-1} for the NO-treated membrane module and 3.23×10^{10} m^{-1} for the control membrane module, indicating that the R_f increase had been reduced by 56% due to the NO treatment. Therefore, the PROLI NONOate treatment could delay the increase of fouling resistance for a fouled MBR membrane.

The effect of continuous NO addition on biofouling

The effect of NO on biofouling when NO donor was added from the beginning of operation of the MBR was also tested. In these experiments, the treated membrane module was exposed to a solution of the NO donor daily for 1 h. The control module was similarly rinsed in the same solution without the NO donor present. In these experiments, the treated membrane module was rinsed in a solution of the NO donor daily for 1 h from the start of the MBR operation. The control module was similarly rinsed in the same solution without the NO donor present. Since the membranes had no biofilm on their surface at the start of the experiment, the change in performance was determined by quantifying changes in both TMP and R_f.

For the control membrane module, the system maintained operation in the low TMP stage (3–15 kPa) for 54 days before entering the TMP jump stage at which time the TMP increased rapidly. In comparison, the NO-treated membrane module was observed to remain in the low TMP stage for 66 days before entering the TMP jump phase (Fig. 2). The average rates of TMP increase were 0.24 kPa day^{-1} for the NO-treated module and 0.294 kPa day^{-1} for the control module in the steady TMP stage (3–15 kPa). When the experiment was terminated at 85 days, the TMP of the control membrane module had increased to 62 kPa while the NO-treated membrane module had only reached to 42 kPa (Fig. 2), representing a 32.3% reduction in TMP due to the NO treatment. The experiment was repeated and, due to time constraints, terminated at 34 days. In this experiment, the control TMP at the end of the experiment was 7 kPa and the PROLI-treated was 5.6 kPa, representing a 20% reduction in the TMP observed for the NO-treated module (Fig. S1). Thus, the NO treatment resulted in a slower TMP increase in MBR.

The R_f profiles for the NO-treated membrane module and control membrane module were similar during the first 46 days of operation. However, after 55 days, the R_f of the control module was higher than that of the NO-treated module (Fig. 2B). At 85 days, the R_f was 7.56×10^9 m^{-1} for the control module and 5.43×10^9 m^{-1} for the NO-treated module, representing a reduction of 28.2% in R_f. This was consistent with the effect of PROLI NONOate treatment on the TMP increase.

The effect of NO treatment on the biofilm bacterial community

To study changes in the microbial communities when NO was used to control fouling from the start of MBR operation, samples were collected at 71 days (20 kPa for NO treatment and 27 kPa for control) and 85 days (42 kPa for NO treatment and 62 kPa for control). The composition of the bacterial community was conserved in the treated and control biofilms on the 2 days tested (71 and 85 days), showing 80% Bray–Curtis similarity (Fig. 3 and Fig. S2). The dominant bacteria in the control and treated biofilms were similar, such as Rhodobacterales, Actinomycetales, Rhodospirillales, Rhizobiales and Sphingobacteriales (Fig. 3A). However, the bacterial community in the NO-treated biofilms at 85 days was closer to the control biofilm community at 71 days (85% similarity) relative to the control biofilm community at 85 days (80% similarity) (Fig. 3B).

There were some minor differences in the bacterial communities in the relative proportions of organisms between the control and treated samples. Five Orders of bacteria, including Thiotrichales, Gemmatimonadales, Xanthomonadales, Rhodocyclales and Myxococcales, had lower abundances in the NO-treated biofilms relative to the untreated control biofilms at both 71 and 85 days (Fig. 4). For example, at 85 days, the Thiotrichales accounted for 2.73% of the bacterial community on the control membrane compared with 0.16% on the NO-treated membranes. This trend was also observed for Gemmatimonadales and Xanthomonadales, which had abundances of 1.12% and 1.16% respectively in control biofilms at 85 days, in comparison to abundances of 0.49% and 0.71% respectively in NO-treated biofilms. These results indicate that these biofilm bacteria may be highly susceptible to dispersal and subsequent removal in the presence of NO.

Additionally, some bacteria, such as Rhizobiales and Actinomycetales, were found to have slightly higher abundances in the NO-treated biofilms in comparison to the control biofilms at both 71 and 85 days (Fig. 4). The Rhizobiales accounted for 13.77% and 15.51% respectively in the NO-treated biofilm community at 71 and 85 days, while their abundances were 11.26% and 11.08% in the control biofilms. Similarly, the abundance of Actinomycetales was also higher in the NO-treated biofilms (25.91% at 71 days and 22.64% at 85 days)

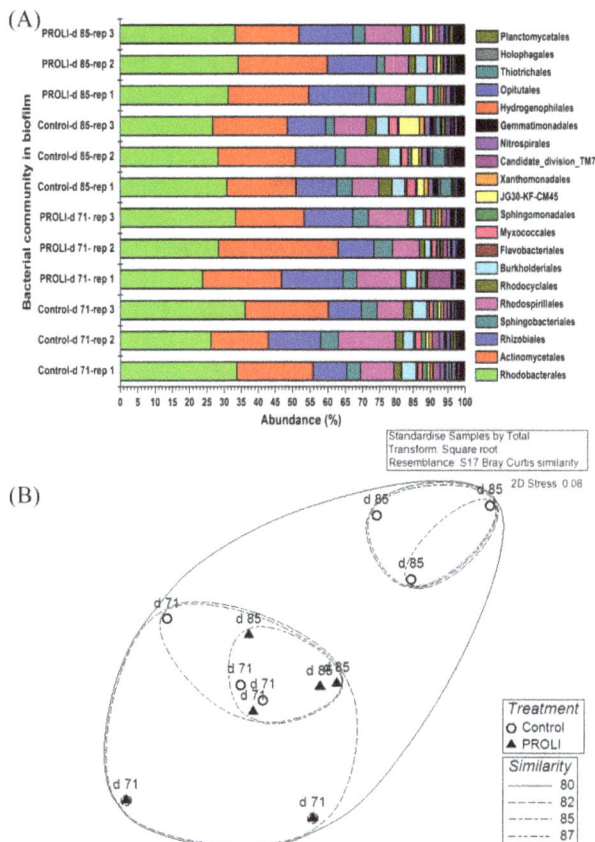

Fig. 3. Bacterial communities on control and PROLI-NONOate-treated membranes. (A) The composition of bacterial communities at 71 days and 85 days. (B) The nonmetric multidimensional scaling (NMDS) plot for biofilm bacterial communities based on the Bray–Curtis similarity. 'rep' indicates the replicate sample. 'PROLI' represents the PROLI NONOate treatment.

than in the control biofilms (20.96% at 71 days and 21.61% at 85 days). This indicated the Rhizobiales and Actinomycetales may not be dispersed by NO as effectively as other bacteria, or that these bacterial may be able to utilize NO for biofilm formation, resulting in relative higher abundances for them in the NO-treated biofilm community.

The effect of NO treatment on the biofilm fungal community

The biofilm fungal community was also compared for the NO-treated and control biofilms at 71 and 85 days (Fig. 5). The dominant fungal groups were conserved in the two types of biofilms and included Eurotiomycetes, Saccharomycetes, Sordariomycetes and unclassified Ascomycota. However, distinct from the bacterial community, the fungal communities distributed unevenly in the replicated biofilm samples, resulting in the lower similarity (32.73% at 71 days and 60.9% at 85 days) between the NO-treated and control biofilms (Fig. 6). Examining the average abundance, the Saccharo-

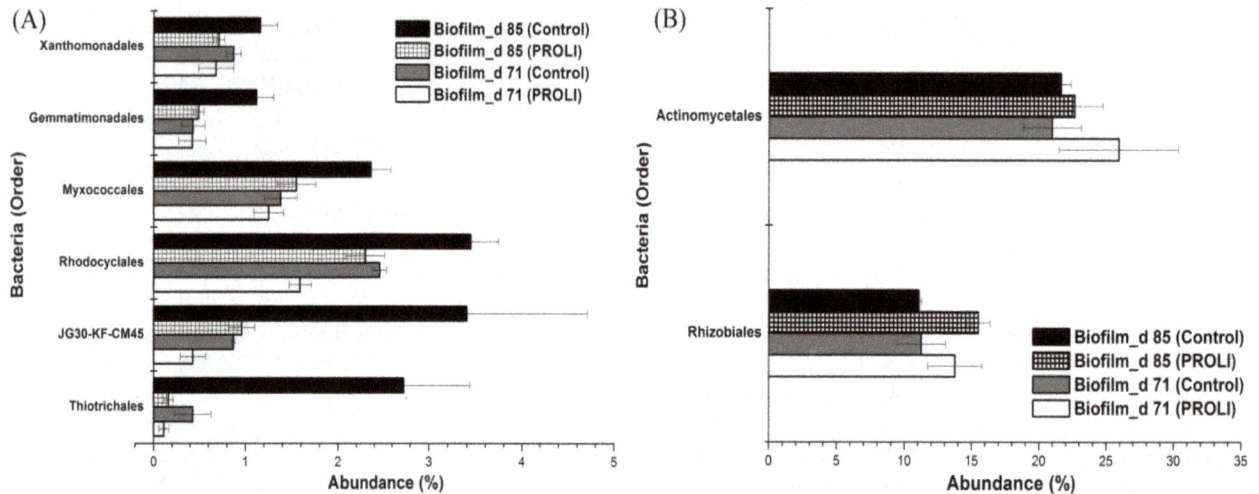

Fig. 4. The change in MBR biofilm communities due to PROLI NONOate treatment. (A) Bacteria that have a lower abundance in the PROLI NONOate-treated biofilms relative to the control biofilms. (B) Bacteria have a higher abundance in the PROLI NONOate-treated biofilms relative to the control biofilms. The abundances were the average values of triplicate samples for the biofilms. The error bars are the standard errors of the mean ($n = 3$ replicate membrane samples).

mycetes and Sordariomycetes became less abundant in the NO-treated biofilms while the Eurotiomycetes and unclassified Ascomycota increased the abundance in the NO-treated biofilms (Fig. 7). Specifically, the Saccharomycetes accounted for 37.7% and 85.4% of the fungal community in control biofilms at 71 days and 85 days respectively but decreased in abundance to 3.6% (71 days) and 57.3% (85 days) in the NO-treated biofilm fungal communities. The Sordariomycetes had the abundance of 31.6% (71 days) and 1.3% (85 days) in the control biofilms. However, this fungal class was not

detected in the NO-treated biofilm at 71 days and had a quite low abundance (0.02%) in the NO-treated biofilms at 85 days. Given that there was a significant reduction in abundance of the Sordariomycetes for the untreated biofilm from 71 days to 85 days, it is difficult to be certain that the reduced percent composition for the NO was a consequence of the NO treatment or whether it was due to other factors. For the other two Classes of fungi, the Eurotiomycetes accounted for 23.92% and 9.5% of the fungal community in the control biofilms at 71 and 85 days respectively, compared with 81% and 31.98% in the NO-treated biofilms. One unclassified class of Ascomycota was also present in the biofilm at a higher abundance in the NO-treated biofilms (11.6% at 71 days, 5.11% at 85 days) relative to the control biofilms (5.21 % at 71 days, 2.41 % at 85 days).

NO treatment results in a reduction of biofilm constituents

In order to determine the effect of NO treatment on the biomass of biofilm, the different biofilm components were quantified by the confocal laser scanning microscopy (CLSM) analysis at the 2 days tested. Four components, including proteins, α-polysaccharides, β-polysaccharides and total cells, were studied. The biovolume of each biofilm component in the NO-treated module was observed to be lower than the corresponding components in the control MBR (Fig. S3). For example, at 71 days, the biovolume of the proteins was 2.1 $\mu m^3 \mu m^{-2}$ on the NO-treated membrane and 3.4 $\mu m^3 \mu m^{-2}$ on the control membrane (Fig. 8), representing a 38.2% reduction as a consequence of NO treatment. Similarly, the biovolumes

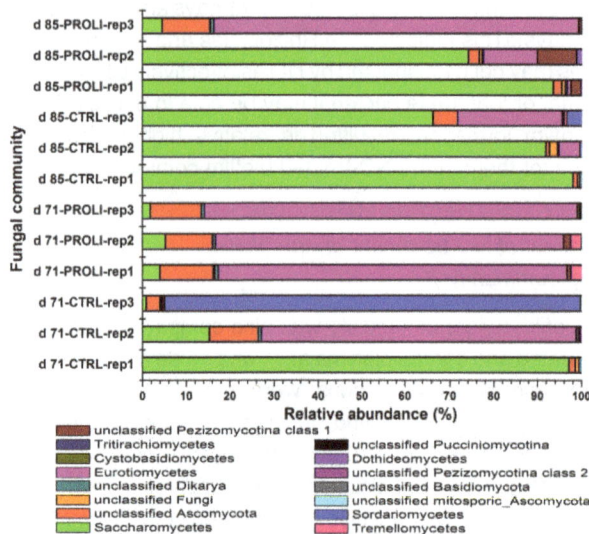

Fig. 5. The fungal community compositions for PROLI NONOate-treated and control biofilms after 71 and 85 days of MBR operation. 'CTRL' and 'PROLI' represent the control and PROLI NONOate-treated biofilms. 'rep' indicates the replicate sample.

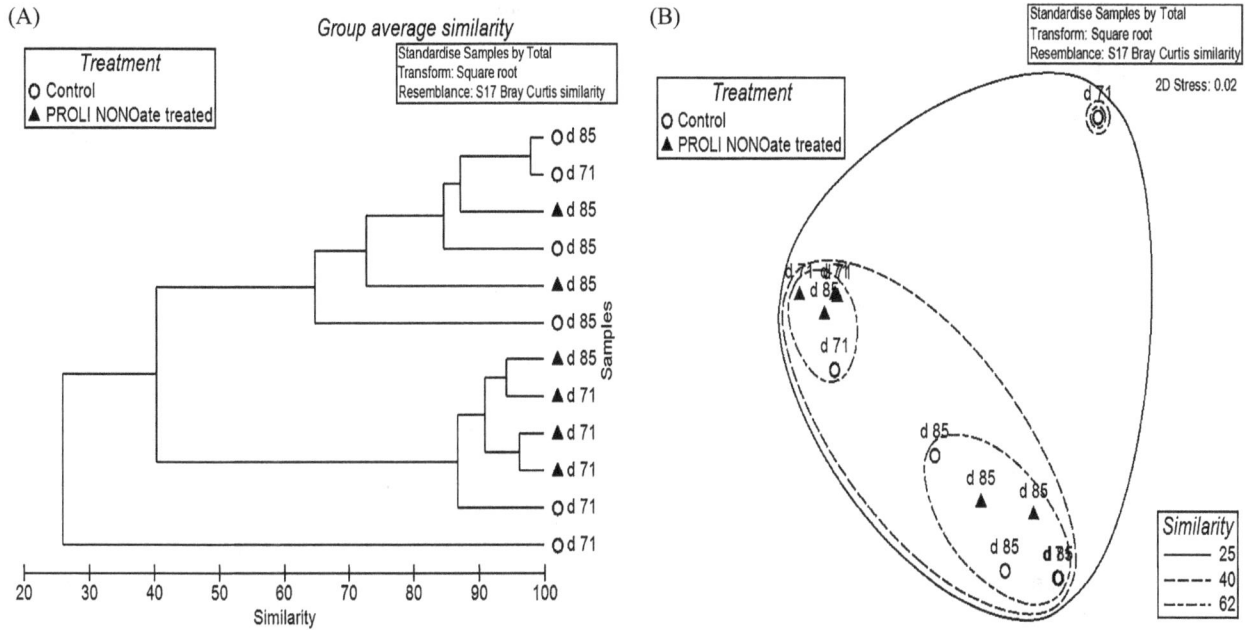

Fig. 6. Comparison of fungal communities for the PROLI NONOate-treated biofilms and control biofilms. (A) The clustering tree for the fungal communities. (B) The nonmetric multidimensional scaling (NMDS) plot for the fungal communities.

for the α-polysaccharides, β-polysaccharides and microbial cells were reduced by 68.1%, 55% and 50% respectively on the NO-treated membranes at 71 days. At 85 days, although the biovolumes of each biofilm component had increased from 71 days in both control and NO-treated modules, they were reduced on the NO-treated membranes. For example, the α-polysaccharide bio-

volume was 1.8 μm³ μm⁻² for the control biofilms compared with 1.5 μm³ μm⁻² in the NO-treated biofilms. The greatest differences observed at 85 days were for the protein and microbial cells, which were 5.3 μm³ μm⁻² and 1.8 μm³ μm⁻² respectively on control membranes compared with 3.3 and 0.7 μm³ μm⁻² on the PROLI NONOate-treated membranes (Fig. 8). Thus, there was a 37.7%

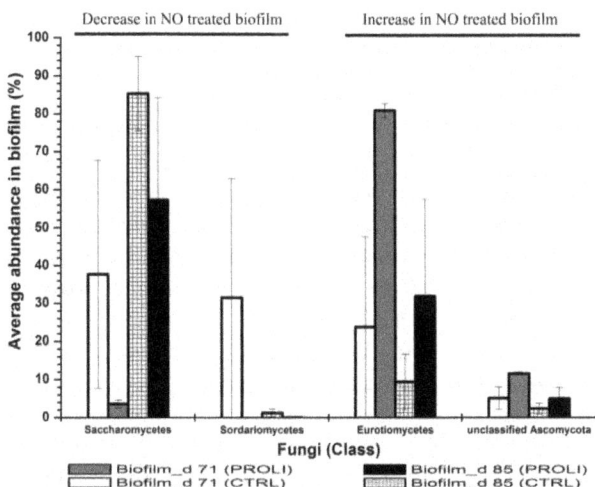

Fig. 7. The variation of fungi in MBR biofilms due to PROLI NONOate treatment. The left two fungi were decreased in abundance in the PROLI NONOate-treated biofilms. The right two fungi were increased in abundance in the PROLI NONOate-treated biofilms. The abundances are the average values of triplicate samples for the biofilms. The error bars are the standard errors of the mean (n = 3 replicate membrane samples).

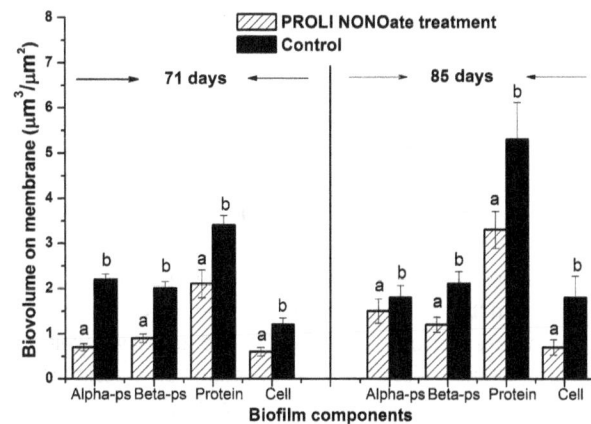

Fig. 8. The biovolumes of biofilm components on membranes as visualized by confocal microscopy for the untreated control and PROLI NONOate-treated MBRs. 'Alpha-ps' and 'Beta-ps' represent the Alpha-polysaccharides and Beta-polysaccharides respectively. The error bars are the standard error of the mean (n = 30). For each PROLI NONOate-treated and control biofilm component at the same day, the bars at different letters (A, B) are significantly different at P < 0.05 (t-test).

reduction in protein and a 66.7% reduction in microorganisms associated with the NO treatment on day 85. Moreover, the amount of DNA extracted on 71 days was 2.7 μg cm^{-2} for the control biofilms and 2.3 μg cm^{-2} for the NO-treated biofilms. At 85 days, 4.2 μg cm^{-2} of DNA was extracted from the control membrane while 3.98 μg cm^{-2} of DNA was extracted from the NO-treated membrane (data not shown). This also suggested that the number of microorganisms on membranes was reduced by NO treatment and was consistent with the image-based assessment of the total biovolume of microorganisms.

Discussion

NO treatment delayed the increase of TMP and fouling resistance

Membrane cleaning with chemical agents, such as sodium hypochlorite, is the most commonly used method to control biofouling (Lee et al., 2013). However, the application of sodium hypochlorite may adversely affect the structure, and hence function, of polymeric membranes (Puspitasari et al., 2010) and the biological function of activated sludge (Lee et al., 2013). It has been demonstrated that sublethal concentrations of NO can induce biofilm dispersal in a range of bacteria (Barraud et al., 2006; Hetrick et al., 2009; Barnes et al., 2013). In this study, NO treatment, utilizing the NO donor PROLI NONOate, was shown to reduce the growth of biofilms formed by mixed species microbial communities in a MBR system, decreasing membrane fouling. After backwashing, established biofilms treated with NO for 37 days showed a 56% reduction of R_f relative to the control membrane module. This indicated that backwashing with NO could partially reduce the growth of established biofilms in MBR. Similarly, exposing the membrane to NO from the start of the experiment resulted in a 32.3% reduction of TMP and a 28.2% reduction of R_f, indicating that the PROLI NONOate treatment could also delay the construction of the new biofilms on membrane. When repeated, the TMP increase was again delayed by 20% after NO treatment for 34 days, showing that effect of NO on TMP and biofouling was reproducible. The delay in TMP increase was supported by the image-based data, which showed that NO treatment resulted in a reduction of biofilm biomass, for both cells (66.7% reduction) as well as macromolecules (e.g. 37.7% reduction for proteins). This finding was consistent with previous results, where a 30% reduction of total microbial biovolume was observed when the biofilms were treated with 80 μM PROLI NONOate for 1 h (Barnes et al., 2013). In combination, these results demonstrate that the proof-of-principle that NO can be used reduce the growth of biofilms in a MBR, delaying the TMP jump and hence increasing the operational time of the MBR.

In previous studies, biological methods based on the molecular signalling have already been demonstrated to control biofouling in MBR systems. For example, it was reported that the Porcine kidney acylase I, which is a quenching enzyme of N-acyl homoserine lactone autoinducers, could prevent biofouling (approximately 26–46% reduction in TMP) in MBR (Yeon et al., 2009a,b; Kim et al., 2013). In this study, NO released from the PROLI NONOate also acts as a biofilm dispersal signal molecule and involves c-di-GMP-based signal transduction (Barraud et al., 2009a). Thus, it is concluded that the signalling-based biological strategies, e.g. NO or anti-QS approaches, may be promising in fouling control in MBRs. However, the NO-based biofouling prevention strategy was not as efficient as the commonly used chemical cleaning strategy, which was reported that treatment with the combination of 3000 mg l^{-1} NaClO and 500 mg l^{-1} NaOH reduced 81.8% of TMP (Li et al., 2011) and treatment with 1000 mg l^{-1} citric acid recovered 90% of the membrane permeability in MBR (Yan et al., 2012). Thus, further experiments are required to improve the biofouling prevention efficiency of NO-based biological strategy, such as optimizing the dosing concentration and regimen of NO-releasing donor or choosing a more stable NO-releasing chemical. Additionally, we have shown that NO can increase the sensitivity of the biofilm to standard cleaning or killing agents, such as hydrogen peroxide or surfactants (Barraud et al., 2006). Therefore, one strategy that might be successful is to induce biofilm dispersal using NO and to include a standard cleaning agent, such as chlorine or caustic treatment to better remove the biofilm.

Effect of NO treatment on the microbial community in biofilm

While the NO treatment was able to delay the increase in biofilm biomass and the associated TMP rise, it was not able to prevent it completely. Characterization of the biofilm communities indicated that the bacterial and fungal communities in the treated and control biofilms were of high similarity, where the dominant bacteria were Rhodobacterales, Actinomycetales, Rhodospirillales, Rhizobiales and Sphingobacteriales and the dominant fungi were Eurotiomycetes, Saccharomycetes and Sordariomycetes. Based on the previous microbial community analysis (data not shown), bacteria including Actinomycetales, Rhodospirillales, Rhizobiales and Sphingobacteriales were the dominant biofilm forming microorganisms at high TMP. In previous studies, the Actinomycetales strains were also isolated from cave biofilms and biofilms on MBR membranes (Kim et al., 2005; Jurado et al., 2009). The Rhodospirillales and Rhizobiales have been reported in the biofilms of urban drinking water distribution system (Liu *et al.*, 2012). The Saccharomycetes, which were the significant sludge community compositions in wastewater treatment systems (Yang *et al.*, 2011), have also been proposed to be important in the biofouling process of MBRs and groundwater treatment facilities (Gillings *et al.*, 2006; Xia *et al.*, 2010). Thus, the treatment with 80 μM PROLI NONOate in this study did not significantly alter the microbial biofilm community composition. However, comparison of the community similarity suggested that the biofilm bacterial community observed for the NO-treated biofilms at 85 days was slightly more similar to the control biofilms at 71 days rather than the control biofilms at 85 days, indicating the development of bacterial community may be delayed by the NO treatment. While it is tempting to speculate that the NO treatment delayed the establishment of a community associated with the maximum TMP, further work is needed to more accurately track the change in community composition across the TMP curve to address this. It should be noted that these data represent relative abundances and not absolute numbers of community members. Thus, it is not possible to determine if the changes in community abundance represent an increase in a specific Operational Taxonomic Unit (OTU) or the lack of dispersal for that organism relative to the other community members.

While the overall biofilm community composition was not strikingly different in the control and treated MBRs, there were still some subtle differences in both the bacterial and fungal communities after NO treatment. The bacterial orders of Thiotrichales, Gemmatimonadales, Xanthomonadales, Rhodocyclales and Myxococcales and fungal classes of Saccharomycetes and Sordariomycetes were reduced in abundance in the PROLI NONOate-treated biofilms. In previous works, the Gemmatimonadales were reported to preferentially form the base structure of biofilm (Besemer *et al.*, 2007). The Xanthomonadales and Rhodocyclales had been identified as biofilm constructors on polyvinyl chloride membranes in MBR (Xia *et al.*, 2010). The Saccharomycetes and Sordariomycetes were primary fungal compositions of biofilm in kitchen and bathroom sinks (Adams *et al.*, 2013). Interestingly, these organisms were reduced in abundance in the NO-treated biofilms. This may suggest that these organisms are particularly sensitive to NO-mediated dispersal and their removal from the biofilm may be important in the observed delay in TMP increase. Therefore, it will be of interest to quantify those specific organisms in future experiments by fluorescence *in situ* hybridization or quantitative PCR to determine their overall relationship to membrane fouling and the TMP rise.

Some bacteria showed an increased abundance in the NO-treated biofilm, such as Rhizobiales and Actinomycetales. Although the NO has been reported to induce dispersal of bacteria from both single and mixed species biofilms (Barraud *et al.*, 2006; 2009b; Barnes *et al.*, 2013), it has also been shown that some bacteria do not disperse in the presence of NO (Barnes *et al.*, 2013) and NO can actually stimulate the biofilm formation of some bacteria, e.g. *Azospirillum brasilense* (Arruebarrena Di Palma *et al.*, 2013) and *Shewanella oneidensis* (Plate and Marletta, 2012). This may explain why these bacteria had a relative higher abundance in PROLI NONOate-treated biofilm. Further investigation is required to determine the specific mechanism for them to have increased abundance in NO-treated biofilm.

In conclusion, the NO donor compound PROLI NONOate showed the potential to control membrane biofouling in MBRs through reducing the production of macromolecules in EPS, delaying the succession of the microbial community and selectively dispersing some microbial groups. While biofouling was not completely prevented, these results are significant as the microbial community of the MBR is highly diverse, comprised here of approximately 103 orders of bacteria and 17 classes of fungi. Further work aimed at optimizing the delivery of NO may improve the overall efficacy of NO in biofouling control. Additionally, alternative NO donor compounds, which exhibit different NO release kinetics, may be better suited for the high organic content environment of the MBR. It has been shown previously that biofilms exposed to NO donor compounds were more susceptible to antimicrobial agents and removal from surfaces with surfactants (Barraud *et al*, 2006). Therefore, NO treatment may be best complemented with traditional biofilm cleaning protocols, e.g. bleaching with low concentration of chlorine, to synergistically remove the biofilms and hence, alleviate the fouling problem.

Experimental procedures

MBR set-up

A laboratory scale submerged MBR was operated to treat artificial synthetic wastewater, composed of 320 mg l^{-1} glucose, 60 mg l^{-1} beef extract, 80 mg l^{-1} peptone, 7 mg l^{-1} KH_2PO_4, 14 mg l^{-1} $MgSO_4 \cdot 7H_2O$, 7.3 mg l^{-1} $FeSO_4 \cdot 7H_2O$ and 90 mg l^{-1} sodium acetate. The TOC was 200 mg l^{-1}. The MBR system was composed of an anoxic sludge tank and an aerobic sludge tank containing two separate membrane modules (Fig. S4). The membrane modules were made with HF Polyvinylidene fluoride (PVDF) membranes (ZeeWeed, GE) and were assembled as a 'curtain' style module. The area of membrane was 565 cm^2 for each membrane modules. For the membrane pieces, one end (the free end) was sealed and hung down into the sludge tank. The other ends of the HF membranes were open and sealed into a chamber that was linked to the suction pump.

Before the experiment, fresh activated sludge was collected from the Ulu Pandan wastewater treatment plant in

Singapore and acclimated in artificial synthetic wastewater for 60 days before the start of the experiment. During MBR operation, the synthetic wastewater from the feedwater tank passed through the anoxic tank and aerobic tank and was degraded by the sludge biomass. The wastewater was then recycled from the aerobic tank to the anoxic tank (1.2 l h^{-1}) to be further degraded. The purified water was subsequently separated from the sludge by the membrane module. The two modules were run with a constant flux of 15 l m^{-2} h^{-1}. The aeration speed in the aerobic tank was 5 l min^{-1}. The MLSS were maintained at 2–4 g l^{-1}. The hydraulic and sludge retention times for the MBR were maintained at approximately 10 h and 25 days respectively. The MBR was run at a room temperature (25–26°C). The parameters, such as membrane flux, TMP, pH, dissolved oxygen and temperature, were monitored and automatically recorded using a data logger and computer. The TOC of the influent and permeate was measured using a multi N/C® 2100s (AnalytikJena).

Application of the NO donor PROLI NONOate to membrane

The effect of NO on membrane biofouling was tested using two distinct approaches. The first approach was to immerse the membrane module into the solution of NO donor compound PROLI NONOate (Cayman Chemicals) from the beginning of the experiment. Another approach was to backwash the membrane module with PROLI NONOate solution after the TMP had increased to 88–90 kPa. All treatments were operated in separated tanks outside of the MBR sludge tank to ensure there was no effect of PROLI NONOate on the sludge microbial community.

For the continuous addition of the PROLI NONOate from the beginning of the MBR operation, the treatment was performed on the membrane modules every 24 h. The NO-treated module and control module were taken out from the sludge tank and treated in two separate beakers. For the NO-treated membrane module, the membrane was immersed in 80 µM PROLI NONOate in diluted synthetic feed water (TOC of 10 mg l^{-1}, pH 8.5–9.0) for 1 h. Dosing concentrations were based on previously published work that used MBR isolates to determine effective concentrations (Barnes et al., 2013). For the control module, the membrane was immersed under the same conditions but without PROLI NONOate. After treatment, the modules were returned to the MBR and operation was continued. The TMP and flux (Jw) were measured immediately in dH$_2$O at room temperature before and after the NO treatment. The hydraulic resistance through the membrane during filtration was calculated using the TMP and flux in pure water based on the formula $R_f = (\Delta P - \Delta \pi)/\mu\ Jw - R_m$ (Chen et al., 2012). The pressure differential (ΔP) and Jw are the measured TMP and flux. The osmotic pressure differential ($\Delta \pi$) was 0 for pure water and μ is the viscosity of water at 25–26°C. R_m is the resistance for the clean membrane module, which was 1.5×10^8 m^{-1} for the PROLI-treated membrane module and 1.1×10^8 m^{-1} for control membrane module respectively.

To test the ability of the NO donor to remove an established fouling layer, the MBR system was operated until both membrane modules had reached the maximum TMP (85–90 kPa) after which time the PROLI NONOate was added. For treat-

ment, the membrane modules were removed from the aerobic sludge tank and immersed into different beakers, which were filled with 500 ml diluted synthetic feed water (TOC of 10 mg l^{-1}, pH 7). For the NO-treated membrane module (45 min exposure to NO per day), one backwashing pump was used to deliver 80 µM PROLI NONOate stock solution, which was dissolved in a 10 mM NaOH solution to prevent the spontaneous release of NO before it reached the membrane. Another backwashing pump, referred to here as the neutralizing pump, was used to deliver a 2.5 mM HCl solution. The ratio of flow rate for the PROLI NONOate backwashing pump and HCl pump was 1:4. The two solutions were mixed directly before they entered the membrane module to adjust the pH to be approximately 7 ($\pm$ 1) and to hence ensure there was no effect of the NaOH or HCl on the biofouling of the membrane. These flow rates and concentrations were verified to produce a neutral pH (data not shown), which allows for NO release from the PROLI NONOate donor. The untreated control module was treated in the same way, but without PROLI NONOate. The total flux for the backwashing was 30 l m^{-2} h^{-1} for both the PROLI-treated and control modules, which was twice the suction flux during the MBR operation.

Fluorescent staining and CLSM observation

Membrane pieces were collected to analyse microbial biomass by CLSM at two time points, 71 and 85 days. Three HF membranes were cut from the free ends of membrane module for each sampling point, immersed in 20 µM SYTO 63 red fluorescent nucleic acid stain solution (Molecular Probes, Invitrogen) and incubated at room temperature in the dark for 30 min to stain the DNA in the microbial cells. The membranes were then removed from the SYTO 63 staining solution and soaked in 500 µl fluorescein-5-isothiocyanate (FITC 'Isomer I', Molecular Probes, Invitrogen) staining solution (1 mg ml^{-1}) for 1 h to image proteins in the biofilm. The membranes were then immersed in freshly prepared 0.2 mg ml^{-1} concanavalin A-tetramethylrhodamine (ConA) (Molecular Probes, Invitrogen) staining solution for 30 min to stain the α-mannopyranosyl and α-glucopyranosyl sugar residues. Finally, the membranes were stained with 1 g l^{-1} calcofluor white solution (Sigma) for 30 min to label the β-D-glucopyranose polysaccharides (McSwain et al., 2005). All staining procedures were performed at 25–26°C in the dark. After staining, the samples were immersed into phosphate-buffered saline for 10 min, twice to remove excess stain.

The stained membrane samples were put onto glass slides for imaging using an inverted CLSM (LSM 710, Carl Zeiss). Four channels were used to image the samples. The calcofluor white was excited at 405 nm and detected at 410–480 nm. The wavelengths used for FITC were 488 nm for excitation and 500–540 nm for emission. The tetramethylrhodamine-conjugated concanavalin A was detected via excitation at 543 nm and emission at 550–600 nm. The SYTO 63 was excited at 633 nm and captured at 650–700 nm (Chen et al., 2007). At every time point, three pieces of membrane were collected to stain. A total of 9–15 images were captured for the triplicate membrane samples to analyse. The quantitative analysis of the three-dimensional

(3D) CLSM images performed using IMARIS (Version 7.3.1, Bitplane) software. The 3D structure of each biofouling component was reconstructed by the 'surface' function in IMARIS. The threshold of fluorescent intensity was adjusted to remove the pixels of background. The biovolume of the fluorescent pixels in each fluorescent channel was calculated by the command 'statistics'. Finally, the biovolume of each biofouling component was standardized as the volume (μm^3) of the specific component per membrane surface area (μm^2). The significance of difference in biofouling components between the PROLI NONOate-treated and control biofilm was calculated by t-test in SPSS (version 16.0). Comparisons were considered significantly different at $P < 0.05$.

DNA extraction

DNA from the biofouling community growing on the HF membranes as well as from the activated sludge in the bulk solution was extracted by a modified CTAB-Polyethylene glycol protocol (Paithankar and Prasad, 1991; Griffiths et al., 2000) at two time points, 71 and 85 days. For each sample point, three independent sludge samples or HF were collected for replicates. The HF membrane pieces were cut into small pieces and put into microfuge tubes containing lysing matrix (MP Biomedicals). Subsequently, 0.5 ml of 5% CTAB lysis solution and 0.5 ml phenol/chloroform/isoamyl alcohol (25:24:1) were added to the tubes. The tubes were placed in a Fast-Prep bead beater (FastPrep-24, M.P. Biomedicals) and shaken using speed setting 5.5 for 30 s. Afterwards, the tubes were centrifuged at 17 000 g for 5 min. The top aqueous layer was transferred to a clean 2 ml tube, RNase was added at a final concentration of 10 μg ml^{-1} and the sample was incubated at 37°C for 30 min. After RNA digestion, 0.5 ml chloroform/isoamyl alcohol (24:1) was added, the samples were vortexed briefly and centrifuged at 17 000 g for 5 mln. The top aqueous layer was transferred into clean 2 ml tubes and mixed well with two volumes of a 30% PEG solution and incubated at 4°C overnight to precipitate the DNA. The following day, the samples were centrifuged at 17 000 g for 15 min and the supernatant was discarded. The DNA pellets were washed with 70% ice-cold ethanol three times and air dried. The DNA pellets were dissolved in DNase and RNase-free distilled water and the concentration was quantified using a NanoDrop spectrophotometer (Thermo Scientific). The aqueous DNA samples were stored at –80°C.

Pyrosequencing and processing of sequence data

The DNA was sequenced using the '454' pyrosequencing platform (Research and Testing Laboratory, TX, USA) targeting bacterial and fungal communities (Handl et al., 2011). The primers selected for the bacterial PCR were Gray28F (5'-GAGTTTGATCNTGGCTCAG-3') and Gray519R (5'-GTNTTACNGCGGCKGCTG-3') (Baker et al., 2003). The primers selected for the fungal PCR were forward funSSUF (5'-TGGAGGGCAAGTCTGGTG-3') and reverse funSSUR (5'-TCGGCATAGTTTATGGTTAAG-3') (Foster et al., 2013). The number of reads for every sample was approximately 3000.

The pyrosequencing data were processed using MOTHUR based on the Costello analysis pipeline (Costello et al., 2009;

Schloss et al., 2009). The sequences were sorted by the barcodes to generate groups of data for the samples. The sequence packets were trimmed and barcodes and primers were removed. Sequences that had poor quality (below the quality score of 25) were removed from the dataset and the size of data packets was reduced to facilitate the analysis through the process of 'unique'. Chimeric sequences were identified and removed using 'chimera.slayer'. The reference was set to be self and the sequences were aligned with the SILVA bacterial 16S and eukaryotic 18S rRNA sequence databases and assigned to taxonomic groups based on the SILVA bacterial and eukaryotes taxonomic reference (Pruesse et al., 2007). The criteria for the sequence classification (identity to the reference sequence) were: species (> 97%), genus (94–97%), family (90–94%), order (85–90%), class (80–85%) and phylum (75–80%) (Lim et al., 2012). Sequences with similarities below these criteria were classified into unidentified groups for each taxonomic rank. Finally, a shared phylotype file was generated through the command 'make.shared' for all the samples. The coverage of phylotypes was calculated based on the phylotypes acquired and the number of sequences pooled. The relationship between the phylotypes and sequences was calculated by MOTHUR using the command of 'summary.single'. All raw DNA sequences were deposited in the GenBank Sequence Read Archive. The accession numbers for the bacterial sequences are SRR1066760–SRR1066769, SRR1066771–SRR1066774, SRR1066776, SRR1066777, SRR1066780, SRR1066783, SRR1066784, SRR1066787, SRR1066788. The accession numbers for the fungal sequence are SRR1067677–SRR1067697.

Phylotype-based community analysis

Phylogenetic trees and nonmetric multidimensional scaling plots were created based on the Bray–Curtis similarity of phylotype compositions in the different groups (Clarke, 1993). The similarity or dissimilarity between the samples or groups was calculated by 'SIMPER' analysis (PRIMER-E, UK) (Clarke, 1993). The contributions of the phylotypes to the similarity or dissimilarity were calculated based on the relative abundances of phylotypes between the samples. The fungi or bacteria which contributed 1% or more to the similarity or dissimilarity were considered to be the key organisms influencing the composition of the community.

Acknowledgements

This research was supported by a research grant (MEWRC651/06/177) from the Environment and Water Industry Programme Office of Singapore. The authors also thank A/Prof Torsten Thomas from the Centre for Marine Bio-Innovation at the University of New South Wales for help in the optimization of the analysis pipeline for pyrosequencing datasets and GE Singapore Water Technology Centre for providing the ZeeWeed membrane.

Conflict of interest

None declared.

References

Adams, R.I., Miletto, M., Taylor, J.W., and Bruns, T.D. (2013) The diversity and distribution of fungi on residential surfaces. *PLoS ONE* **8:** e78866.

Arruebarrena Di Palma, A., Pereyra, C.M., Moreno Ramirez, L., Xiqui Vázquez, M.L., Baca, B.E., Pereyra, M.A., *et al.* (2013) Denitrification-derived nitric oxide modulates biofilm formation in *Azospirillum brasilense. FEMS Microbiol Lett* **338:** 77–85.

Baker, G.C., Smith, J.J., and Cowan, D.A. (2003) Review and re-analysis of domain-specific 16S primers. *J Microbiol Methods* **55:** 541–555.

Barnes, R.J., Bandi, R.R., Wong, W.S., Barraud, N., McDougald, D., Fane, A., *et al.* (2013) Optimal dosing regimen of nitric oxide donor compounds for the reduction of *Pseudomonas aeruginosa* biofilm and isolates from wastewater membranes. *Biofouling* **29:** 203–212.

Barraud, N., Hassett, D.J., Hwang, S.H., Rice, S.A., Kjelleberg, S., and Webb, J.S. (2006) Involvement of nitric oxide in biofilm dispersal of *Pseudomonas aeruginosa. J Bacteriol* **188:** 7344–7353.

Barraud, N., Schleheck, D., Klebensberger, J., Webb, J.S., Hassett, D.J., Rice, S.A., *et al.* (2009a) Nitric oxide signaling in *Pseudomonas aeruginosa* biofilms mediates phosphodiesterase activity, decreased cyclic di-GMP levels, and enhanced dispersal. *J Bacteriol* **191:** 7333–7342.

Barraud, N., Storey, M.V., Moore, Z.P., Webb, J.S., Rice, S.A., and Kjelleberg, S. (2009b) Nitric oxide-mediated dispersal in single- and multi-species biofilms of clinically and industrially relevant microorganisms. *Microb Biotechnol* **2:** 370–378.

Barraud, N., Kelso, M.J., Rice, S.A., and Kjellberg, S. (2014) Nitric oxide: a key mediator of biofilm dispersal with applications in infectious diseases. *Curr Pharm Des* **21:** 31–42.

Besemer, K., Singer, G., Limberger, R., Chlup, A. K., Hochedlinger, G., Hodl, I., *et al.* (2007). Biophysical controls on community succession in stream biofilms. *Applied and Environmental Microbiology* **73:** 4966–4974.

Chen, J., Zhang, M., Wang, A., Lin, H., Hong, H., and Lu, X. (2012) Osmotic pressure effect on membrane fouling in a submerged anaerobic membrane bioreactor and its experimental verification. *Bioresour Technol* **125:** 97–101.

Chen, M.Y., Lee, D.J., Tay, J.H., and Show, K.Y. (2007) Staining of extracellular polymeric substances and cells in bioaggregates. *Appl Microbiol Biotechnol* **75:** 467–474.

Chen, X., Suwarno, S.R., Chong, T.H., McDougald, D., Kjelleberg, S., Cohen, Y., *et al.* (2013) Dynamics of biofilm formation under different nutrient levels and the effect on biofouling of a reverse osmosis membrane system. *Biofouling* **29:** 319–330.

Clarke, K.R. (1993) Non-parametric multivariate analyses of changes in community structure. *Aust J Ecol* **18:** 117–143.

Costello, E.K., Lauber, C.L., Hamady, M., Fierer, N., Gordon, J.I., and Knight, R. (2009) Bacterial community variation in human body habitats across space and time. *Science* **326:** 1694–1697.

Estrela, C., Estrela, C.R., Barbin, E.L., Spanó, J.C.E., Marchesan, M.A., and Pécora, J.D. (2002) Mechanism of action of sodium hypochlorite. *Braz Dent J* **13:** 113–117.

Foster, M.L., Dowd, S.E., Stephenson, C., Steiner, J., and Suchodolski, J.S. (2013) Characterization of the fungal microbiome (mycobiome) in fecal samples from dogs. *Vet Med Int* **2013:** 2013:658373. doi: 10.1155/2013/658373. Epub 2013 Apr 23.

Gillings, M.R., Holley, M.P., and Selleck, M. (2006) Molecular identification of species comprising an unusual biofilm from a groundwater treatment plant. *Biofilms* **3:** 19–24.

Griffiths, R.I., Whiteley, A.S., O'Donnell, A.G., and Bailey, M.J. (2000) Rapid method for coextraction of DNA and RNA from natural environments for analysis of ribosomal DNA- and rRNA-based microbial community composition. *Appl Environ Microbiol* **66:** 5488–5491.

Handl, S., Dowd, S.E., Garcia-Mazcorro, J.F., Steiner, J.M., and Suchodolski, J.S. (2011) Massive parallel 16S rRNA gene pyrosequencing reveals highly diverse fecal bacterial and fungal communities in healthy dogs and cats. *FEMS Microbiol Ecol* **76:** 301–310.

Hetrick, E.M., Shin, J.H., Paul, H.S., and Schoenfisch, M.H. (2009) Anti-biofilm efficacy of nitric oxide-releasing silica nanoparticles. *Biomaterials* **30:** 2782–2789.

Jurado, V., Kroppenstedt, R.M., Saiz-Jimenez, C., Klenk, H.P., Mouniee, D., Laiz, L., *et al.* (2009) *Hoyosella altamirensis* gen. nov., sp. nov., a new member of the order Actinomycetales isolated from a cave biofilm. *Int J Syst Evol Microbiol* **59:** 3105–3110.

Kim, K.K., Park, H.Y., Park, W., Kim, I.S., and Lee, S.T. (2005) *Microbacterium xylanilyticum* sp. nov., a xylan-degrading bacterium isolated from a biofilm. *Int J Syst Evol Microbiol* **55:** 2075–2079.

Kim, M., Lee, S., Park, H.D., Choi, S.I., and Hong, S. (2012) Biofouling control by quorum sensing inhibition and its dependence on membrane surface. *Water Sci Technol* **66:** 1424–1430.

Kim, S.R., Oh, H.S., Jo, S.J., Yeon, K.M., Lee, C.H., Lim, D.J., *et al.* (2013) Biofouling control with bead-entrapped quorum quenching bacteria in membrane bioreactors: physical and biological effects. *Environ Sci Technol* **47:** 836–842.

Le Clech, P., Jefferson, B., Chang, I.S., and Judd, S.J. (2003) Critical flux determination by the flux-step method in a submerged membrane bioreactor. *J Membr Sci* **227:** 81–93.

Lee, D.Y., Li, Y.Y., Noike, T., and Cha, G.C. (2008) Behavior of extracellular polymers and bio-fouling during hydrogen fermentation with a membrane bioreactor. *J Membr Sci* **322:** 13–18.

Lee, E.-J., Kwon, J.-S., Park, H.-S., Ji, W.H., Kim, H.-S., and Jang, A. (2013) Influence of sodium hypochlorite used for chemical enhanced backwashing on biophysical treatment in MBR. *Desalination* **316:** 104–109.

Li, J., Yu, D., and Wang, D. (2011) Experimental test for high saline wastewater treatment in a submerged membrane bioreactor. *Desalination Water Treat* **36:** 171–177.

Lim, S., Kim, S., Yeon, K.M., Sang, B.I., Chun, J., and Lee, C.H. (2012) Correlation between microbial community structure and biofouling in a laboratory scale membrane bioreactor with synthetic wastewater. *Desalination* **287:** 209–215.

Liu, R., Yu, Z., Zhang, H., Yang, M., Shi, B., and Liu, X. (2012) Diversity of bacteria and mycobacteria in biofilms of two urban drinking water distribution systems. *Can J Microbiol* **58:** 261–270.

McSwain, B., Irvine, R., Hausner, M., and Wilderer, P. (2005) Composition and distribution of extracellular polymeric substances in aerobic flocs and granular sludge. *Appl Environ Microbiol* **71:** 1051–1057.

Madaeni, S.S., Saedi, S., Rahimpour, F., and Zereshki, S. (2009) Optimization of chemical cleaning for removal of biofouling layer. *Chem Prod Process Model* **4:** 1934-2659, doi: 10.2202/1934-2659.1309, May 2009.

Martin, K.J., Bolster, D., Derlon, N., Morgenroth, E., and Nerenberg, R. (2014) Effect of fouling layer spatial distribution on permeate flux: a theoretical and experimental study. *J Membr Sci* **471:** 130–137.

Meng, F., Chae, S.R., Drews, A., Kraume, M., Shin, H.S., and Yang, F. (2009) Recent advances in membrane bioreactors (MBRs): membrane fouling and membrane material. *Water Res* **43:** 1489–1512.

Paithankar, K.R., and Prasad, K.S.N. (1991) Precipitation of DNA by polyethylene-glycol and ethanol. *Nucleic Acids Res* **19:** 1346–1346.

Plate, L., and Marletta, M.A. (2012) Nitric oxide modulates bacterial biofilm formation through a multicomponent cyclic-di-GMP signaling network. *Mol Cell* **46:** 449–460.

Pruesse, E., Quast, C., Knittel, K., Fuchs, B.M., Ludwig, W., Peplies, J., *et al.* (2007) SILVA: a comprehensive online resource for quality checked and aligned ribosomal RNA sequence data compatible with ARB. *Nucleic Acids Res* **35:** 7188–7196.

Puspitasari, V., Granville, A., Le-Clech, P., and Chen, V. (2010) Cleaning and ageing effect of sodium hypochlorite on polyvinylidene fluoride (PVDF) membrane. *Sep Purif Technol* **72:** 301–308.

Schloss, P.D., Westcott, S.L., Ryabin, T., Hall, J., Hartmann, M., Hollister, E., *et al.* (2009) Introducing mothur: open-source, platform-independent, community-supported software for describing and comparing microbial communities. *Appl Environ Microbiol* **75:** 7537–7541.

Tan, K., and Obendorf, S.K. (2007) Development of an antimicrobial microporous polyurethane membrane. *J Membr Sci* **289:** 199–209.

Van Nieuwenhuijzen, A.F., Evenblij, H., Uijterlinde, C.A., and Schulting, F.L. (2008) Review on the state of science on membrane bioreactors for municipal wastewater treatment. *Water Sci Technol* **57:** 979–986.

Veerasamy, D., Supurmaniam, A., and Nor, Z.M. (2009) Evaluating the use of *in-situ* ultrasonication to reduce fouling during natural rubber skim latex (waste latex) recovery by ultrafiltration. *Desalination* **236:** 202–207.

Wang, Z., Wu, Z., Yin, X., and Tian, L. (2008) Membrane fouling in a submerged membrane bioreactor (MBR) under sub-critical flux operation: membrane foulant and gel layer characterization. *J Membr Sci* **325:** 238–244.

Williams, M.D., and Pirbazari, M. (2007) Membrane bioreactor process for removing biodegradable organic matter from water. *Water Res* **41:** 3880–3893.

Xia, S., Li, J., He, S., Xie, K., Wang, X., Zhang, Y., *et al.* (2010) The effect of organic loading on bacterial community composition of membrane biofilms in a submerged polyvinyl chloride membrane bioreactor. *Bioresour Technol* **101:** 6601–6609.

Yan, X., Bilad, M.R., Gerards, R., Vriens, L., Piasecka, A., and Vankelecom, I.F.J. (2012) Comparison of MBR performance and membrane cleaning in a single-stage activated sludge system and a two-stage anaerobic/aerobic (A/A) system for treating synthetic molasses wastewater. *J Membr Sci* **394–395:** 49–56.

Yang, C., Zhang, W., Liu, R., Li, Q., Li, B., Wang, S., *et al.* (2011) Phylogenetic diversity and metabolic potential of activated sludge microbial communities in full-scale wastewater treatment plants. *Environ Sci Technol* **45:** 7408–7415.

Yeon, K.M., Cheong, W.S., Oh, H.S., Lee, W.N., Hwang, B.K., Lee, C.H., *et al.* (2009a) Quorum sensing: a new biofouling control paradigm in a membrane bioreactor for advanced wastewater treatment. *Environ Sci Technol* **43:** 380–385.

Yeon, K.M., Lee, C.H., and Kim, J. (2009b) Magnetic enzyme carrier for effective biofouling control in the membrane bioreactor based on enzymatic quorum quenching. *Environ Sci Technol* **43:** 7403–7409.

Supporting information

Additional Supporting Information may be found in the online version of this article at the publisher's web-site:

Fig. S1. The TMP profile for the PROLI NONOate-treated and control membrane module. The TMP values are the daily average transmembrane pressures. The vertical dashed line splits the TMP profile into the low pressure phase (left side) and the rapid TMP increase phase (right side).

Fig. S2. The clustering tree of the bacterial communities for the PROLI NONOate-treated and control biofilms at 71 and 85 days. The clustering tree was constructed based on the average Bray–Curtis similarity. The labels 'control' and 'PROLI' represent the control biofilms and PROLI NONOate-treated biofilms respectively.

Fig. S3. Representative 3D confocal laser scanning microscopic (CLSM) images of biofilm matrix components and cells on the PROLI NONOate-treated and control membranes. The α-polysaccharides (A) are shown in red, the β-polysaccharides (B) are shown in blue, the proteins (C) are shown in green and the microorganisms (D) are shown in purple. All images are top down projects of 3D reconstructions of the biofilm. The total magnification for the images was 630×.

Fig. S4. Schematic drawing of the internal submerged MBR.

12

A lithotrophic microbial fuel cell operated with pseudomonads-dominated iron-oxidizing bacteria enriched at the anode

Thuy Thu Nguyen,[1] Tha Thanh Thi Luong,[1†]
Phuong Hoang Nguyen Tran,[1†] Ha Thi Viet Bui,[1,2]
Huy Quang Nguyen,[1,3] Hang Thuy Dinh,[4]
Byung Hong Kim[5,6,7] and Hai The Pham[1,2*]

[1]Research group for Physiology and Applications of Microorganisms (PHAM group) at Center for Life Science Research, Departments of
[2]Microbiology and [3]Biochemistry, Faculty of Biology, Vietnam National University – University of Science, Nguyen Trai 334, Thanh Xuan, Hanoi, Vietnam.
[4]Laboratory of Microbial Ecology, Institute of Microbiology and Biology, Vietnam National University, Xuan Thuy 144, Cau Giay, Hanoi, Vietnam.
[5]Korea Institute of Science and Technology, Hwarangno 14-gil, 5 Seongbuk-gu, Seoul 136-791, Korea.
[6]Fuel Cell Institute, National University of Malaysia, Bangi 43600 UKM, Selangor, Malaysia.
[7]School of Municipal and Environmental Engineering, Harbin Institute of Technology, 73 Huanghe Road, Nangang District, Harbin 150090, China.

Summary

In this study, we attempted to enrich neutrophilic iron bacteria in a microbial fuel cell (MFC)-type reactor in order to develop a lithotrophic MFC system that can utilize ferrous iron as an inorganic electron donor and operate at neutral pHs. Electrical currents were steadily generated at an average level of 0.6 mA (or 0.024 mA cm^{-2} of membrane area) in reactors initially inoculated with microbial sources and operated with 20 mM Fe^{2+} as the sole electron donor and 10 ohm external resistance; whereas in an uninoculated reactor (the control), the average current level only reached 0.2 mA (or 0.008 mA cm^{-2} of membrane area). In an inoculated MFC, the generation of electrical currents was correlated with increases in cell density of bacteria in the anode suspension and coupled with the oxidation of ferrous iron. Cultivation-based and denaturing gradient gel electrophoresis analyses both show the dominance of some *Pseudomonas* species in the anode communities of the MFCs. Fluorescent in-situ hybridization results revealed significant increases of neutrophilic iron-oxidizing bacteria in the anode community of an inoculated MFC. The results, altogether, prove the successful development of a lithotrophic MFC system with iron bacteria enriched at its anode and suggest a chemolithotrophic anode reaction involving some *Pseudomonas* species as key players in such a system. The system potentially offers unique applications, such as accelerated bioremediation or on-site biodetection of iron and/or manganese in water samples.

*For correspondence. E-mail phamthehai@vnu.edu.vn; hai.phamthe@gmail.com;

†Both authors contributed equally to the research.
Funding Information This research is funded by Vietnam National Foundation for Science and Technology Development (NAFOSTED) under Grant No. 106.03-2012.06. It also received support from Korea Institute of Science and Technology (KIST) IRDA Alumni Program and International Foundation for Science (IFS – Sweden).

Introduction

The research interest in microbial fuel cells (MFCs) has increased recently, due to their unique property of exploiting microbial activity to generate electricity from energy-storing substances. In MFCs, microorganisms act as biocatalysts to convert chemical energy comprised in electron donors to electrical energy (Allen and Bennetto, 1993; Logan *et al.*, 2006). These systems can also be modified (and assisted with energy) to become microbial electrolysis cells, in which hydrogen or other substances can be produced (Logan *et al.*, 2006; Rozendal *et al.*, 2006; Rabaey and Rozendal, 2010). 'Bioelectrochemical systems' is therefore a broad sense term to designate all kinds of these systems (Rabaey *et al.*, 2007).

Up to now, the majority of MFC researches have been focused on optimization of the device for the recovery of energy from biomass (mostly in waste) or from light or for bioremediation or the production of future clean energy (Rosenbaum *et al.*, 2010; Lovley and Nevin, 2011; Wang and Ren, 2013). However, due to some performance-limiting factors, including the microbial activity, the electron transfer process, the internal resistance of the device and particularly the cathode reaction rate (Pham *et al.*, 2006; Kim *et al.*, 2007), the maximum power output (per volume unit) of an MFC is still limited. Moreover, scale-up difficulties also hindered the realization of this technology in the field of energy recovery (Rozendal *et al.*, 2008; Cheng *et al.*, 2014). Therefore, currently, many MFC researches are being directed towards exploiting the special characteristics of MFCs for environmental bioremediation or for biosynthesis and the development of biosensors (Kim *et al.*, 2007; Rabaey and Rozendal, 2010; Lovley and Nevin, 2011; Arends and Verstraete, 2012).

There have been MFCs utilizing various types of substrates, including a wide range of soluble or dissolved complex organic matter (Pant *et al.*, 2010). Most substrates tested in MFCs are indeed artificial or real wastewaters containing different kinds of compounds (Rabaey *et al.*, 2007; Pant *et al.*, 2010). There are also MFCs operated with single substances such as acetate, formate or Acid orange 7, etc. (Lee *et al.*, 2003; Ha *et al.*, 2008; Fernando *et al.*, 2014). It is common that most substrates (electron donors) used in MFC systems so far are organic (Pant *et al.*, 2010), meaning that bacteria in these systems are heterotrophic. There have been only few reports about the use of an inorganic electron donor, such as sulfide, in a MFC (Rabaey *et al.*, 2006) but as for sulfide, it can only be used in the presence of another organic electron donor, i.e. acetate. Little is known about whether an 'inorganic' MFC that utilizes metal ions can actually function.

In principle, the development of such an 'inorganic' MFC operated with metal ions such as ferrous ions should be feasible because there exist a group of bacteria that can oxidize ferrous ions to gain energy – the chemolithotrophic iron-oxidizing bacteria (or iron bacteria) (Cullimore and McCann, 1978; Hedrich *et al.*, 2011). Taxonomically, these bacteria are classified into several groups but most of them belong to proteobacteria (Hedrich *et al.*, 2011). The acidophilic iron-oxidizing proteobacteria, such as *Acidithiobacillus ferrooxidans*, were probably considered typical iron bacteria (Hedrich *et al.*, 2011). However, there are also phototrophic iron bacteria, neutrophilic iron bacteria that respire on nitrate or even neutrophilic aerobic iron bacteria, including some *Pseudomonas* species (Sudek *et al.*, 2009; Hedrich *et al.*, 2011).

In this study, we attempted to enrich neutrophilic iron bacteria in a MFC in order to develop an iron-oxidizing MFC system that can operate at neutral pHs, as this condition will be convenient for practical applications. Such a lithotrophic MFC can be not only scientifically interesting but also promisingly used as a biosensor detecting iron or as a bioremediation means to remove iron or other metal pollutants from water.

Results

Generation of electrical currents in MFCs fed with ferrous iron as the sole electron donor – an indication of the enrichment of iron-oxidizing bacteria

Several modified National Centre for Biotechnology Education (UK) (NCBE)-type MFC reactors were set up, inoculated and operated with a modified M9 medium containing only Fe^{2+} (20 mM) as the sole electron donor at the anode. Within the first 2 days of operation, all of the reactors already began to generate electrical currents (Fig. 1). After 2 weeks of operation, the electrical currents of the reactors were steady. The generation of electrical currents while being fed with Fe^{2+} as the only electron donor is the first evidence of the function of the reactors as MFCs and of the enrichment of iron-oxidizing bacteria.

Differences in the levels of current generation could be clearly observed ($P < 0.05$) between an MFC that was initially inoculated with a microbial source (an inoculated MFC) and a MFC that was not (the control). At the steady state, with 20 mM Fe^{2+} supplied into the anode compartment, the average currents of an inoculated MFC was 0.6 ± 0.11 mA (equivalent to 0.024 ± 0.0044 mA cm^{-2} membrane area) while that of the control was only 0.2 mA (equivalent to 0.008 mA cm^{-2} membrane area) (Fig. 1). The amount of coulombs produced by the former was even six times as much as that produced by the latter (Fig. S2). The differences of inoculated MFCs versus the control imply that the generation of current in an inoculated MFC is due to electroactive bacteria that might be feasibly enriched from the initial

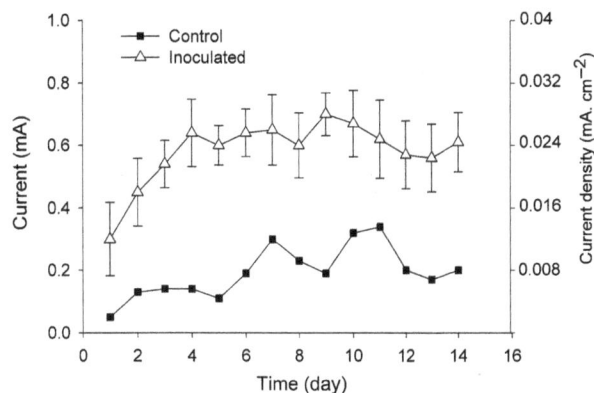

Fig. 1. Typical patterns of the generations of electrical currents by an inoculated MFC and the control (uninoculated but not abiotic) during the enrichment period. The MFCs were operated with ferrous iron as the only electron donor in the anodes (see *Experimental procedures*) and with a 10 ohm external resistor, at 25°C. Each inoculated MFC was inoculated with the mud from a natural stream suspected to contain iron bacteria. The control was not inoculated with any microbial source at the beginning. Each data point is an average current per batch on the corresponding day, generated by the corresponding MFC(s).

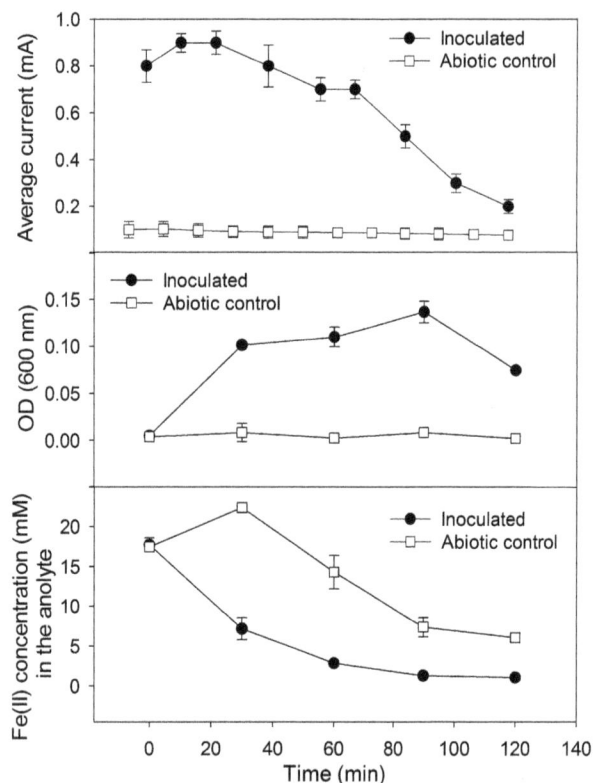

Fig. 2. Typical patterns of changes of the generated current (top), the optical density at the wavelength of 600 nm of the anode suspension (center), and the concentration of ferrous iron in the anolytes (bottom) during an operational batch of a MFC inoculated with bacteria in comparison with those of the abiotic control. The MFCs were operated with a 10 ohm external resistor at 25°C. Error bars represent standard deviations.

microbial source. Such a source probably allows the selection from a large microbial community for electrochemically active bacteria that can use Fe^{2+} as the electron donor. Thus in the control, without an initial inoculum, such bacteria might not be enriched. The current generated by the control, although at low levels, might be due partially to plain chemical reactions (as described later) and partially to the activity of contaminating bacteria from the surroundings. These bacteria might gradually adapt to the anode conditions but their electrochemical activity might not be competent enough.

The generation of electricity in relation to the microbial activity in the MFCs

To prove that the generation of electricity in the MFCs was not due to plain chemical reactions, an abiotic control with the anode compartment sterilized (see *Experimental procedures*) was tested with Fe^{2+}. Under such conditions [with optical density (OD) (600 nm) values being approximately 0 – Fig. 2], when only

abiotically chemical reactions could occur, the current generated in an operational batch was very limited (0.1 ± 0.02 mA) and distinctively much lower than that of an inoculated MFC (under biotic conditions) (Fig. 2). Noticeably, the current generated by an abiotic control, although reaching certain levels right after the supply of Fe^{2+}, rapidly decreased down to near the bottom level during a batch. Together with a generated current that remained at significant levels for a long time in an inoculated MFC, increases in the number of bacterial cells [reflected by OD (600 nm) values] were also observed at some points (Fig. 2). However, it seems that under both abiotic and biotic conditions, the trends of change of the Fe(II) concentration were similar (Fig 2). These results proved that abiotically chemical oxidation of ferrous ions does occur in the anode of our reactors, but the electrode itself was probably not an electron acceptor for this abiotic oxidation, resulting in little electricity generated by the abiotic control. Probably, ferrous ions were abiotically oxidized by oxygen diffused from the cathode through the membrane, as shown elsewhere (Pham *et al.*, 2004). Only in a MFC inoculated with bacteria, a significant electrical current could be generated (0.6 ± 0.11 mA versus 0.1 ± 0.02 mA of the abiotic control). It could be the result of efficient interactions of the enriched bacteria with the anodic electrode (the microbial activity) – a property that the abiotic control does not have. This also implies that the consortium of bacteria that was enriched in an inoculated MFC can oxidize Fe^{2+} before transferring electrons to the electrode. Ferric precipitate was also observed more in the anode compartment of an inoculated MFC than in that of the abiotic control.

All the results reported above suggest that our inoculated MFCs were successfully developed and functioned as MFCs that generate electricity upon oxidizing ferrous ion, due to the electrochemical activity of the microbes enriched at the anodes.

Culturable bacteria in the anode of iron-oxidizing MFCs

Culturable bacteria in the anode suspensions of an inoculated MFC and the control (that was not uninoculated) were grown and isolated on Winogradsky medium to find potential iron bacteria (Starosvetsky *et al.*, 2008). The number and types of strains isolated from the inoculated MFC were different from those from the control (Fig. 3). Only two isolates could be cultivated from the anode of the control while six isolates could be obtained from the anode of the inoculated MFC. Colony plating and counting results showed the high presence frequence of isolate FC 2.5 in the microbial community of the studied inoculated MFC (Fig. 3). Results of 16S rDNA sequence analyses showed that strain FC 2.5 was close to *Pseudomonas*

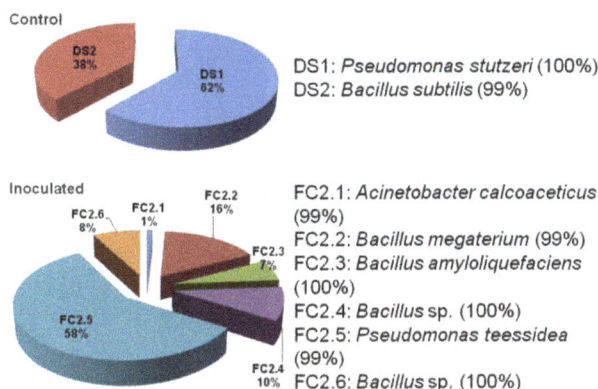

Control

DS2 38%
DS1 62%

DS1: *Pseudomonas stutzeri* (100%)
DS2: *Bacillus subtilis* (99%)

Inoculated

FC2.6 8%　FC2.1 1%　FC2.2 16%
FC2.3 7%
FC2.5 58%
FC2.4 10%

FC2.1: *Acinetobacter calcoaceticus* (99%)
FC2.2: *Bacillus megaterium* (99%)
FC2.3: *Bacillus amyloliquefaciens* (100%)
FC2.4: *Bacillus* sp. (100%)
FC2.5: *Pseudomonas teessidea* (99%)
FC2.6: *Bacillus* sp. (100%)

Fig. 3. The levels of presence frequence (expressed as percentage per total culturable colonies) of the bacterial isolates from an inoculated MFC and the control. Isolation was done on solid Winogradsky medium. Isolates named with 'DS' were from the control, while those named with 'FC2' were from an inoculated MFC. Notes on the right indicate the proposed taxonomic identification of the corresponding isolates based on observations of their cell and colony morphology and analyses of their 16S rDNA sequences. The percentage of similarity between the 16S rDNA sequence of an isolate and that of the proposed species was shown in the brackets next to the corresponding note.

teessidea (99% sequence homology). Also based on 16S rDNA sequence analyses, some other strains could be identical to *Bacillus* sp. (FC 2.3 and FC 2.6) and other species such as *Acinetobacter* sp. (FC 2.1). These results are interesting because none of those isolates are related to popularly-known iron bacteria.

Anode bacterial communities of iron-oxidizing MFCs

In order to analyse the bacterial community at the anode of each MFC, total DNA of bacterial cells in the anode suspension was extracted, together with total DNA of those scraped off from the electrode surface. Surprisingly, in all cases, very little DNA was obtained from the electrode surfaces (data not shown), indicating that bacteria did not occupy the anode surfaces. This result is also consistent with the fluorescent in situ hybridization (FISH) results reported later (Fig. 5).

Since no DNA was obtained from the electrode surfaces, only the bacterial compositions of the anode suspensions of the MFCs were analysed and compared by denaturing gradient gel electrophoresis (DGGE) (Fig. 4). DGGE patterns clearly showed that the anode communities enriched with and without a microbial source are significantly different and change with time in different manners. As can be seen in Fig. 4, in the control, there was a community that was established and became stable already after the first week of operation (with a rate of change of ~ 0% week^{-1}). That community appeared to consist of a number of species but no species seemed to dominate. Band sequence analyses showed that the

presence of many *Pseudomonas* species was evident. In an inoculated MFC, in contrast, the microbial community still changed after the first week of operation (with a rate of change of ~ 73 ± 12% week^{-1}) and only showed a steady state after 2 weeks (still with a rate of change of ~ 20 ± 10% week^{-1}). Moreover, there appeared some species that dominate the community. Based on band sequence analyses, these species were suspected to be *Pseudomonas* sp., *Geobacter* sp. and *Bacillus* sp. (Fig. 4).

In order to assess the presence of iron bacteria, particularly neutrophilic iron-oxidizing ones, in the anode of an iron-oxidizing MFC, FISH analyses of anode surface samples and anode suspension samples of another inoculated MFC were carried out using probe PS 1. FISH images (Fig. 5) showed a significant level of PS 1 signal (white dots) reflecting the presence of iron bacteria in the anode suspension of that inoculated MFC. Analyses of these images by Image J revealed that the proportion of iron bacteria in the anode suspension of that MFC was 36.5% of the total bacteria, while that in the inoculum was only 25%. This indicates a significant increase in the quantity of iron bacteria in the anode suspension after the enrichment period. On the other hand, it is interesting to note that few iron bacteria, even few bacteria, were

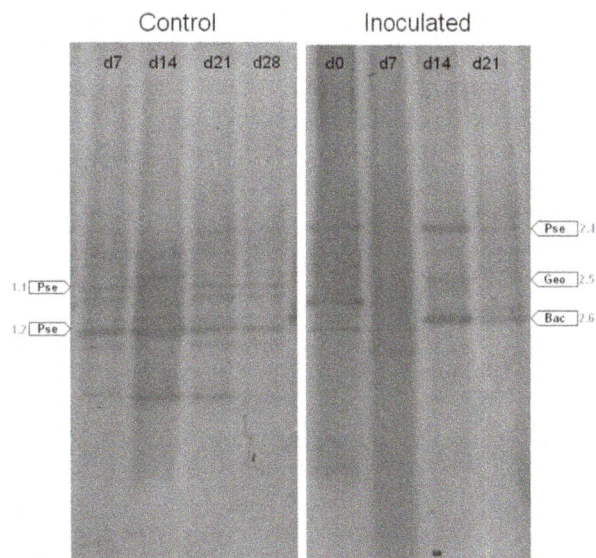

Control　　　　Inoculated

d7　d14　d21　d28　　d0　d7　d14　d21

Pse 2.1
Geo 2.5
Bac 2.6

1.1 Pse
1.2 Pse

Fig. 4. DGGE analysis of the anodic bacterial communities of the MFCs at different time points during the enrichment period. The note on each lane of a gel indicates the moment the sample was taken (for example, d7 = at the 7th day). A sample at d0 indeed resembles the corresponding inoculum. The note on each white arrow indicates the genus, 16S rDNA sequence of which has the highest similarity (> 98%) to the DNA sequence of the corresponding band on the gel (based on BLAST analysis). Pse = *Pseudomonas* sp., Geo = *Geobacter* sp., Bac = *Bacillus* sp. (The note next to each arrow is the assigned number of the corresponding band). The DGGE was repeated three times with three replicates of each sample. As the results of these repetitions were absolutely similar, only typical patterns were shown here.

| Inoculum (natural mud) | Anodic suspension | Anode surface |

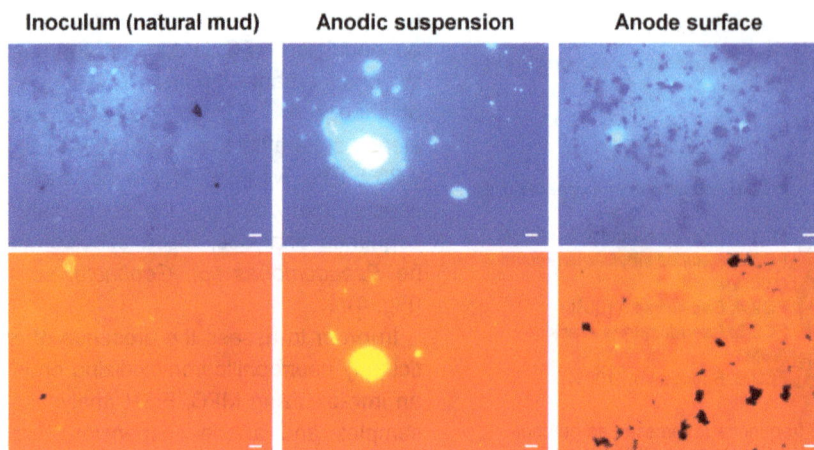

Fig. 5. FISH analyses of iron bacteria in the inoculum and in the anode suspension as well as on the anode surface of an inoculated MFC. Upper are images created from fluorescent signals of DAPI, with white dots showing the presence of all bacteria in the samples. Lower are images created from fluorescent signals of Cy3 attached to the probe PS1, with white dots showing the presence of iron bacteria in the samples. Bars, 5 μm.

present on the anode surface. These results suggest that iron bacteria are mostly present and active in the anode suspension but not on the electrode surface.

Discussion

The successful development of iron-oxidizing MFCs

In this study, MFCs that can generate electricity upon utilizing ferrous iron as the electron donor have been experimented and shown to function. As mentioned, our results show that with a proper inoculum (i.e. collected from a site with a high chance to contain iron bacteria), a bacterial community can be enriched in the anode of such a reactor and responsible for the generation of electricity coupled to the oxidation of Fe^{2+}. Indications of the successful enrichment of such a consortium include: (i) enrichment period current patterns that resemble those in other types of MFCs, (ii) the significant generation of electricity only when bacteria are present (Fig. 2) ($P < 0.05$) and (iii) the detection of bacterial communities including iron bacteria based on DGGE and FISH analyses.

Normally, the required time for an electrochemically active community to establish and stabilize is around 2 weeks (Kim *et al.*, 2003; Rabaey and Verstraete, 2005) although shorter enrichment time (e.g. 5 days) has been reported (Ishii *et al.*, 2014). The slow growth of iron bacteria at neutral pHs due to the lower Fe^{2+}/Fe^{3+} redox potential (Hedrich *et al.*, 2011) might be also an explanation for this relatively long enrichment time. The pattern of current generation of an inoculated MFC in this study in the first 2 weeks of operation is similar to that during the enrichment period of a typical MFC (Kim *et al.*, 2004; Rabaey and Verstraete, 2005). During enrichment, shifts

in the composition of the anode community (of an inoculated MFC) (Fig. 5) indicate a selection process in which only bacteria that can well adapt to the anode conditions (by being able to utilize Fe^{2+} and interact with the electrode) become dominant. This was also supported by the increased quantity of iron bacteria after enrichment, as demonstrated by the FISH results (Fig. 4). Similar community shifts during enrichment to finally shape a working electroactive community were also observed in other MFC systems (Aelterman *et al.*, 2006; Pham *et al.*, 2009). A noticeable point about our MFCs is that unlike other systems, they are not operated with organic matter as fuel, and thus require an anode community that contains not only electrochemically active bacteria but also chemolithotrophic ones living on ferrous ions. Our study demonstrates that the development of such an iron-oxidizing MFC system is feasible. There have been reports on MFC systems using iron-containing compounds or ferric iron reducing bacteria or even ferrous iron oxidizing ones at the cathodes (ter Heijne *et al.*, 2006; 2007) but there has been no similar study on MFCs utilizing ferrous iron as the fuel.

The role of the microbial source for inoculation

It is clearly shown in this study that an initial microbial source is essential for the establishment of a final working community that oxidizes Fe^{2+} and transfer electrons to the anode. Moreover, it appears that a natural source might enable the enrichment of a working community that is stable and performs well. The enrichment and stabilization of a working community from a natural microbial source may take longer but this is definitely not a critical matter. Indeed, most well-performing and stable open-system MFCs are operated with mixed cultures, enriched from natural microbial sources (Kim *et al.*, 2004; Logan and Regan, 2006). The fact that an iron-oxidizing MFC can function well with an anode consortium enriched from a natural source also means that practical development of this type of MFC is straightforward.

The correlation between the composition of an anode community and its performance

The differences in the composition between the microbial communities, as shown by both solid-medium growth and DGGE results, might account for the differences in their performance. The bacterial community of an inoculated MFC is different from that of the control (the uninoculated but not abiotic), particularly in the aspect that the former is dominated by some species while the latter is not. This definitely has some links with the differences in their performance: the former performs distinctively better than the latter. The correlation between the composition of an anode community and its performance has been reported previously (Pham *et al.*, 2008a; 2009).

It should be also noted from both the solid medium growth results and the DGGE results that the number of species in the anode community of each MFC is small (less than 10), i.e. the community is basically not very diverse. This could be explained by the poor nutrient conditions at the anode of each MFC, forcing the community to carry out chemoautotrophic or chemolithotrophy-associated metabolism to survive. Such conditions definitely select for a community so specialized to adapt that it contains only a limited number of species. Indeed, unique communities were observed in MFC systems operated with specific substrates such as formate or acetate (Lee *et al.*, 2003; Ha *et al.*, 2008).

Hypothesis about the electron transfer mechanism in an iron-oxidizing MFC

Our results, altogether, reveal several striking facts. First, both solid-medium growth results and DGGE results indicate the presence of *Pseudomonas* species in the anode communities of all the MFCs and their dominance in an inoculated MFC, which is a well-performing MFC. Second, bacteria detected by probe PS 1, supposed to be neutrophilic iron bacteria, became increased in quantity in the anode community of a well-performing MFC, as shown by FISH results. Third, bacteria, including iron bacteria, could be found in the anode suspensions of the MFCs but hardly detected on the anode surfaces. These facts lead us to a hypothesis that iron-oxidizing bacteria are present in the anode communities of our iron-oxidizing MFCs but they do not directly transfer electrons to the electrode. In the studies on direct electron transfer to electrodes (either via outer membrane proteins or

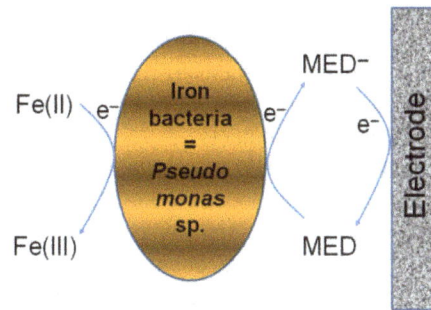

Fig. 6. The hypothesized mechanism of electron transfer occurring at the anode of the iron-oxidizing MFC in this study. MED: mediator.

nanowires), bacteria could always be observed on the electrode surfaces (Kim *et al.*, 1999; Reguera *et al.*, 2005). Only bacteria that can indirectly transfer electrons via chemicals or self-produced mediators are present and active in the anode suspension (Allen and Bennetto, 1993; Rabaey *et al.*, 2005; Pham *et al.*, 2008b). *Pseudomonas* species have been well known as electrochemically active heterotrophic bacteria that can self-produce electron mediators to reduce an electrode upon oxidizing organic matter (Rabaey *et al.*, 2004; 2005). However, some *Pseudomonas* species have been reported to be able to chemolithotrophically metabolize Fe^{2+} at neutral pHs (Straub *et al.*, 1996; Sudek *et al.*, 2009). Considering those facts and the fact that *Pseudomonas* species are dominant in the anode of all the MFCs, our hypothesis for the anode electron transfer in our iron-oxidizing MFC is as follows (Fig. 6): Some *Pseudomonas* species could be actually the dominant 'neutrophilic iron bacteria' that can oxidize Fe^{2+} and transfer electrons to the anodic electrode via their self-produced mediators. In an inoculated MFC, the enriched anode consortium might contain these specialized *Pseudomonas* species that dominate and enable the MFC to function well, generating remarkable currents. In the control (uninoculated but not abiotic), probably some *Pseudomonas* cells that somewhat can do the same 'job' from surroundings could invade the anode but their activities might not be specific or efficient enough to enable a significant generation of electricity.

If our hypothesis is true, it is probable that the probe PS 1 used in the FISH experiments might unspecifically hybridize DNAs or RNAs of *Pseudomonas* species and thus could also detect *Pseudomonas* species instead of the targeted 'neutrophilic iron bacteria'. PS1 was, in fact, designed upon aligning the 16S rDNA sequences of various groups of neutrophilic iron bacteria, particularly those of *Leptothrix* group (Siering and Ghiorse, 1997). However, it was not certain that this probe did not give positive results when tested with *Pseudomonas* species.

The presence of *Geobacter* sp. in the anode community of an inoculated MFC is interesting yet remains questioned because several *Geobacter* sp. are known electroactive bacteria that transfer electrons to the electrode by means of direct contacts (e.g. conductive pili) (Lovley and Nevin, 2011) while few bacteria were found on the anode surface of our inoculated MFC. The presence of *Bacillus* sp., which are heterotrophic Grampositive bacteria that have not been shown to have electrochemical activities, is not either interpretable. Probably, in our iron-oxidizing MFCs, *Geobacter* and *Bacillus* species only act as opportunistic bacteria that take advantage of the output from iron oxidation autotrophy.

Potential applications of iron-oxidizing MFCs in this study

It is noticeable that with the anode community dominated by neutrophilic iron-oxidizing bacteria, most possibly *Pseudomonas* species, our MFC system can be operated at pHs around 7. This enables a convenient operation and handling of the MFC. These MFC systems could be used for an accelerated bioremediation of iron in water samples, as the soluble ferrous salts could be converted to insoluble ferric salts by the activity of the anode electroactive iron-oxidizing community. They can also be potentially used as on-site biosensors to detect and monitor iron or manganese (as iron bacteria also oxidize Mn) in water samples, because of the correlation between their generated current and the concentration of Fe^{2+}. This would be a substantial environmental application, if one considers the use of such an on-site biosensor in remote areas (in Vietnam for instance) where ground water is used as the major water source.

Overall, the results of this research showed that it is feasible to develop a lithotrophic MFC with an anode enriched with an iron-oxidizing electroactive bacterial consortium that can function well at neutral pHs. The key role of some *Pseudomonas* species as lithotrophic neutrophilic iron bacteria is probably the most convincing explanation for the anode electron transfer in such a MFC system. The MFC, with its unique properties, would offer potential applications in bioremediation or biomonitoring of iron in water. However, further studies are required to realize these potentials.

Experimental procedures

Fabrication and operation of MFC reactors

The MFC reactors in this study were fabricated following the NCBE model (Allen and Bennetto, 1993) with some modifications (Fig. S1). Each reactor consisted of two big polyacrylic frames (12 cm × 12 cm × 2 cm) and two small poly-acrylic rectangle-holed frames of anode and cathode compartments (8 cm × 8 cm × 1.5 cm). The dimension of each rectangle hole on each small frame was 5 cm × 5 cm and thus each compartment had the dimension of 5 cm × 5 cm × 1.5 cm. Each compartment was filled in with graphite granules (3–5 mm in diameter) (Xilong Chemical, China), used as the electrode material and packed enough so that the granules well contacted each other and a graphite rod (5 mm in diameter) (Xilong Chemical) to collect the electrical current. This rod penetrated the big frame of each compartment via a drilled hole (5 mm in diameter) and stuck outside. The gaps between the rod and the big frame were sealed up by epoxy glue to ensure that the compartment is closed. Also, for this purpose, rubber gaskets were placed between the poly-acrylic parts when the reactor was assembled. A 6 cm × 6 cm Nafion 117 membrane (Du Pont, USA) was used to separate the two compartments of each reactor. Each reactor was assembled using nuts and bolts penetrating holes at four corners of each big frame. Anode and cathode graphite rods were connected to crocodile clamps and through wires to an external resistor of 10 ohm and to a multimeter. Such a low resistance should allow the generation of higher current levels (Gil *et al.*, 2003).

For the influent and effluent (of anolyte or catholyte), two holes (5 mm in diameter) were created on the big frame of each compartment and PVC pipes were sealed to them. The anode influent pipe was inserted with a three-way connector before connected via a drip chamber to a bottle containing modified M9 medium (0.44 g KH2PO4 l^{-1}, 0.34 g K2HPO4 l^{-1}, 0.5 g NaCl l^{-1}, 0.2 g MgSO4.7H2O l^{-1}, 0.0146 g CaCl2 l^{-1}, pH 7) (Clauwaert *et al.*, 2007). The reactors were operated in batch mode at room temperature (22 ± 3°C) (unless otherwise stated). Before a batch, the M9 medium bottle was sterilized, cooled and purged with nitrogen (Messer, Vietnam) for 30–60 min to minimize the amount of oxygen, the potential competitor with the anode to accept electrons. To start a batch, a FeCl$_2$ solution (the source of ferrous ions) was syringed, together with a trace element solution (with the recipe following Clauwaert *et al.*, 2007), into the anode compartment of each MFC through the three-way connector on the anode influent pipe. The supplied volume and the concentration of the FeCl$_2$ solution were calculated so that the final concentration of Fe^{2+} in the anolyte will be as desired. The volume of the trace element solution was also calculated so that its final proportion in the anolyte was 0.1% (v/v). Subsequently, sterilized and nitrogen-purged M9 medium was sucked from the containing bottle, with a syringe, and pumped into the anode compartment, also through the three-way connector. The volume of the pumped-in medium was calculated such that half of the anolyte was replaced (approximately 10 ml). This replacement also helped remove a part of ferric precipitate formed in the anode compartment. Finally, a NaHCO$_3$ solution (the carbon source) was supplied into the anode compartment, in a similar manner, such that its final concentration in the anolyte was 2 g l^{-1} (Clauwaert *et al.*, 2007). This sequence of supplying the components of the anolyte ensures that ferrous carbonate precipitate was not formed (experimentally checked, data not shown). The cathode compartment of each MFC reactor contained only a buffer solution (0.44 g KH2PO4 l^{-1}, 0.34 g K2HPO4 l^{-1}, 0.5 g NaCl l^{-1}). At the beginning of each batch, this catholyte was renewed completely. During a batch, the cathode compartment was aerated, through the cathode influent pipe, with an air pump (model SL-2800, Silver Lake, China) to supply oxygen, the final electron acceptor. The aeration rate was adjusted to be slightly above 50 ml min^{-1} to ensure that the catholyte was air-saturated (Pham *et al.*, 2005) but did not evaporate fast. A batch of operation for a reactor was timed from the moment right after the anolyte was replaced until when the current dropped down to the baseline (usually about 2 h). Each reactor was operated at least three batches per day (with 1 h being the interval between two consecutive batches) and left standby during the nighttime. (This mode of operation did not affect the stability in the performance of the reactors.)

Measurement and calculation of electrical parameters

A digital multimeter (model DT9205A+, Honeytek, Korea) was used to measure the voltage between the anode and the cathode of each MFC. Electrical parameters [current I(A), voltage U(V) and resistance R(Ω)] were measured and/or calculated according to Aelterman and colleagues (2006) and Logan and colleagues (2006). Unless otherwise stated, all the values of average currents and charges reported in this study were the results of at least three repetitions.

Inoculation and enrichment procedures

After assembled, all the reactors were double-checked to ensure no leakages and bad electrical connections occurring. Several MFC reactors were set up in this study. One reactor [the (biotic) control] was not initially inoculated with any microbial source but operated in the same manner as other reactors and thus could be contaminated with microbes from the environment. Furthermore, in order to prove that the generation of electricity in the MFCs was not due to plain chemical reactions, an abiotic control was used. The abiotic control was a reactor of the same MFC type, with the anode compartment (including the electrode) sterilized (at 121°C, 1 atm, for 20 min) and subsequently tested with Fe^{2+} for only 2 h right after assemblage. Three other reactors, designated as the inoculated MFCs, were inoculated with a bacterial source (an inoculum) from a natural mud taken from a brownish water stream at the depth of 20 cm underneath the stream bottom, in Ung Hoa, Hanoi, Vietnam.

The inoculation was carried out as follows: In the first 3 days, the inoculum was daily supplemented into the anode compartment of each reactor (except the control) and the reactors were operated with 20 mM of Fe^{2+}. The inoculum was prepared by mixing 1 ml of sterile M9 medium with the pellet (after centrifuged at 4000× *g*, for 5 min) of 2 ml of the original bacterial source (the mud). After day 3, the reactors were operated without supplementation of inocula.

During the enrichment period (the first 4 weeks), all the MFC reactors were operated in the manner mentioned above with 20 mM of Fe^{2+} supplied into each anode compartment and the generation of electricity was monitored. Samples from their anolytes (1 ml each) were daily taken and preserved at 4°C (for later microbiological analyses) or at −20°C (for molecular analyses).

Measurement of bacteria density and ferrous iron

The OD of bacterial cells in each anode suspension sample was measured at 600 nm using a UV/VIS spectrophotometer (BioMate 3S, Thermo Scientific, USA). Before measurement, one volume of the sample was pretreated with 1/50 volume of 25% HCl solution to prevent the formation of ferrous precipitates that might interfere the OD signal.

The concentration of Fe^{2+} in an anolyte sample was measured by the phenanthroline assay (Braunschweig *et al.*, 2012). In short, 5 µl of each sample was diluted 60 times before being treated with 6 µl of 25% HCl solution as mentioned above. The resulting solution was centrifuged (4000× *g*, 10 min) and the supernatant was added with 30 µl of 5.2 M ammonium acetate solution, 12 µl of 21 mM phenanthroline solution and 255 µl of distilled water. After about 30 min, the 490 nm absorbance value of that final solution was determined using the mentioned spectrophotometer. That value was used to calculate the concentration of Fe^{2+} in the sample, based on a pre-determined calibration line.

Isolation and morphology analyses of bacteria from the anodes of the MFCs

After the enrichment period (i.e. when the current generations of the MFCs were steady), culturable bacteria in anode suspension samples of an inoculated MFC and the (biotic) control were analysed.

Bacteria in an anode suspension sample of a MFC were plated for isolation and colony counting on Winogradsky medium (0.5 g KH2PO4 l^{-1}, 0.5 g NaNO3 l^{-1}, 0.2 g CaCl2 l^{-1}, 0.5 g MgSO4.7H2O l^{-1}, 0.5 g NH4NO3 l^{-1}, 6 g Ammonium Ferric Citrat l^{-1}; 16 g agar l^{-1}) (Starosvetsky *et al.*, 2008) by dilution method. Colonies of each isolate were counted from at least three plates per dilution level. Isolates were subcultured and preserved in Luria–Bertani (LB) medium.

Cells of isolated strains were fixed and Gram stained following standard procedures (Madigan *et al.*, 2004) and observed under a light microscopy (Carl-Zeiss, Germany).

The judgment for overlapping isolates was based on colony morphology as well as cell morphology analyses.

Molecular analyses

Preparation of samples: Anode suspension samples from the MFCs were used as such. To prepare an anode surface sample, particles on 1 cm^2 of the electrode surface were scrapped off using a sterile razor and suspended in 1 ml of a pH 7 buffer solution (0.44 g KH2PO4 l^{-1}, 0.34 g K2HPO4 l^{-1}, 0.5 g NaCl l^{-1}).

DNA extraction and polymerase chain reaction (PCR)-DGGE: Samples were centrifuged (4000× *g*, 10 min) and the pellets were used for DNA extraction. Total DNA of a sample was extracted using standard methods (Boon *et al.*, 2000). DNA was quantified based on UV absorption at 260 nm. 16S rRNA gene fragments were amplified with the primers P63F (5′-CAGGCCTAACACATGCAAGTC-3′) and P1378R (5′-CGGTGTGTACAAGGCC CGGGAACG-3′). These fragments were used as the templates to amplify 200 bp fragments with the primers P338fGC and P518R (Muyzer *et al.*,

1993). These 200 bp fragments were subjected to DGGE with a denaturing gradient ranging from 45% to 60% (Boon et al., 2002) to analyse the compositions of the bacterial communities in the samples. The rate of change of a community was calculated as the percentage of change in the corresponding DGGE pattern (Marzorati et al., 2008).

Bands of interest on DGGE gel were cut off from the gel and spliced into small pieces using a sterile razor. The small gel pieces were subsequently suspended in 50 µl of deionized water for 24 h at 4°C to allow DNA to elute. This eluted DNA was used as the template to amplify again the DNA fragment corresponding to each band. The PCR products were purified with ExoSAP – IT kit (Affymetric, USA) before submitted to Integrated ADN Technologies – IDT (Singapore) for DNA sequencing.

For 16S rDNA analysis of the isolates, each isolate was cultured in LB broth for 24 h before the cells were harvested for DNA extraction (following the same procedure). From extracted DNAs, 16S rRNA gene fragments were amplified with the primers P63F and P1378R. PCR products were similarly purified and sequenced.

The analysis of DNA sequences and homology searches were completed with standard DNA sequencing programs and the BLAST server of the National Center for Biotechnology Information using the BLAST algorithm (Altschul et al., 1990).

FISH

After the enrichment, when the electrical generation was stable, samples from the anode suspension and anode surface of an inoculated MFC were taken and prepared as mentioned. Samples were fixed as follows: 1 ml of each sample was mixed with 0.1 ml of phosphate buffer saline (PBS) (10×, pH7). The mixture was supplemented with 1 ml of 37% (v/v) formaldehyde solution and kept at 4°C for at least 30 min. This mixture was subsequently centrifuged (5000× g, 5 min) and the pellet was washed with PBS. After washing, this cells-containing pellet was suspended in a solution containing 50% ethanol and 50% PBS solution. DNA hybridization was carried out as follows using probe PS1 (5′-ACGGUAGAGGAGCAAUC -3′) specific for neutrophilic iron-oxidizing bacteria (Siering and Ghiorse, 1997): 20 µl of each fixed sample was diluted in 2 ml of sterile deionized water and applied on a polycarbonate membrane and allow to naturally dry. Two microlitre of the probe solution (containing 5 ng µl⁻¹ of PS1 bound with the fluorescent-emitting compound Cy3 and supplied by IDT) were mixed with 18 µl of hybridization buffer (including 180 µl of 5 M NaCl ml⁻¹; 20 µl of 1 M Tris-HCl ml⁻¹; 350 µl of formamide ml⁻¹, and 1 µl of 10% SDS ml⁻¹), and chilled on ice, in the dark. This mixture was dropped onto the polycarbonate membrane carrying the already dried sample. The membrane was subsequently incubated in a hybridization chamber at 46°C for 1.5–3 h. After that, this sample-carrying membrane was washed with washing buffer (including 800 µl of 5M NaCl, 1 ml of 1M Tris/HCl, 0.5 ml of 0.5M EDTA, 50 µl of 10% SDS and water in a total volume of 50 ml) and heated at 48°C for 15 min. Next, the membrane was washed in sterile water for several seconds before wiped up with sterile tissue. Each sample-carrying membrane was treated with 50 µl of 4,6-diamidino-

2-phenylindole (DAPI) solution (1 µg ml⁻¹) for 3 min in the dark and quickly washed with sterile water and with 80% ethanol for several seconds. Finally, after drying, the samples were observed under a fluorescent microscope (model Axiostar Plus, Carl-Zeiss, Germany) using a specialized filter (552 nm) for Cy3 signals and a blue/cyan filter (460 nm) for DAPI signals. Images of samples were captured by a camera connected to the microscope. In these images, white dots corresponding to signals from stained cells were calculated by using Image J. Thus, the quantity of white dots from DAPI signals (x) indicated the number of total bacteria while that from Cy3 signals (y) indicated the number of PS1-probed bacteria. Therefore, in each sample, the proportion (%) of iron bacteria could be calculated as $y/x*100$.

Data analysis

All the experiments, unless otherwise stated, were repeated three times. Data were analysed using basic statistical methods by using Microsoft Excel: differences in data were evaluated by t-Test analysis; errors among replicates were expressed in the form of standard deviations.

Acknowledgement

This research is funded by Vietnam National Foundation for Science and Technology Development (NAFOSTED) under grant number 106.03-2012.06. It also received support from Korea Institute of Science and Technology (KIST) IRDA Alumni Program and International Foundation for Science (IFS – Sweden). The authors assure that there is no conflict of interest from any other party regarding the content of this paper.

References

Aelterman, P., Rabaey, K., Pham, H.T., Boon, N., and Verstraete, W. (2006) Continuous electricity generation at high voltages and currents using stacked microbial fuel cells. *Environ Sci Technol* **40**: 3388–3394.

Allen, R.M., and Bennetto, H.P. (1993) Microbial fuel cells – electricity production from carbohydrates. *Appl Biochem Biotechnol* **39**: 27–40.

Altschul, S.F., Gish, W., Miller, W., Myers, E.W., and Lipman, D.J. (1990) Basic local alignment search tool. *J Mol Biol* **215**: 403–410.

Arends, J.B., and Verstraete, W. (2012) 100 years of microbial electricity production: three concepts for the future. *Microb Biotechnol* **5**: 333–346.

Boon, N., Goris, J., De Vos, P., Verstraete, W., and Top, E.M. (2000) Bioaugmentation of activated sludge by an indigenous 3-chloroaniline-degrading *Comamonas testosteroni* strain, I2gfp. *Appl Environ Microbiol* **66**: 2906–2913.

Boon, N., De Windt, W., Verstraete, W., and Top, E.M. (2002) Evaluation of nested PCR-DGGE (denaturing gradient gel electrophoresis) with group-specific 16S rRNA primers for the analysis of bacterial communities from different wastewater treatment plants. *Fems Microbiol Ecol* **39**: 101–112.

Braunschweig, J., Bosch, J., Heister, K., Kuebeck, C., and Meckenstock, R.U. (2012) Reevaluation of colorimetric iron determination methods commonly used in geomicro-biology. *J Microbiol Methods* **89:** 41–48.

Cheng, S., Ye, Y., Ding, W., and Pan, B. (2014) Enhancing power generation of scale-up microbial fuel cells by optimizing the leading-out terminal of anode. *J Power Sources* **248:** 931–938.

Clauwaert, P., Rabaey, K., Aelterman, P., De Schamphelaire, L., Ham, T.H., Boeckx, P., *et al.* (2007) Biological denitrification in microbial fuel cells. *Environ Sci Technol* **41:** 3354–3360.

Cullimore, D.R., and McCann, A.E. (1978) The identification, cultivation and control of iron bacteria in ground water. In *Aquatic Microbiology.* Skinner, F.A., and Shewan, J.M. (eds.). London, UK: Academic Press, pp. 219–261.

Fernando, E., Keshavarz, T., and Kyazze, G. (2014) Complete degradation of the azo dye Acid Orange-7 and bioelectricity generation in an integrated microbial fuel cell, aerobic two-stage bioreactor system in continuous flow mode at ambient temperature. *Bioresour Technol* **156:** 155–162.

Gil, G.-C., In-Seop, C., Byung Hong, K., Mia, K., Jae-Kyung, J., Hyung Soo, P., and Hyung Joo, K. (2003) Operational parameters affecting the performance of a mediator-less microbial fuel cell. *Biosens Bioelectron* **18:** 327–334.

Ha, P.T., Tae, B., and Chang, I.S. (2008) Performance and bacterial consortium of microbial fuel cell fed with formate. *Energy Fuels* **22:** 164–168.

Hedrich, S., Schlomann, M., and Johnson, D.B. (2011) The iron-oxidizing proteobacteria. *Microbiology* **157:** 1551–1564.

ter Heijne, A., Hamelers, H.V.M., de Wilde, V., Rozendal, R.A., and Buisman, C.J.N. (2006) A bipolar membrane combined with ferric iron reduction as an efficient cathode system in microbial fuel cells. *Environ Sci Technol* **40:** 5200–5205.

ter Heijne, A., Hamelers, H.V.M., and Buisman, C.J.N. (2007) Microbial fuel cell operation with continuous biological ferrous iron oxidation of the catholyte. *Environ Sci Technol* **41:** 4130–4134.

Ishii, S.I., Suzuki, S., Norden-Krichmar, T.M., Phan, T., Wanger, G., Nealson, K.H., *et al.* (2014) Microbial population and functional dynamics associated with surface potential and carbon metabolism. *ISME J* **8:** 963–978.

Kim, B.H., Kim, H.J., Hyun, M.S., and Park, D.H. (1999) Direct electrode reaction of Fe(III)-reducing bacterium, Shewanella putrefaciens. *J Microbiol Biotechnol* **9:** 127–131.

Kim, B.H., Chang, I.S., Gil, G.C., Park, H.S., and Kim, H.J. (2003) Novel BOD (biological oxygen demand) sensor using mediator-less microbial fuel cell. *Biotechnol Lett* **25:** 541–545.

Kim, B.H., Park, H.S., Kim, H.J., Kim, G.T., Chang, I.S., Lee, J., and Phung, N.T. (2004) Enrichment of microbial community generating electricity using a fuel-cell-type electrochemical cell. *Appl Microbiol Biotechnol* **63:** 672–681.

Kim, B.H., Chang, I.S., and Gadd, G.M. (2007) Challenges in microbial fuel cell development and operation. *Appl Microbiol Biotechnol* **76:** 485–494.

Lee, J.Y., Phung, N.T., Chang, I.S., Kim, B.H., and Sung, H.C. (2003) Use of acetate for enrichment of electrochemically active microorganisms and their 16S rDNA analyses. *FEMS Microbiol Lett* **223:** 185–191.

Logan, B.E., and Regan, J.M. (2006) Electricity-producing bacterial communities in microbial fuel cells. *Trends Microbiol* **14:** 512–518.

Logan, B.E., Hamelers, B., Rozendal, R., Schrorder, U., Keller, J., Freguia, S., *et al.* (2006) Microbial fuel cells: methodology and technology. *Environ Sci Technol* **40:** 5181–5192.

Lovley, D.R., and Nevin, K.P. (2011) A shift in the current: new applications and concepts for microbe-electrode electron exchange. *Curr Opin Biotechnol* **22:** 441–448.

Madigan, M.T., Martinko, J., and Parker, J. (2004) *Brock Biology of Microorganisms.* New York, NJ, USA: Pearson Education.

Marzorati, M., Wittebolle, L., Boon, N., Daffonchio, D., and Verstraete, W. (2008) How to get more out of molecular fingerprints: practical tools for microbial ecology. *Environ Microbiol* **10:** 1571–1581.

Muyzer, G., de Waal, E.C., and Uitterlinden, A. (1993) Profiling of complex microbial populations using denaturing gradient gel electrophoresis analysis of polymerase chain reaction-amplified genes coding for 16S rRNA. *Appl Environ Microbiol* **59:** 695–700.

Pant, D., Van Bogaert, G., Diels, L., and Vanbroekhoven, K. (2010) A review of the substrates used in microbial fuel cells (MFCs) for sustainable energy production. *Bioresour Technol* **101:** 1533–1543.

Pham, H., Boon, N., Marzorati, M., and Verstraete, W. (2009) Enhanced removal of 1,2-dichloroethane by anodophilic microbial consortia. *Water Res* **43:** 2936–2946.

Pham, T.H., Jang, J.K., Chang, I.S., and Kim, B.H. (2004) Improvement of cathode reaction of a mediatorless microbial fuel cell. *J Microbiol Biotechnol* **14:** 324–329.

Pham, T.H., Jang, J.K., Moon, H.S., Chang, I.S., and Kim, B.H. (2005) Improved performance of microbial fuel cell using membrane-electrode assembly. *J Microbiol Biotechnol* **15:** 438–441.

Pham, T.H., Rabaey, K., Aelterman, P., Clauwaert, P., De Schamphelaire, L., Boon, N., and Verstraete, W. (2006) Microbial fuel cells in relation to conventional anaerobic digestion technology. *Eng Life Sci* **6:** 285–292.

Pham, T.H., Boon, N., Aelterman, P., Clauwaert, P., De Schamphelaire, L., Rabaey, K., and Verstraete, W. (2008a) High shear enrichment improve the performance of the anodophillic microbial consortium in a microbial fuel cell. *Microb Biotechnol* **1:** 487–496.

Pham, T.H., Boon, N., Aelterman, P., Clauwaert, P., De Schamphelaire, L., Vanhaecke, L., *et al.* (2008b) Metabolites produced by *Pseudomonas* sp. enable a Gram-positive bacterium to achieve extracellular electron transfer. *Appl Microbiol Biotechnol* **77:** 1119–1129.

Rabaey, K., and Rozendal, R.A. (2010) Microbial electrosynthesis – revisiting the electrical route for microbial production. *Nat Rev Microbiol* **8:** 706–716.

Rabaey, K., and Verstraete, W. (2005) Microbial fuel cells: novel biotechnology for energy generation. *Trends Biotechnol* **23:** 291–298.

Rabaey, K., Boon, N., Siciliano, S.D., Verhaege, M., and Verstraete, W. (2004) Biofuel cells select for microbial consortia that self-mediate electron transfer. *Appl Environ Microbiol* **70:** 5373–5382.

Rabaey, K., Boon, N., Hofte, M., and Verstraete, W. (2005) Microbial phenazine production enhances electron transfer in biofuel cells. *Environ Sci Technol* **39:** 3401–3408.

Rabaey, K., Van de Sompel, K., Maignien, L., Boon, N., Aelterman, P., Clauwaert, P., *et al.* (2006) Microbial fuel cells for sulfide removal. *Environ Sci Technol* **40:** 5218–5224.

Rabaey, K., Rodriguez, J., Blackall, L.L., Keller, J., Gross, P., Batstone, D., *et al.* (2007) Microbial ecology meets electrochemistry: electricity-driven and driving communities. *ISME J* **1:** 9–18.

Reguera, G., McCarthy, K.D., Mehta, T., Nicoll, J.S., Tuominen, M.T., and Lovley, D.R. (2005) Extracellular electron transfer via microbial nanowires. *Nature* **435:** 1098–1101.

Rosenbaum, M., He, Z., and Angenent, L.T. (2010) Light energy to bioelectricity: photosynthetic microbial fuel cells. *Curr Opin Biotechnol* **21:** 259–264.

Rozendal, R.A., Hamelers, H.V.M., Euverink, G.J.W., Metz, S.J., and Buisman, C.J.N. (2006) Principle and perspectives of hydrogen production through biocatalyzed electrolysis. *Int J Hydrogen Energy* **31:** 1632–1640.

Rozendal, R.A., Hamelers, H.V.M., Rabaey, K., Keller, J., and Buisman, C.J.N. (2008) Towards practical implementation of bioelectrochemical wastewater treatment. *Trends Biotechnol* **26:** 450–459.

Siering, P.L., and Ghiorse, W.C. (1997) Development and application of 16S rRNA-targeted probes for detection of iron- and manganese-oxidizing sheathed bacteria in environmental samples. *Appl Environ Microbiol* **63:** 644–651.

Starosvetsky, J., Starosvetsky, D., Pokroy, B., Hilel, T., and Armon, R. (2008) Electrochemical behaviour of stainless steels in media containing iron-oxidizing bacteria (IOB) by corrosion process modeling. *Corros Sci* **50:** 540–547.

Straub, K.L., Benz, M., Schink, B., and Widdel, F. (1996) Anaerobic, nitrate-dependent microbial oxidation of ferrous iron. *Appl Environ Microbiol* **62:** 1458–1460.

Sudek, L.A., Templeton, A.S., Tebo, B.M., and Staudigel, H. (2009) Microbial ecology of Fe (hydr)oxide mats and basaltic rock from Vailulu'u Seamount, American Samoa. *Geomicrobiol J* **26:** 581–596.

Wang, H., and Ren, Z.J. (2013) A comprehensive review of microbial electrochemical systems as a platform technology. *Biotechnol Adv* **31:** 1796–1807.

Supporting information

Additional Supporting Information may be found in the online version of this article at the publisher's web-site:

Fig. S1. Picture of a MFC reactor in this study. The reactor was assembled from polyacrylic frames and contains two compartments separated by a Nafion 117 membrane. Graphite granules were used as the electrode matrixes connecting with graphite rods at both compartments. Red wires connect an external circuit with the cathode while black wires connect it with the anode. Plastic tubes are for the influent and the effluent of purged air at the cathode and for the influent and the effluent of 'fuel' solution at the anode. The device is operated with an external resistor of 10 ohm at 25°C.

Fig. S2. Per-batch coulombic amounts generated by the MFCs in this study. Each coulombic amount (Q) is calculated as: $Q = I \times t$; whereas I is the current intensity (A) and t is the average time of a batch (from the feeding moment till the moment the current decreases by 95% of the maximum steady current).

Appendix S1. DNA sequences in this study

13

NextGen microbial natural products discovery

Claudia Schmidt-Dannert, Department of Biochemistry, Molecular Biology and Biophysics, University of Minnesota, 1479 Gortner Avenue, St. Paul, MN 55108, USA.

Small-molecule secondary metabolites isolated from microorganisms and plants provide the chemical scaffolds of a large fraction of today's pharmaceuticals. Evolutionary forces shaped the molecular complexity of these natural products that contribute to the exquisite binding of these compounds to biological targets. Starting with the discovery of penicillin by Fleming, we have seen a rapid increase in the discovery and production of natural products and derivatives thereof as antibiotics and other drugs. But once the 'easy to access' bioactive compounds have been isolated, the drug discovery pipeline slowed down beginning in the 1990s. Pharmaceutical companies turned away from natural products as screening programmes led to the rediscovery of known structures and development of structurally complex natural products into drugs using synthetic methods proved to be challenging and too expensive if no reliable biological sources were available. Considering the urgent need for the development of new drugs to combat multidrug-resistant pathogens and overcome long-term side-effects and/or reduction in effectiveness of current drugs, unlocking nature's treasure trough of small-molecule chemodiversity will be crucial for next-generation drug development (Gerwick and Moore, 2012; Basmadjian *et al.*, 2014; Genilloud, 2014).

Driven by advances in sequencing, gene synthesis, bioinformatics and metabolomics, the natural products discovery process is beginning to undergo a major transformation – away from the tedious isolation, screening and dereplication process to *in silico*-based bioprospecting approaches that seek to eventually transform genomic information directly into biosynthetic outputs (Lewis, 2013; Deane and Mitchell, 2014). The explosion in the number of available microbial genome sequences has given us a glance at the hidden natural product biosynthetic capacity of these organisms. Based on known sequence information for enzymes involved in synthesizing, e.g. the scaffolds of bioactive polyketides, non-ribosomal peptides or terpenes, numerous gene clusters (fungi) and operons (bacteria) can be identified in microbial genomes that are silent and for which no secondary metabolite products have been identified. This also includes many well-studied natural products producers such as *Streptomyces* and *Aspergillus* strains that express only a subset of their secondary metabolome under typical laboratory growth conditions (Brakhage, 2013; Doroghazi and Metcalf, 2013; Lim and Keller, 2014; Rebets *et al.*, 2014).

Our sequencing capacity is outpacing – by orders of magnitude – our ability to identify natural products gene cluster and most importantly, translate this sequence information into screenable molecules. The number of sequenced microbial genome sequences is rapidly approaching 5000 sequenced genomes, of which a large majority is bacterial genomes with only a few hundred fungal genomes available. With this large number of sequences available, the question becomes 'How does one most effectively search this vast sequence space for interesting natural products pathways?' One approach commonly used is to focus on a few groups of bacteria of fungi known to produce bioactive natural products and comprehensively identify within their genomes natural products biosynthetic operons or gene cluster, and then target the most diverse biosynthetic gene cluster for characterization. In many cases, products of target gene clusters are not produced at all or only at very low levels under laboratory growth conditions, requiring gene cluster activation either through exogenous stimuli or manipulation of genetic control elements which may be strain specific and a laborious undertaking. In the case that a strain is genetically tractable, gene disruption can then be used to specifically characterize biosynthetic gene functions. This 'reverse discovery' approach has been quite successfully used in genome-driven bioprospecting for a number of natural products identified in bacteria and some filamentous fungi (Lewis, 2013; Deane and Mitchell, 2014; Jensen *et al.*, 2014).

Such 'reverse discovery strategies', however, are limited to microorganisms that can be cultivated in the laboratory and that can be genetically manipulated, leaving out enormous biosynthetic diversity found in unculturable microbial species such as many higher fungi (see below) and from complex microbial ecosystems. Recent work has shown that metagenomic libraries from microbial ecosystems can be successfully arrayed and screened for large biosynthetic gene clusters of interest based on homology to conserved regions of known biosynthetic genes such non-ribosomal peptide synthases or polyketide synthases (Owen *et al.*, 2013). Fungi have a tremendous capacity for natural products biosynthesis, yet only a relatively small fraction of its large biodiversity has been explored so far. Natural products pathways have mostly been characterized from a relatively small subset of Ascomycota, including filamentous fungi like *Aspergillus*, *Penicillium* and *Fusarium* that are genetically tractable and can be readily cultured in the laboratory (Lazarus *et al.*, 2014). Basidiomycota, including the mushroom-forming fungi, have received almost no attention so far, despite the fact that they may have a quite distinct arsenal of natural products (Quin *et al.*, 2014). Genome surveys of the few hundred genomes in Joint Genome Institute's Fungal Genomics database shows that we have barely scratched the surface of the biosynthetic potential encoded in the small number of sequences genomes that represent a minuscule fraction of the fungal diversity. Major reasons for the slow progress in characterizing the secondary metabolome of many fungi (especially many Basidiomycota) is that they are frequently hard to work with: laboratory growth may be slow or not possible and genetic tools so readily available for bacteria and filamentous fungi are largely absent.

The future of natural products and drug discovery will be greatly influenced by how quickly the scientific community can develop strategies that will enable us to move away from the slow approaches for pathway identification and characterization that depend on first the growth of the producer organism and it then being genetically tractable to some extent. Instead, we should take full advantage of rapid and affordable whole genome sequencing, RNAseq and DNA synthesis where we can move rapidly from *in silico* biosynthetic pathway identification into a high-throughput synthetic biology workflow with the concurrent analytical profiling of heterologously assembled expression libraries. Implementation of such an *in silico* to natural products discovery platform begins with the accurate identification and structural annotation of biosynthetic pathways and genes in genomic data. A number of bioinformatics tools have been developed for genomic bioprospecting (Weber, 2014), but these tools rely on algorithms trains with hidden Markov models derived from known biosynthetic genes. These models need to be expanded to capture a larger biosynthetic diversity. Coding information will then be directly used to synthesize corresponding genetic constructs suitable for high-throughput pathway assembly which could be done using already existing synthetic biology methods (Cobb *et al.*, 2014).

Precise structural gene annotations will be essential for such an envisioned high-throughput synthetic biology workflow that relies on gene synthesis and assembly. From our own experience, we know that gene annotations in the genomes of many fungi are incorrect. Basidiomycota genes are very intron rich and many small intron/exons are incorrectly predicted using available models. Deep RNA sequencing of a cross-section of microbial species (fungi and bacteria) that can be grown in the lab will be crucial to develop algorithms for accurate structural annotation. High-resolution transcriptomics analysis of diverse species will enable the construction of gene co-expression networks built on physical distance to seed genes that are frequently associated with natural products biosynthetic pathways (e.g. cytochrome P450s, group transferases, transporters) could be a means for the discovery of novel pathways and sequences for broader *in silico* searchers. Considering that microbial secondary metabolite pathways are typically clustered and gene expression is co-regulated, such network analysis will be a powerful method for accurate delineation of biosynthetic gene clusters, including satellite clusters and split super-cluster pathways known in fungi. Finally, we may need to develop more than the common *Escherichia coli* and *Saccharomyces cerevisiae* host platforms for high-throughput refactoring and functional expression of pathways from a variety of sources to overcome for example potential co-factor, precursor limitations, product toxicity or the ability to express very large gene cluster. Considering the fast pace at which progress has and continues to be made in genomics and synthetic biology and also new methods being developed for compound screening and identification through high-resolution mass spectrometry (Krug and Muller, 2014), we should be optimistic that genomics-driven natural products drug discovery has bright future.

Acknowledgement

The author's research in natural products biosynthesis has been supported by the National Institutes of Health Grant GM080299.

References

Basmadjian, C., Zhao, Q., Bentouhami, E., Djehal, A., Nebigil, C.G., Johnson, R.A., *et al.* (2014) Cancer wars: natural products strike back. *Front Chem* **2:** 20.

Brakhage, A.A. (2013) Regulation of fungal secondary metabolism. *Nat Rev Microbiol* **11:** 21–32.

Cobb, R.E., Ning, J.C., and Zhao, H. (2014) DNA assembly techniques for next-generation combinatorial biosynthesis of natural products. *J Ind Microbiol Biotechnol* **41:** 469–477.

Deane, C.D., and Mitchell, D.A. (2014) Lessons learned from the transformation of natural product discovery to agenome-driven endeavor. *J Ind Microbiol Biotechnol* **41:** 315–331.

Doroghazi, J.R., and Metcalf, W.W. (2013) Comparative genomics of actinomycetes with a focus on natural product biosynthetic genes. *BMC Genomics* **14:** 611.

Genilloud, O. (2014) The re-emerging role of microbial natural products in antibiotic discovery. *Antonie Van Leeuwenhoek* **106:** 173–188.

Gerwick, W.H., and Moore, B.S. (2012) Lessons from the past and charting the future of marine natural products drug discovery and chemical biology. *Chem Biol* **19:** 85–98.

Jensen, P.R., Chavarria, K.L., Fenical, W., Moore, B.S., and Ziemert, N. (2014) Challenges and triumphs to genomics-based natural product discovery. *J Ind Microbiol Biotechnol* **41:** 203–209.

Krug, D., and Muller, R. (2014) Secondary metabolomics: the impact of mass spectrometry-based approaches on the discovery and characterization of microbial natural products. *Nat Prod Rep* **31:** 768–783.

Lazarus, C.M., Williams, K., and Bailey, A.M. (2014) Reconstructing fungal natural product biosynthetic pathways. *Nat Prod Rep* **31:** 1339–1347.

Lewis, K. (2013) Platforms for antibiotic discovery. *Nat Rev Drug Discov* **12:** 371–387.

Lim, F.Y., and Keller, N.P. (2014) Spatial and temporal control of fungal natural product synthesis. *Nat Prod Rep* **31:** 1277–1286.

Owen, J.G., Reddy, B.V., Ternei, M.A., Charlop-Powers, Z., Calle, P.Y., Kim, J.H., and Brady, S.F. (2013) Mapping gene clusters within arrayed metagenomic libraries to expand the structural diversity of biomedically relevant natural products. *Proc Natl Acad Sci USA* **110:** 11797–11802.

Quin, M.B., Flynn, C.M., and Schmidt-Dannert, C. (2014) Traversing the fungal terpenome. *Nat Prod Rep* **31:** 1449–1473.

Rebets, Y., Brotz, E., Tokovenko, B., and Luzhetskyy, A. (2014) Actinomycetes biosynthetic potential: how to bridge in silico and in vivo? *J Ind Microbiol Biotechnol* **41:** 387–402.

Weber, T. (2014) In silico tools for the analysis of antibiotic biosynthetic pathways. *Int J Med Microbiol* **304:** 230–235.

14

Culture-dependent and culture-independent characterization of potentially functional biphenyl-degrading bacterial community in response to extracellular organic matter from *Micrococcus luteus*

Xiao-Mei Su,[1] Yin-Dong Liu,[1]
Muhammad Zaffar Hashmi,[1] Lin-Xian Ding[2] and
Chao-Feng Shen[1]*
[1]*Department of Environmental Engineering, College of Environmental and Resource Sciences, Zhejiang University, Hangzhou 310058, China.*
[2]*College of Geography and Environmental Science, Zhejiang Normal University, Jinhua 321004, China.*

Summary

Biphenyl (BP)-degrading bacteria were identified to degrade various polychlorinated BP (PCB) congers in long-term PCB-contaminated sites. Exploring BP-degrading capability of potentially useful bacteria was performed for enhancing PCB bioremediation. In the present study, the bacterial composition of the PCB-contaminated sediment sample was first investigated. Then extracellular organic matter (EOM) from *Micrococcus luteus* was used to enhance BP biodegradation. The effect of the EOM on the composition of bacterial community was investigated by combining with culture-dependent and culture-independent methods. The obtained results indicate that *Proteobacteria* and *Actinobacteria* were predominant community in the PCB-contaminated sediment. EOM from *M. luteus* could stimulate the activity of some potentially difficult-to-culture BP degraders, which contribute to significant enhancement of BP biodegradation. The potentially difficult-to-culture bacteria in response to EOM addition were mainly *Rhodococcus* and *Pseudomonas* belonging to G*ammaproteobacteria* and *Actinobacteria* respectively. This study provides new insights into exploration of functional difficult-to-culture bacteria with EOM addition and points out broader BP/PCB degrading, which could be employed for enhancing PCB-bioremediation processes.

Introduction

Polychlorinated biphenyls (PCBs) are toxic persistent pollutants that threaten both natural ecosystem and human health (Zanaroli *et al.*, 2010). *In situ* PCB bioremediation have aroused increasing concern because they are less expensive and more environmentally sound than conventional methods (Leigh *et al.*, 2006). A large number of bacteria have been identified and shown to cometabolize PCBs through the biphenyl (BP) catabolic pathway (Petrić *et al.*, 2007). Indeed, research on aerobic PCB biodegradation bacteria isolated so far has mainly focused on BP-utilizing bacteria. Numerous phylogenetically diverse BP-utilizing bacteria that have the capability to transform several PCB congeners have been isolated (Abraham *et al.*, 2002; Pieper, 2005). However, until recently, the full-scale bacterial remediation of PCB-contaminated environment performed not very well, because the activity and capability of BP/PCB-degrading strains soon decreased when exposed to natural environment. Although bioaugmentation of sites with degradative bacteria have been unsuccessful application in the field, efforts to stimulate indigenous BP/PCB-degrading bacteria have been promising (Leigh *et al.*, 2006).

In the natural environment, it is very common for bacteria to survive under a wide variety of stress conditions by entering a 'viable but non-culturable' (VBNC) state, in which cells are intact and alive but fail to normally grow on the routine bacteriological media (Oliver, 2010). The high toxicity of PCBs and their low bioavailability exerted significant stress on indigenous microorganisms. It should be interesting to find out the survival and activity

*For correspondence. E-mail ysxzt@zju.edu.cn, purple@zju.edu.cn

Funding Information We gratefully acknowledge the financial support provided by the Fundamental Research Funds for the Central Universities (2014QNA6012), the National Natural Science Foundation of China (41271334), the National High Technology Research and Development Program of China (2012AA06A203) and Zhejiang Provincial Natural Science Foundation of China (LR12D01001).

of bacteria exposed to such adverse conditions (Chávez et al., 2006). Although numerous BP/PCB-degrading bacteria have been isolated and studied, there are more phylogenetically diverse bacteria detected in the environment using molecular tools (Macedo et al., 2007). Undoubtedly, using conventional plate separation methods, only a small fraction of BP/PCB-degrading bacteria existed in nature can be obtained. Furthermore, it is common knowledge that artificial mixed cultures consisting of purified cultivable isolates from enrichment cultures are less efficient in BP/PCB degradation than mixed cultures (Mikesková et al., 2012; Uhlik et al., 2012). One reason for this discrepancy is that there are abundant VBNC or uncultured bacteria possessing BP/PCB-degrading abilities in mixed cultures (Tang et al., 2013). Hence, the resuscitation and stimulation of potential BP/PCB-degrading indigenous bacteria is crucial for in situ bioremediation of PCBs.

The most exciting development in reaction of VBNC bacteria is the role of extracellular protein secreted by Micrococcus luteus, known as a resuscitation-promoting factor (Rpf), which has been shown to promote the resuscitation and growth of high $G + C$ Gram-positive organisms (Mukamolova et al., 1998; Su et al., 2013b). It is worth noting that these organisms contain well-known BP/PCB degraders, including Rhodococcus, Arthrobacter, Bacillus and Microbacterium (Pieper, 2005; Petrić et al., 2007). In addition, Ding and Yokota (2010) indicated that Rpf could also enhance the culturability of several other Gram negative. Hence, Rpf is capable of culturing difficult-to-culture bacteria involved in the BP/PCB degradation process and stimulating the activity of indigenous bacterial communities in PCB-contaminated environment. Mukamolova and colleagues (2006) indicated that Rpf control the culturability of several bacteria because of its muralytic activity. And at least two additional extracellular proteins in the M. luteus culture supernatant were also found to possess the same muralytic activity as Rpf protein. On the other hand, the recombinant Rpf protein presented lower activity than the native Rpf protein (purified from M. luteus culture supernatant), and both of them were prone to lose its activity after storage at 4°C for 1 week (Mukamolova et al., 2006). By analysing on the disadvantage of Rpf protein, extracellular organic matter (EOM) from M. luteus offers an attractive and cost-effective alternative additive for resuscitating and stimulating VBNC or difficult-to-culture bacteria, as well as enhancing the activity and degrading capability of indigenous bacteria in PCB-contaminated area.

The bacterial composition of the PCB-contaminated sediment sample was first investigated in the present study. Above all, to test the hypothesis that EOM from M. luteus was a feasible and effective additive for enhancing bacterial BP-degradation capability, the effect of EOM on BP biodegradation and bacterial community was assessed. Specifically, the degradation abilities, bacterial composition and abundance of enrichment cultures with different experimental treatment were investigated. In this work, we aimed to identify whether EOM could resuscitate the BP-degrading potential of VBNC or difficult-to-culture bacteria by isolating pure cultures unique to the enrichment culture with EOM addition. To our best knowledge, this is the first attempt to reveal the effect of EOM from M. luteus on BP-degrading bacterial communities using the combination of Illumina high-throughput sequencing and culture-dependent methodology.

Results and discussion

Bacterial composition in the PCB-contaminated sediment

Illumina high-throughput sequencing was performed to determine the diversity and composition of the bacterial communities in the sediment sample. The top 10 dominant phyla/genera are shown in Fig. 1. As shown in Fig. 1A, of the 10 major phyla, the most predominant phylum was Proteobacteria, which represented 31.3% of the total bacteria. Actinobacteria, Chloroflexi and Firmicutes were the subdominant groups, constituting around 50% of the total population. At the genus level (Fig. 1B), the four most abundant genera, which belonged to Rhodococcus, Achromobacter, Mycobacterium and Dechloromonas, accounted for 18.8% of all sequences. The proportion of sequences assigned as unclassified constituted highly 31.3% of the total population. The composition of the bacterial community of the sediment sample identified in this study is in accordance with previous studies (Fukuda et al., 1998; Macedo et al., 2007; Uhlik et al., 2009), which are rather common abundant bacteria in PCB-contaminated environments adapted for biodegradation of the pollutants. To improve bioremediation practices, it is important to understand the composition of bacterial community in long-term PCB-contaminated sites (Petrić et al., 2011). These results indicate that genera Rhodococcus and Achromobacter with outstanding degradation abilities were widespread in long-term PCB-contaminated sediment, in which the total of 20 PCB congeners (from one to six chlorine atoms) concentration averaged 29.16, and the standard deviation was 0.72 µg g^{-1}. In addition, it would be interesting to find that majority indigenous bacteria cannot be cultivated. Indeed, yet-to-be cultured bacteria probably account for the majority of the degradation activity among indigenous bacteria (Uhlik et al., 2012; Tang et al., 2013).

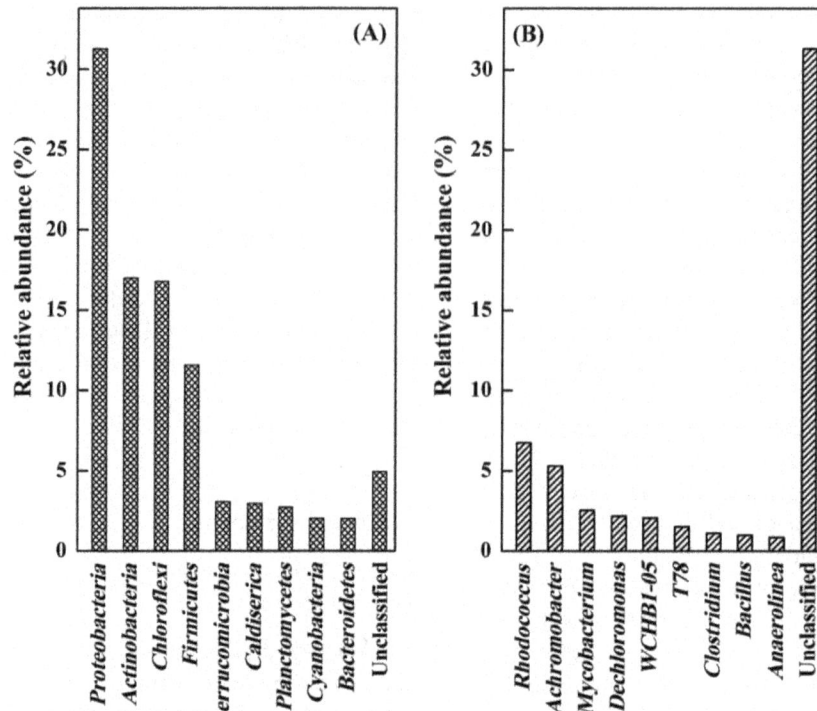

Fig. 1. Taxonomic composition of bacterial community in the PCB-contaminated sediment sample. (A) Taxonomic affiliations at the phylum level. (B) Taxonomic affiliations at the genus level.

Effect of EOM on BP degradation

Based on the BP-degradation efficiency and cell growth (OD$_{600}$), the effect of EOM on the BP tolerance concentration, BP-degrading curve and cell growth was investigated. The degradation of BP by enrichment samples under various BP concentrations (from 500 to 4500 mg l^{-1}) is detailed in Fig. 2. As shown in Fig. 2, the BP-degradation efficiency of all the groups decreased with elevated BP concentrations. When BP concentration was increased from 500 to 3500 mg l^{-1}, the BP-degradation efficiency decreased from 99.9% to 68.3% in treatment group (TG), whereas it decreased from 65.5% to 12.4% in blank group (BG). Under a BP concentration of 4000 mg l^{-1}, BP-degradation efficiency reached 59.5% in TG, whereas the efficiency was 29.6% and 7.6% in CG and BG respectively. Similar to the trend of variations in BP-degradation efficiency, cell growth decreased with elevated BP concentrations in the CG and BG. However, the values of OD$_{600}$ in TG underwent an increasing trend, reaching its maximum value of 2.63 at 2000 mg l^{-1}, and then decreased to 0.58 at BP concentration of 4500 mg l^{-1}. Meanwhile, under a BP concentration of 4000 mg l^{-1}, the value of OD$_{600}$ in TG was observed to be around fivefold and ninefold greater than that observed in control group (CG) and BG. Meanwhile, Fig. 3 depicts the changes in the BP-degradation curves

and cell growth among the three groups. As shown in Fig. 3, the BP-degradation efficiency and cell growth were significantly enhanced in TG. After 60 h, the BP-degradation efficiency in TG reached up to 97.4%, whereas the efficiency was less than 60.3% in CG and 47.2% in BG. The overall trend in cell growth coincided with the degradation rate, and significant increases in the value of OD$_{600}$ were recorded. These results indicated that TG with EOM addition maintained higher degradation efficiency and faster cell growth and was more tolerant to increasing BP concentration. Specifically, the BP-degradation efficiencies of TG were higher than those of CG, suggesting that EOM performed better than autoclaved EOM in enhancing BP degradation. The better performance of TG could be attributed to the positive effects of some proteins in EOM on cell growth and BP-degradation efficiency, because autoclaved EOM had the same constituents as EOM except proteins. In addition, the BP-degradation efficiencies of CG were higher than BG, suggesting that some polysaccharides of EOM may also play a role in enhancing BP degradation.

In previous studies, the resuscitation and stimulation function of proteins in the culture supernatants of *M. luteus* has been tentatively verified (Ding, 2004; Su et al., 2013b). This study further indicated that EOM from *M. luteus* as an efficient additive could significantly enhance the performance of BP biodegradation. Recent

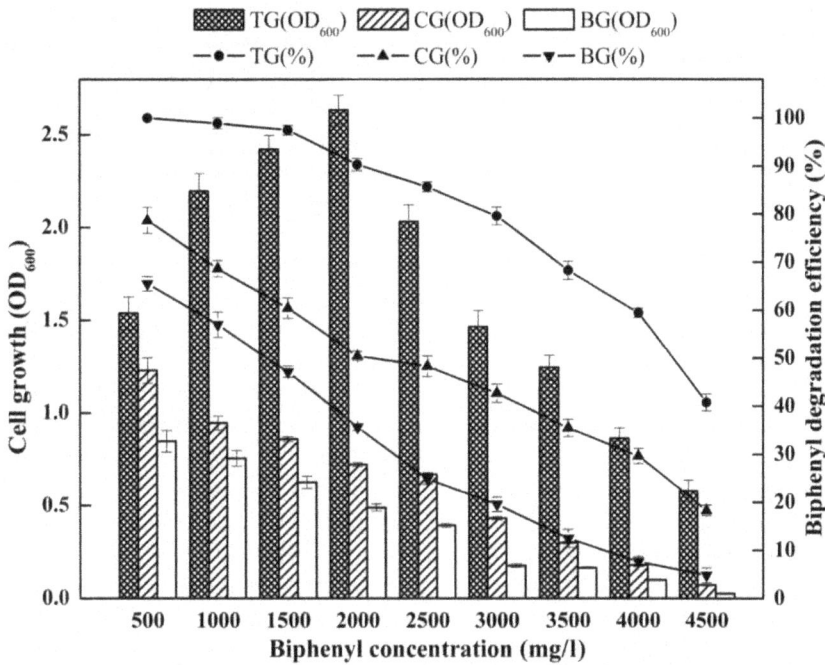

Fig. 2. Comparison of tolerance concentration of biphenyl among different groups of enrichment cultures. TG: addition of EOM; CG: addition of autoclaved EOM; BG: without EOM. Error bars indicate the standard deviations of triplicate samples.

advances in PCB bioremediation processes have focused on selecting BP-degrading enrichment cultures to obtain BP/PCB-degrading community and identifying the key members involved in this process using culture-independent approaches (Cámara *et al.*, 2004; Uhlik

et al., 2012). Nevertheless, most of bacteria and enrichment cultures, which were obtained with high removal rates in the laboratory, showed exceptionally slow removal rates *in situ* (Abraham *et al.*, 2002). Therefore, the challenge for PCB-bioremediation is to enhance

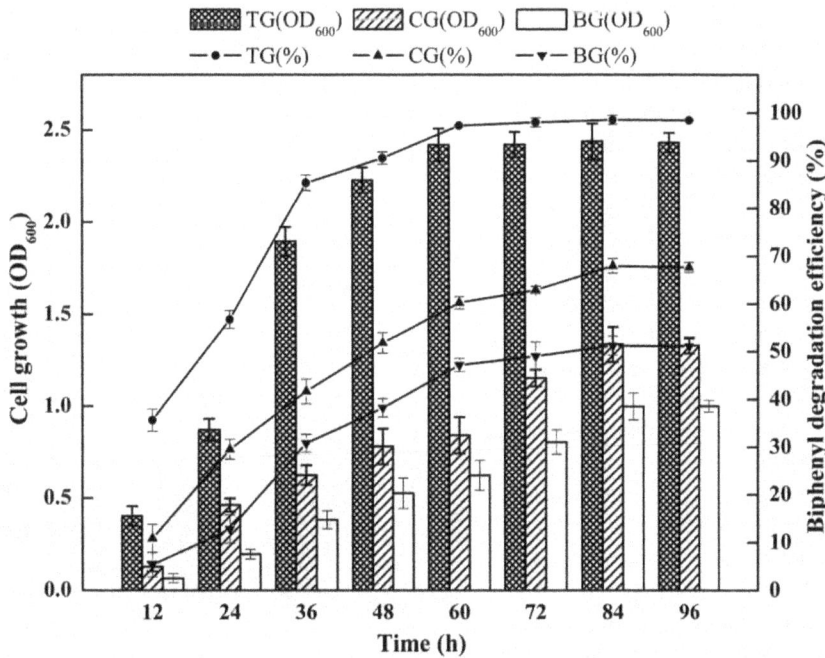

Fig. 3. Comparison of biphenyl-degrading curve among different enrichment cultures at a concentration of 1500 mg l⁻¹. TG: addition of EOM; CG: addition of autoclaved EOM; BG: without EOM. Error bars indicate the standard deviations of triplicate samples.

Table 1. Alpha diversity of samples.

	OTUs	Chao 1 richness estimation	Shannon diversity
Sediment sample	5623	23 187	9.86
Enrichment sample TG	2272	10 561	5.60
Enrichment sample CG	2175	9803	5.02
Enrichment sample BG	1981	9259	4.87

the activities of bacterial communities involved in biodegradative processes and develop approaches to realize the full potential of BP/PCB degraders (Gomes et al., 2013). Clearly, culture-dependent studies can focus only on a limited number of bacteria that are not likely to represent the bacterial diversity in the environment, as there are abundant of VBNC or difficult-to-culture bacteria that are important players in bioremediation existed ubiquitously in PCB-contaminated (de Cárcer et al., 2007; Zanaroli et al., 2010). Although culture-independent molecular techniques can investigate the diversity of bacteria potentially responsible for ecologically relevant processes, to elucidate bacteria related function and genotype, it will be necessary to recover them and study their microbiology in pure cultures (Uhlik et al., 2012; Su et al., 2013b). Thus, EOM from M. luteus can be employed for culturing difficult-to-culture bacteria and for exploiting the potential environmental functions of VBNC or difficult-to-culture bacteria, which will demonstrate highly valuable in future bioremediation strategies.

Effect of EOM on bacterial diversity and composition

Illumina high-throughput sequencing was adopted to investigate bacterial community diversity and richness among sediment sample and enrichment cultures (TG, CG and BG) (Table 1). In order to compare the bacterial species richness, operational taxonomic units (OTUs) were estimated by Chao1 estimator at a distance level of 3%. The total number of OTUs in TG (2272) with EOM addition was the largest among the enrichment cultures. And BG (1981) contained the least OTUs amount, indicating that TG with EOM addition exhibited the greatest bacterial richness. Furthermore, the bacterial phylotype richness can also be reflected using Shannon diversity index, which is generally used to demonstrate species diversity in a bacterial community, and it accounts for both evenness and abundance of the species present (Ma et al., 2013). Considering the fact that Shannon index of TG (5.60) was much higher than that of CG (5.02) and BG (4.87), it could be inferred that the enriched OTUs in TG community distributed more evenly than those in CG and BG. These results suggested that bacterial community diversity and richness increased with EOM addition.

The effective bacterial sequences in the TG, CG and BG were all assigned to corresponding taxonomies. Figure 4 shows the taxonomic compositions at the class level for the three enrichment cultures, demonstrating a relative abundance greater than 0.1% in at least one enrichment culture. In total, 10 identified class were observed; Bataproteobacteria, Actinobacteria and Gammaproteobacteria were the dominant class, which were consistent with previous studies on bacterial community composition in polycyclic aromatic hydrocarbon and PCB exposed environments (Stach and Burns, 2002; Petrić et al., 2011). However, the relative abundance of each dominant class among the three enrichment cultures was distinct. Especially, the relative abundance of Bataproteobacteria, Actinobacteria and Gammaproteobacteria were 27.6%, 39.3% and 24.3% in TG, and 58.0%, 15.6% and 18.1% in BG. Actinobacteria were the most dominant bacterial community in TG instead of Bataproteobacteria in CG and BG. Thus, it could be inferred that after EOM addition, Actinobacteria and Gammaproteobacteria were greatly enriched. Moreover, a small fraction of class Bacteroidia accounting for less than 0.1% of total community showed a similar trend, which the order of abundance was TG > CG > BG. The Bacteroidia class is believed to be involved in the degradation of aromatic compounds (Xu et al., 2012). Petrić and colleagues (2011) had previously reported that Actinobacteria and Bacteroides were the predominant phyla in bioremediation of PCB-contaminated soil. In view of these results, it could be concluded that EOM from M. luteus greatly affects the composition and abundance of bacterial communities closely related to BP/PCB degradation.

To further explore the variation in bacterial community with EOM addition, bacterial abundance was also analysed more specifically at the genus level. As Fig. 5

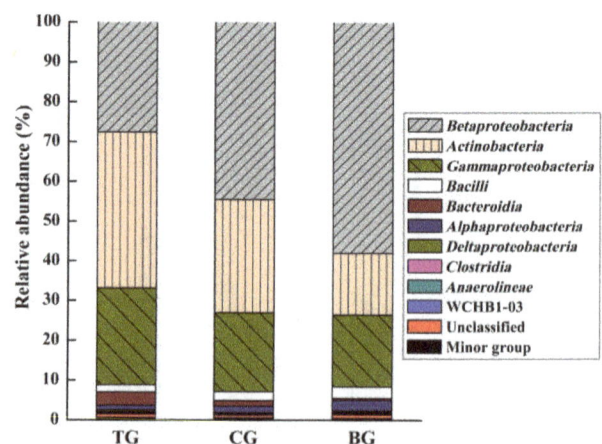

Fig. 4. Taxonomic composition of bacterial communities at the class level in enrichment cultures. TG: addition of EOM; CG: addition of autoclaved EOM; BG: without EOM.

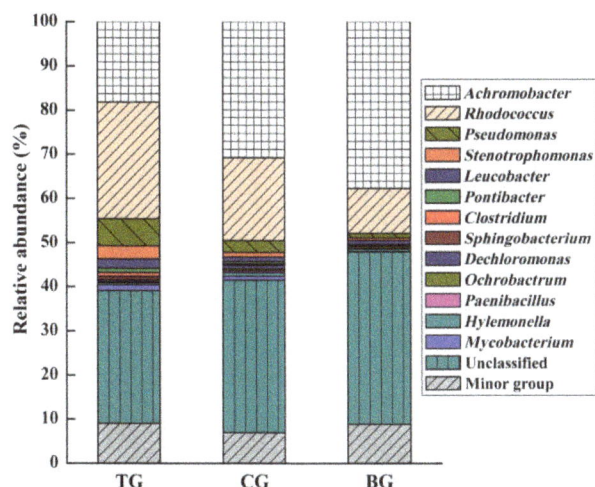

Fig. 5. Taxonomic composition of bacterial communities at the genus level in enrichment cultures. TG: addition of EOM; CG: addition of autoclaved EOM; BG: without EOM.

illustrates, a relative abundance greater than 0.1% in at least one enrichment culture is summarized. Other genera were grouped into minor groups. At the genus level, the majority of sequences were affiliated to five types, such as *Achromobacter*, *Rhodococcus*, *Pseudomonas*, *Stenotrophomonas* and unclassified bacteria. And the composition of TG with EOM addition was also reflected to be dramatically different. The genus *Rhodococcus* dominated TG (26.4% of total reads) community, followed by *Achromobacter* (18.2%), *Pseudomonas* (6.1%) and *Stenotrophomonas* (3.1%). However, the genus *Achromobacter* was the most dominant bacterial community, accounting for 30.8% and 37.69% of the total reads in CG and BG community respectively. Notably, the order of abundance for the genera *Rhodococcus*, *Pseudomonas* and *Stenotrophomonas* was TG > CG > BG. The discrepancy between TG and CG could be attributed to the function of some proteins in EOM. Compared with BG, CG with autoclaved EOM had higher relative abundance of the three genera which may be attributable to the stimulation function of polysaccharides in EOM. Overall, well-known PCB degraders (*Rhodococcus*, *Pseudomonas* and *Stenotrophomonas*) were greatly abundant after EOM addition (Macedo *et al.*, 2007; Correa *et al.*, 2010). The results in this study is consistent with earlier reports that Rpf could promote the resuscitation and growth of not only high G + C Gram-positive organisms, but also several other Gram-negative organisms (Mukamolova *et al.*, 2006; Oliver, 2010; Su *et al.*, 2013a).

It was interesting to observe that a large number of unclassified sequences (30.1–38.9% of total reads) were found in the enrichment cultures, suggesting that a wide variety of novel species or yet-to-be cultured bacteria may

inhabit complex BP enrichment cultures communities. The results indicated that some uncultured or VBNC bacteria frequently detected in PCB-contaminated sites and BP enrichment cultures may be highly correlated to the functions and performances of BP/PCB biodegradation process. This is in agreement with previous investigation on degradative bacterial communities in BP/PCB-contaminated environments (Abraham *et al.*, 2002; Uhlik *et al.*, 2012). Indeed, the traditionally isolated genera of bacteria possessing the capability for pollutants degradation represent only a small fraction of the total diversity in nature. In this study, bacteria in the enrichment culture were exposed to high concentrations of BP and progressive nutrient depletion, both of which may initiate the VBNC state. Tang and colleagues (2013) found that many clones related to BP/PCB degradation were uncultured bacteria. Furthermore, *Actinobacteria* as the predominance of the phylum in the BP/PCB-degrading community process was prone to enter VBNC state in which they have significantly reduced metabolic activity and lose culturability (Keep *et al.*, 2006; Mukamolova *et al.*, 2006). In addition, Mukamolova and colleagues (2006) had previously shown that the culturability of several *Actinobacteria* is controlled by Rpf. And Rpf homologues are widespread throughout the *Actinobacteria*. Meanwhile, Schroeckh and Martin (2006) found that Rpf could resuscitate *Actinobacteria* and was a useful tool for isolating new actinobacterial species. Typically, *Rhodococcus* is known to enter into a VBNC state quite rapidly under adverse conditions, such as *R. rhodochrous* and *R. fascians* (Shleeva *et al.*, 2002). The results supported the hypothesis that the positive effects of EOM on BP-degrading capability enhancement was mainly caused by its ability to resuscitate and stimulate VBNC bacteria, especially for BP/PCB-degrader *Rhodococcus*-like populations. Therefore, it may be inferred that EOM from *M. luteus* containing the Rpf and Rpf-like proteins provides a useful approach for isolating novel species or yet-to-be cultured bacteria and realizing the full potential of BP/PCB degraders.

Isolation and phylogenetic analysis of bacteria

In order to further verify the resuscitating function of EOM, the three enrichment cultures (TG, CG and BG) were also investigated using a culture-dependent method. Colonies that were unique in TG, with no counterpart in CG and BG, were isolated using a modified most probable number method. Eleven unique strains in TG (TG1–TG11) were isolated on mineral salts agar plates. Phylogenetic analysis was based on 16S rDNA sequences. Closely matching representatives were determined by a Blast search at GenBank. Figure 6 presents an overview of the 11 unique strains based phylogenetic tree, generated by including

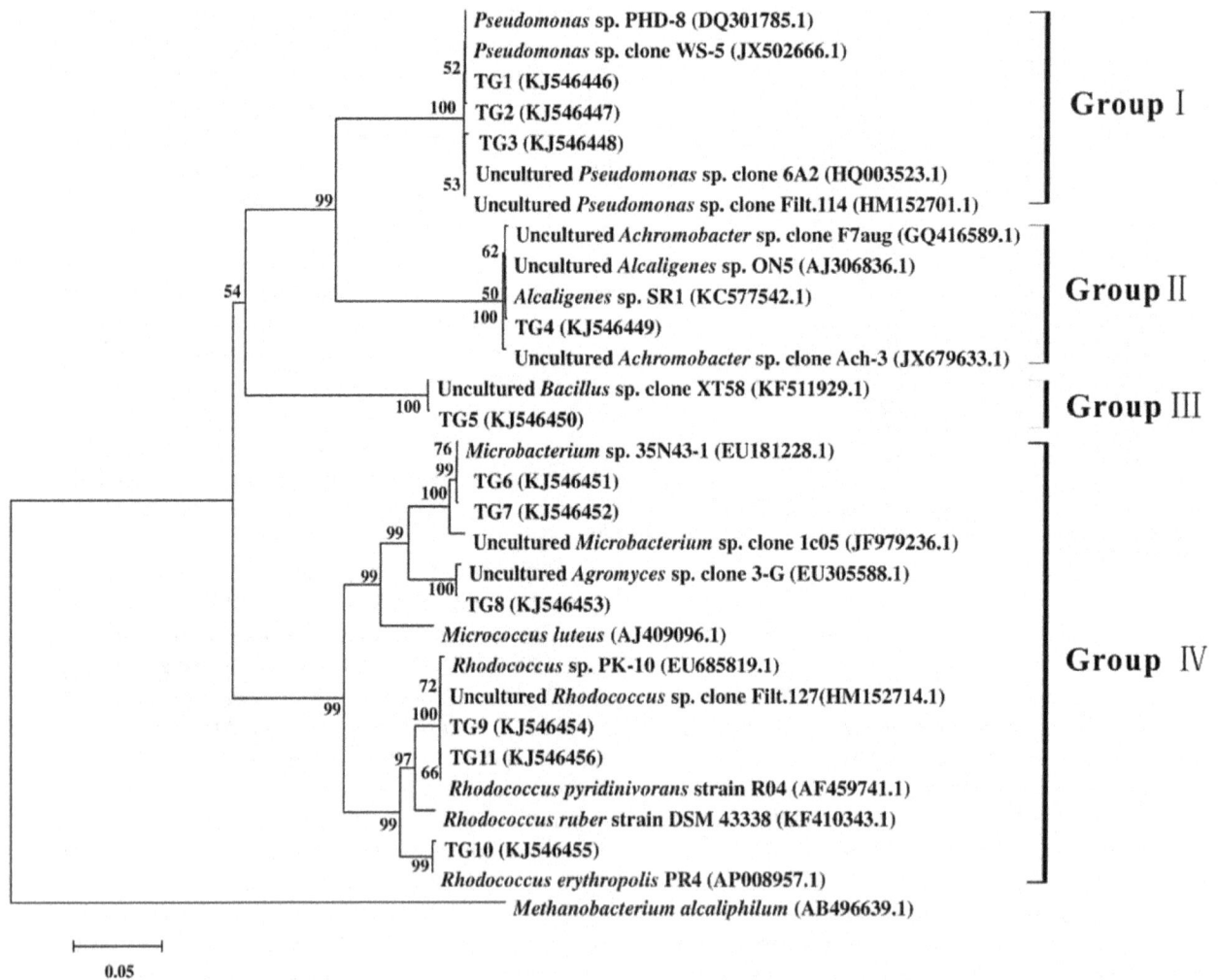

Fig. 6. Phylogenetic tree of bacterial 16S rRNA gene sequences, including eleven isolates unique to enrichment culture with EOM addition and 18 of their most similar GenBank sequences. Bootstrap values (> 50%) are showed at branch points. The bar represents a sequence distance of 0.1.

representative members. Overall, at the class level, strains belonging exclusively to TG can be broadly divided into four subgroups. The most dominant classified subgroup was *Actinobacteria* (subgroups IV), accounting for 54.5% (6/11) of the total strains. G*ammaproteobacteria* (subgroups I) was the subdominant group, which constituted 27.2% of the total strains. *Bataproteobacteria* (subgroups II) and *Bacilli* (subgroups III) only contributed to about 18% of the total population. The results are consistent with the investigations on variation of community composition in TG analysed by Illumina high-throughput sequencing. And the analyses based on the culture-dependent method further confirmed that EOM could resuscitate functional bacteria for BP degradation, especially the *Actinobacteria*.

Moreover, it was worth noting that strains TG6–TG11 belonging to genera *Microbacterium*, *Agromyces* and *Rhodococcus* are highly homologous with *M. luteus*.

Furthermore, it is evident that strain TG9 represents a novel species of the genus *Rhodococcus*, for which the name *Rhodococcus biphenylivorans* sp. nov. is proposed (Su *et al.*, 2015). In addition, VBNC state of the strain TG9^T in response to low temperature and oligotrophic nutrients was verified (data not shown). This is agreement with previous reports that Rpf could promote the resuscitation and growth of Gram-positive organisms (Keep *et al.*, 2006; Su *et al.*, 2013a). By contrast, strains TG1–TG4 belonging to genera *Pseudnomonas* and *Achromobacter* are Gram-negative bacteria, which were well supported by the observation that Rpf also resuscitated the growth of several other Gram-negative bacteria (Ding and Yokota, 2010; Su *et al.*, 2013b). Additionally, these bacteria unique to TG were closely related to uncultured bacteria, which are known for their catabolic diversity in degradation of aromatic compounds (Zanaroli *et al.*, 2010; Xu *et al.*, 2012).

In view of these results, it could be inferred that after EOM addition, functional difficult-to-culture bacteria belonging to genera *Microbacterium*, *Agromyces*, *Rhodococcus*, *Pseudnomonas* and *Achromobacter* recovered their culturability.

Application prospect of EOM from M. luteus

In the present study, the influence of EOM from *M. luteus* on the BP-degradation capability and composition of bacterial community in PCB-contaminated sediment was first evaluated by combining with culture-dependent and culture-independent methods. The obtained results provided evidence that EOM from *M. luteus* could significantly enhance the BP-degrading capability, which could be attributed to enrichment of some potentially VBNC or difficult-to-culture BP/PCB degraders. In addition, it was evident that the mainly functional bacteria in response to EOM addition were *Actinobacteria*, which were prone to enter into VBNC state and showed a low degrading activity in long-term PCB-contaminated sediment.

Recent advances in the bioremediation of PCB-contaminated sites have focused on the development of ways to stimulate the activities of indigenous BP/PCB-degrading community (Leigh *et al.*, 2006; Gomes *et al.*, 2013). However, the application of PCB bioremediation is still inefficient and not well established up to now. It has been known that diverse BP-utilizing bacteria have the capability to transform several PCB congeners by *bph* encoded BP pathway (Cámara *et al.*, 2004). And numerous studies have focused on the BP-degrading bacterial community and BP degradation pathway in order to explore a competitive advantage of PCB degraders and establish optimized PCB-bioremediation processes (Pieper, 2005; Uhlik *et al.*, 2009). However, highly efficient BP/PCB-degrading community in the laboratory experiments showed lower efficiency and survived poorly when these cultures were inoculated in PCB-contaminated sites. Thus, it will be of critical importance to stimulate functional bacteria to enhance their BP/PCB degrading capabilities for *in situ* bioremediation. The present study demonstrated that EOM from *M. luteus* is an efficient additive which can significantly enhance BP biodegradation by recovering and stimulating the potentially functional BP-degrading community. It provided new insight into the exploration of potentially functional bacteria for enhancing in situ bioremediation, which could be inferred that the addition of EOM to PCB-contaminated areas holds great potential for the efficient and cost-effective bioremediation of PCB-contaminated environments.

Conclusions

The obtained results suggest that *Proteobacteria* and *Actinobacteria* were two predominant classes in long-term PCB-contaminated sediment. EOM from *M. luteus* enhanced the performance of BP biodegradation, which could be attributed to stimulation of the growth and activity of some potentially BP/PCB-degraders. Illumina high-throughput sequencing and culture-dependent methods indicated that the genera of *Rhodococcus* and *Pseudomonas* which were related to BP/PCB-degradation were greatly abundant after EOM addition, and potentially difficult-to-culture bacteria in response to EOM addition were mainly *Actinobacteria*. This study provides new insights into the identity of as-yet uncultured and unclassified bacteria actively metabolizing BP/PCB with EOM addition, and points out broader BP/PCB-degrading community which could be employed in bioremediation of sites.

Experimental procedures

Preparation of EOM from M. luteus

Micrococcus luteus IAM 14879 (= NCIMB 13267) used in this study had previously been described (Mukamolova *et al.*, 1998; Ding, 2004). The pure culture was inoculated at 30°C on a rotary shaker at 160 r.p.m. for 36 h in modified lactate minimal medium (LMM) of Kaprelyants and Kell (1992). Then, the pre-culture was grown under the same culture conditions until the cells reached the stationary phase. The obtained fermentation broth was centrifuged (8000 r.p.m., 15 min) to separate the cells and then filtered through a 0.22 μm filter to remove floating cells. Finally, EOM was obtained and stored at −20°C for further experiments. The pH and redox potential (Eh) of EOM were 8.8 and −106 mV respectively. The concentration of total carbon, nitrogen, phosphorus and sulphur of EOM were 867 8, 270.3, 99.8 and 42.1 mg l^{-1} respectively. The main components of EOM were polysaccharides and proteins, and their concentrations were 405.7 and 25.1 mg l^{-1} respectively. Especially, Rpf protein was dominated in the proteins of EOM (Su *et al.*, 2014).

Enrichment and cultivation

A PCB-contaminated sediment sample was obtained from a river very close to the e-waste recycling site in Luqiao Town of Taizhou City, which has been involved in e-waste disassembly for nearly 30 years (Shen *et al.*, 2008; Chen *et al.*, 2010; 2012). The sediment (4%, w/v) was incubated in a mineral salts medium (Su *et al.*, 2013b), which BP (reagent grade, Sigma-Aldrich) was added as the sole carbon and energy source at an initial concentration of 500 mg l^{-1}. Cultures were cultivated in conical flasks at 30°C on a rotary shaker at 180 r.p.m. Initial enrichments were transferred into fresh medium with 5% (v/v) inoculum after 5 days of cultivation. The BP concentration was increased in steps of 500 mg l^{-1} until a final concentration of 2000 mg l^{-1} was reached. Throughout the entire enrichment period, an equal amount EOM (10%, v/v), autoclaved EOM (sterilized at 121°C for 15 min) and LMM were added to TG, CG and BG respectively. Finally, the obtained enrichment samples were subjected to three different treatments, including TG, CG and BG.

BP degradation of enrichment cultures

The BP degradation capability of the enrichment samples (TG, CG and BG) were assessed by investigating the BP-tolerance concentrations and BP-degradation curve of the samples described elsewhere (Su *et al.*, 2013b). Briefly, after cultivating for 64 h at 30°C on a rotary shaker, BP-tolerance concentrations were investigated with the concentration of BP varying from 500 to 4500 mg l^{-1} in steps of 500 mg l^{-1}. And the BP-degradation and cell growth curve was depicted at 12 h intervals within 96 h under a BP concentration of 1500 mg l^{-1}. Lastly, the cell growth (OD_{600}) and BP-degradation efficiency of each experimental culture were measured as previously described (Su *et al.*, 2013b). All of the experiments were performed in triplicate, and standard deviation (SD) was calculated by SPSS software (version 18.0, Chicago, IL, USA) for analysis of statistical significance of the triplicate samples.

DNA extraction

After lyophilization, DNA extraction of sediment sample was carried out using a beating method (FastDNATM SPIN Kit for Soil, Bio101, USA) following the manufacturer's protocol. For the enrichment samples and pure bacterial cultures, DNA extraction was performed using the EZ-10 spin column genomic DNA miniprep kit (Bio Basic, Canada) according to the manufacturer's instructions. The DNA extracts were stored at –20°C for further analysis.

Illumina high-throughput sequencing

Illumina high-throughput sequencing was performed to determine the diversity and composition of the bacterial communities in the sediment and enrichment samples (Capodicasa *et al.*, 2009). PCR amplifications were conducted with the 515F/806R primer set that amplifies the V4 region of the 16S rRNA gene. The reverse primer contained a 6 bp error-correcting barcode unique to each sample. DNA was amplified following the protocol described by (Magoč and Salzberg, 2011). Sequencing was conducted on an Illumina MiSeq platform. Sequences were analysed with the Quantitative Insights Into Microbial Ecology software package and UPARSE pipeline (Wang *et al.*, 2007), and were assigned to OTUs at 97% similarity. All sequences have been deposited in GenBank short-read archive (SRA: SRS632100).

Isolation and phylogenetic analysis of bacteria

The enrichment cultures TG, CG and BG were diluted respectively in 10-fold steps. Then serial dilutions (from 10^2-fold to 10^8-fold) was plated on mineral-salts agar plates with BP as the carbon source. Colonies that were unique in the TG, with no counterpart in the CG and BG, were selected. All isolates were purified and stored at 4°C for further identification. Genomic DNA was extracted from the pure bacterial

cultures and then was amplified by PCR using primers 8F/1541R as described previously (Su *et al.*, 2013b). Phylograms were constructed by the neighbour-joining method with 1000 replicate trees in the MEGA6 computer software program (Tamura *et al.*, 2013). All sequences reported in this study had been submitted to NCBI GenBank under accession numbers KJ546446-KJ546456.

Conflict and interest

None declared.

References

Abraham, W.R., Nogales, B., Golyshin, P.N., Pieper, D.H., and Timmis, K.N. (2002) Polychlorinated biphenyl-degrading microbial communities in soils and sediments. *Curr Opin Microbiol* **5**: 246–253.

Cámara, B., Herrera, C., González, M., Couve, E., Hofer, B., and Seeger, M. (2004) From PCBs to highly toxic metabolites by the biphenyl pathway. *Environ Microbiol* **6**: 842–850.

de Cárcer, D.A., Martín, M., Karlson, U., and Rivilla, R. (2007) Changes in bacterial populations and in biphenyl dioxygenase gene diversity in a polychlorinated biphenyl-polluted soil after introduction of willow trees for rhizoremediation. *Appl Environ Microbiol* **73**: 6224–6232.

Capodicasa, S., Fedi, S., Carnevali, M., Caporali, L., Viti, C., Fava, F., and Zannoni, D. (2009) Terminal-restriction fragment length polymorphism analysis of biphenyl dioxygenase genes from a polychlorinated biphenyl-polluted soil. *Res Microbiol* **160**: 742–750.

Chávez, F.P., Gordillo, F., and Jerez, C.A. (2006) Adaptive responses and cellular behaviour of biphenyl-degrading bacteria toward polychlorinated biphenyls. *Biotechnol Adv* **24**: 309–320.

Chen, L., Yu, C., Shen, C., Zhang, C., Liu, L., Shen, K., *et al.* (2010) Study on adverse impact of e-waste disassembly on surface sediment in East China by chemical analysis and bioassays. *J Soil Sediment* **10**: 359–367.

Chen, L., Yu, C., Shen, C., Cui, J., Chen, C., and Chen, Y. (2012) Occurrence of (anti) estrogenic effects in surface sediment from an E-waste disassembly region in East China. *Bull Environ Contam Toxicol* **89**: 161–165.

Correa, P.A., Lin, L., Just, C.L., Hu, D., Hornbuckle, K.C., Schnoor, J.L., and Van Aken, B. (2010) The effects of individual PCB congeners on the soil bacterial community structure and the abundance of biphenyl dioxygenase genes. *Environ Int* **36**: 901–906.

Ding, L.X. (2004) Studies on the isolation of viable but non-culturable bacteria and the phylogenetic analysis of the genus *Aquaspirillum*. Tokyo: The University of Tokyo.

Ding, L.X., and Yokota, A. (2010) *Curvibacter fontana* sp. nov., a microaerobic bacteria isolated from well water. *J Gen Appl Microbiol* **56**: 267–271.

Fukuda, M., Shimizu, S., Okita, N., Seto, M., and Masai, E. (1998) Structural alteration of linear plasmids encoding the genes for polychlorinated biphenyl degradation in *Rhodococcus* strain RHA1. *Antonie Van Leeuwenhoek* **74**: 169–173.

Gomes, H.I., Dias-Ferreira, C., and Ribeiro, A.B. (2013) Overview of in situ and ex situ remediation technologies for PCB-contaminated soils and sediments and obstacles for full-scale application. *Sci Total Environ* **445–446:** 237–260.

Kaprelyants, A., and Kell, D. (1992) Rapid assessment of bacterial viability and vitality by rhodamine 123 and flow cytometry. *J Appl Microbiol* **72:** 410–422.

Keep, N.H., Ward, J.M., Robertson, G., Cohen-Gonsaud, M., and Henderson, B. (2006) Bacterial resuscitation factors: revival of viable but non-culturable bacteria. *Cell Mol Life Sci* **63:** 2555–2559.

Leigh, M.B., Prouzová, P., Macková, M., Macek, T., Nagle, D.P., and Fletcher, J.S. (2006) Polychlorinated biphenyl (PCB)-degrading bacteria associated with trees in a PCB-contaminated site. *Appl Environ Microbiol* **72:** 2331–2342.

Ma, J., Wang, Z., Yang, Y., Mei, X., and Wu, Z. (2013) Correlating microbial community structure and composition with aeration intensity in submerged membrane bioreactors by 454 high-throughput pyrosequencing. *Water Res* **47:** 859–869.

Macedo, A.J., Timmis, K.N., and Abraham, W.R. (2007) Widespread capacity to metabolize polychlorinated biphenyls by diverse microbial communities in soils with no significant exposure to PCB contamination. *Environ Microbiol* **9:** 1890–1897.

Magoč, T., and Salzberg, S.L. (2011) FLASH: fast length adjustment of short reads to improve genome assemblies. *Bioinformatics* **27:** 2957–2963.

Mikesková, H., Novotný, Č., and Svobodová, K. (2012) Interspecific interactions in mixed microbial cultures in a biodegradation perspective. *Appl Microbiol Biotechnol* **95:** 861–870.

Mukamolova, G.V., Kaprelyants, A.S., Young, D.I., Young, M., and Kell, D.B. (1998) A bacterial cytokine. *Proc Natl Acad Sci USA* **95:** 8916–8921.

Mukamolova, G.V., Murzin, A.G., Salina, E.G., Demina, G.R., Kell, D.B., Kaprelyants, A.S., and Young, M. (2006) Muralytic activity of *Micrococcus luteus* Rpf and its relationship to physiological activity in promoting bacterial growth and resuscitation. *Mol Microbiol* **59:** 84–98.

Oliver, J.D. (2010) Recent findings on the viable but nonculturable state in pathogenic bacteria. *FEMS Microbiol Rev* **34:** 415–425.

Petrić, I., Hršak, D., Fingler, S., Vončina, E., Ćetković, H., Begonja Kolar, A., and Udiković Kolić, N. (2007) Enrichment and characterization of PCB-degrading bacteria as potential seed cultures for bioremediation of contaminated soil. *Food Technol Biotech* **45:** 11–20.

Petrić, I., Bru, D., Udiković-Kolić, N., Hršak, D., Philippot, L., and Martin-Laurent, F. (2011) Evidence for shifts in the structure and abundance of the microbial community in a long-term PCB-contaminated soil under bioremediation. *J Hazard Mater* **195:** 254–260.

Pieper, D.H. (2005) Aerobic degradation of polychlorinated biphenyls. *Appl Microbiol Biotechnol* **67:** 170–191.

Schroeckh, V., and Martin, K. (2006) Resuscitation-promoting factors: distribution among actinobacteria, synthesis during life-cycle and biological activity. *Antonie Van Leeuwenhoek* **89:** 359–365.

Shen, C.F., Huang, S.B., Wang, Z.J., Qiao, M., Tang, X.J., Yu, C.N., *et al.* (2008) Identification of Ah receptor agonists in soil of e-waste recycling sites from Taizhou area in China. *Environ Sci Technol* **42:** 49–55.

Shleeva, M., Bagramyan, K., Telkov, M., Mukamolova, G.V., Young, M., Kell, D.B., and Kaprelyants, A.S. (2002) Formation and resuscitation of 'non-culturable' cells of *Rhodococcus rhodochrous* and *Mycobacterium tuberculosis* in prolonged stationary phase. *Microbiology* **148:** 1581–1591.

Stach, J.E., and Burns, R.G. (2002) Enrichment versus biofilm culture: a functional and phylogenetic comparison of polycyclic aromatic hydrocarbon-degrading microbial communities. *Environ Microbiol* **4:** 169–182.

Su, X.M., Chen, X., Hu, J.X., Shen, C.F., and Ding, L.X. (2013a) Exploring the potential environmental functions of viable but non-culturable bacteria. *World J Microbiol Biotechnol* **29:** 2213–2218.

Su, X.M., Ding, L.X., Hu, J.X., Shen, C.F., and Chen, Y.X. (2013b) A novel approach to stimulate the biphenyl-degrading potential of bacterial community from PCBs-contaminated soil of e-waste recycling sites. *Bioresour Technol* **146:** 27–34.

Su, X.M., Liu, Y.D., Hashmi, M.Z., Hu, J.X., Ding, L.X., Wu, M., and Shen, C.F. (2015) *Rhodococcus biphenylivorans* sp. nov., a polychlorinated biphenyl-degrading bacterium. *Antonie Van Leeuwenhoek* **107:** 55–63.

Su, X.M., Liu, Y.D., Hu, J.X., Ding, L.X., and Shen, C.F. (2014) Optimization of protein production by *Micrococcus luteus* for exploring pollutant-degrading uncultured bacteria. *Springerplus* **3:** 117.

Tamura, K., Stecher, G., Peterson, D., Filipski, A., and Kumar, S. (2013) MEGA6: molecular evolutionary genetics analysis version 6.0. *Mol Biol Evol* **30:** 2725–2729.

Tang, X.J., Chen, C., Chen, L.T., Cui, J.L., Yu, C.N., Chen, L., *et al.* (2013) Bacterial communities of polychlorinated biphenyls polluted soil around an e-waste recycling workshop. *Soil Sediment Contam* **22:** 562–573.

Uhlik, O., Jecna, K., Mackova, M., Vlcek, C., Hroudova, M., Demnerova, K., *et al.* (2009) Biphenyl-metabolizing bacteria in the rhizosphere of horseradish and bulk soil contaminated by polychlorinated biphenyls as revealed by stable isotope probing. *Appl Environ Microbiol* **75:** 6471–6477.

Uhlik, O., Wald, J., Strejcek, M., Musilova, L., Ridl, J., Hroudova, M., *et al.* (2012) Identification of bacteria utilizing biphenyl, benzoate, and naphthalene in long-term contaminated soil. *PLoS ONE* **7:** e40653.

Wang, Q., Garrity, G.M., Tiedje, J.M., and Cole, J.R. (2007) Naive Bayesian classifier for rapid assignment of rRNA sequences into the new bacterial taxonomy. *Appl Environ Microbiol* **73:** 5261–5267.

Xu, M., Chen, X., Qiu, M., Zeng, X., Xu, J., Deng, D., *et al.* (2012) Bar-coded pyrosequencing reveals the responses of PBDE-degrading microbial communities to electron donor amendments. *PLoS ONE* **7:** e30439.

Zanaroli, G., Balloi, A., Negroni, A., Daffonchio, D., Young, L.Y., and Fava, F. (2010) Characterization of the microbial community from the marine sediment of the Venice lagoon capable of reductive dechlorination of coplanar polychlorinated biphenyls (PCBs). *J Hazard Mater* **178:** 417–426.

15

Production of lactic acid using a new homofermentative *Enterococcus faecalis* isolate

Mohan Raj Subramanian[1†], Suvarna Talluri[1] and Lew P. Christopher[1,2*]

[1]*Center for Bioprocessing Research and Development,*
[2]*Department of Civil and Environmental Engineering,*
South Dakota School of Mines and Technology, Rapid City, SD 57701, USA.

Summary

Lactic acid is an intermediate-volume specialty chemical for a wide range of food and industrial applications such as pharmaceuticals, cosmetics and chemical syntheses. Although lactic acid production has been well documented, improved production parameters that lead to reduced production costs are always of interest in industrial developments. In this study, we describe the production of lactic acid at high concentration, yield and volumetric productivity utilizing a novel homofermentative, facultative anaerobe *Enterococcus faecalis* CBRD01. The highest concentration of 182 g lactic acid l^{-1} was achieved after 38 h of fed-batch fermentation on glucose. The bacterial isolate utilized only 2–13% of carbon for its growth and energy metabolism, while 87–98% of carbon was converted to lactic acid at an overall volumetric productivity of 5 g l^{-1} h^{-1}. At 13 h of fermentation, the volumetric productivity of lactate production reached 10.3 g l^{-1} h^{-1}, which is the highest ever reported for microbial production of lactic acid. The lactic acid produced was of high purity as formation of other metabolites was less than 0.1%. The present investigation demonstrates a new opportunity for enhanced production of lactic acid with potential for reduced purification costs.

*For correspondence. E-mail lew.christopher@sdsmt.edu

Funding Information Financial support by the Center for Bioprocessing Research & Development (CBRD) at the South Dakota School of Mines & Technology (SDSM&T), the South Dakota Board of Reagents (SD BOR), the South Dakota Governor's Office for Economic Development (SD GOED), and the US Air Force Research Laboratory (AFRL) is gratefully acknowledged.

Introduction

Lactic acid ($C_3H_6O_3$, MW 90.08) is a chiral, three-carbon carboxylic acid recognized as GRAS (generally regarded as safe) by the US Food and Drug Administration because of its unique physicochemical properties and the diverse applications in food and other chemical industries (Narayanan *et al.*, 2004). The world production of lactic acid in 2009 was 258 000 tons, with a projected growth of 329 000 tons by year 2015, and 367 000 tons by 2017 (Global Industry Analysts Inc, 2011). The estimated consumption of lactic acid in the US alone is 30 million pounds per year with a potential annual demand of 5.5 billion pounds, and the global consumption of lactic acid is expected to increase rapidly in the near future (Bajpai, 2013). This growth would mainly come from the lactic acid consumption in chemical applications. For example, the use of polylactic acid and ethyl lactate is expected to expand 19% per year (Jarvis, 2001; Garlotta, 2002; Södergård and Stolt, 2010). With a current price of $1.5/kg for a 88% purity food-grade product (www.icispricing.com), lactic acid has the potential to become a very large volume, commodity-chemical intermediate (Bozell and Petersen, 2010).

Lactic acid can be produced chemically using petrochemical feedstocks such as lactonitrile (Narayanan *et al.*, 2004). However, the chemical synthesis produces a racemic mixture of D(−)-lactic acid and L(+)-lactic acid, which is not suitable for the synthesis of polylactic acid, as the optical purity of lactic acid is one of the major factors that determine the physical properties of its polymer. In contrast, about 90% of current commercial lactic acid is obtained via biological fermentation of sugars (Hofvendahl and Hahn-Hagerdal, 2000; Zhou *et al.*, 2003). Microbial fermentation of sugars produces stereospecific D(−)-lactic acid or L(+)-lactic acid depending on the strains used. For example, *Lactobacillus*, *Bacillus*, *Rhizopus*, *Streptococcus* and *Enterococcus* produce L(+)-lactic acid, while *Leuconostoc* and *La. vulgaricus* produce D(−)-lactic acid (Park *et al.*, 2010). L(+)-lactic acid is the preferred component of many food and industrial applications that is currently produced via biological fermentation utilizing lactic acid bacteria or fungi such as *Rhizopus*. Some recombinant yeast strains have also been used to produce lactic acid from various carbon feedstocks (Porro *et al.*, 2008). However, the yield and productivities of fungal and yeast strains are very low compared with lactic acid bacteria. In addition, the mycelial morphology of fungal strains affects the viscosity of the fermentation medium and can cause blockage during fermentation (Maas *et al.*, 2006; Zhang *et al.*, 2007; Ilmen *et al.*, 2013). Therefore, these organisms require further gene manipulations and process development for the industrial production of lactic acid.

Recently, we have isolated a bacterial strain with high potential for lactic acid production. This strain was identified as *Enterococcus faecalis* and partially characterized. Subsequently, its full genome was sequenced, and the isolate *E. faecalis* CBRD01 was deposited with the American Type Culture Collection with a patent deposition designation PTA-12846 (unpublished results). Here we describe the abilities of *E. faecalis* CBRD01 to produce L(+)-lactic acid from glucose at high yields, titres and productivity with a minimum by-product formation at both flask and bioreactor scales.

Results

Morphology and biochemical characteristics of isolate CBRD01

Enterococcus faecalis CBRD01 belongs to Gram-positive elongated Cocci with cell size of approximately 1.2–1.5 µm. CBRD01 was able to grow anaerobically under dark conditions at temperature between 30 and 45°C, with a temperature optimum of 37°C. CBRD01 was able to utilize trehalose, lactose, ribose, saccharose, melizitose, cellobiose, mannose, mannitol, sorbitol and inositol (Table 1). The strain did not show oxidase, catalase, alkaline phosphatase, β-galactosidase, aminopeptidase and urease activities. The cellular fatty acid compositional analysis indicated that the strain CBRD01 had a match with *E. faecalis* (data not shown).

Production of lactic acid by E. faecalis *CBRD01 at flask scale*

In order to examine the production of lactic acid, the strain *E. faecalis* CBRD01 was cultured in 100 ml serum bottles with 50 ml of LA5 medium containing three different concentrations of glucose (28.79, 56.13 and 110.22 mM) at 37°C and 150 r.p.m. in an orbital incubator shaker under anaerobic condition. The cells were inoculated at 0.1 ± 0.01 Optical Density at 600 nm (OD600), and fermentation was continued for 12 h. The results revealed that a maximum yield of lactate was obtained at 110.22 mM glucose (Table 2). On average, 27–34 mM of glucose was consumed regardless of the initial concentrations of glucose. However, a reverse relationship was observed between the initial glucose concentration and glucose consumption after 12 h of incubation. The

Table 1. Biochemical characterization of isolate CBRD01.

Characteristics	Reaction
Shape	Cocci (elongated)
Size (in diameter)	1.2–1.5
Gram-reaction	+
Aminopeptidase activity	−
Potassium hydroxide test	−
Oxidase activity	−
Catalase activity	−
Acid from	
Glucose	+
Trehalose	+
Mannitol	+
Raffinose	−
Lactose	+
Ribose	+
Saccharose	+
Arabinose	−
Melibiose	−
Sorbitol	+
Melezitose	+
L-Rhamnose	−
Cellobiose	+
Mannose	+
Inositol	+
Aldehyde dehydrogenase activity	+
Urease activity	−
Voges Proskauer test	+
β-Galactosidase activity	−
Alkaline phosphatase activity	−
Growth at 45°C	+
Growth at 50°C	−
pH optimum	7.0

+, positive reaction; −, negative reaction.

glucose consumption at lower starting concentrations (28.79 and 56.13 mM) was 93% and 60% respectively. This was 3.3- and 2.1-fold, respectively, higher than that at 100.22 mM glucose (28%). The lactate concentration varied from 49 to 61 mmol l^{-1}. In theory, 1 mol of glucose yields a maximum of 2 mol of lactate under anaerobic condition. Accordingly, the glucose consumption at 110.22 mM glucose concentration was 30.55 mM. The lactate titre was 60.21 mmol l^{-1} after 12 h, thus yielding 1.97 mol mol^{-1} glucose, which is equivalent to 98.54% of the theoretical maximum. However, it should be noted that lactate production ceased after 12 h, although the experiment was carried out for 24 h.

Production of lactic acid by E. faecalis *CBRD01 in batch process at bioreactor scale*

To investigate the potential of *E. faecalis* CBRD01 for lactic acid production at bioreactor scale, the strain was cultured under controlled conditions at a constant pH of 7.0 ± 0.2 in a 1 l glass jar bioreactor (DASGIP) with a working volume of 0.5 l for 24 h. Results are shown in Fig. 1. The initial glucose concentration was 94 mM. In 24 h of the bioreactor culture, *E. faecalis* CBRD01

Table 2. Lactate production from different glucose concentrations by *E. faecalis* CBRD01 under anaerobic batch fermentation.

Starting glucose concentration (mM)	Maximum specific growth rate μ_{max} (h^{-1})[a]	Glucose consumed (mM)[b]	Glucose uptake rate (mmol g^{-1} cdw h^{-1})[b]	Lactate produced[c] (mmol l^{-1})[b]	Lactate production rate (mmol g^{-1} cdw h^{-1})[b]	Lactate yield (mol mol^{-1} glucose)[b]	Lactate yield (%)[b]
28.79	0.59	26.87	19.92	49.12	36.42	1.83	91.39
56.13	0.59	33.91	23.89	60.84	42.86	1.79	89.71
110.22	0.64	30.55	21.22	60.21	41.83	1.97	98.54

a. Estimated between 0 and 3 h.
b. Estimated for 12 h.
c. Optically pure L(+)-lactic acid.

produced 172.8 mmol lactate l^{-1}, which is 2.87-fold higher than the titre obtained from the flask culture. The biomass production reached a maximum of 0.64 g l^{-1} at 12 h (from 0.07 g l^{-1} at 0 h) and then decreased to 0.62 g l^{-1} at 24 h (Fig. 1A). Both the specific glucose uptake rate and specific lactate production rate peaked at 12 h of cultivation (Fig. 1B), with the former being 0.57% of the latter. At 12 h, a maximum volumetric productivity of 10 mmol^{-1} l^{-1} h^{-1} was attained (Fig. 1C). Therefore, to better understand the behaviour of CBRD01, it was of interest to discuss its fermentation profile in two phases (phase 1, 0–12 h; phase 2, 12–24 h).

The total glucose consumption at 24 h was 94.06 mmol l^{-1} (Table 3). Although the lactate yield in the first half of cultivation (0–12 h) was 88%, it improved to 98% in the second (12–24 h), resulting in an overall yield of 91.86 % of theoretical maximum of glucose. The μ_{max}, estimated between 0 and 3 h, was 0.59 h^{-1}. The maximum specific production rate (q_{max}) of lactate, estimated between 0 and 12 h (phase 1), was 66.72 mmol g cdw^{-1} h^{-1}. In phase 2 (12–24 h), q_{max} decreased to 15 mmol g cdw^{-1} h^{-1}.

Lactate was the major product of the fermentation process with *E. faecalis* CBRD01. Other metabolites, such as acetate and formate, were produced in quantities of less than 4% in the 1st phase and less than 0.6% in the 2nd phase (Table 4). Analysis of the carbon material balance revealed that 101.16% of glucose carbon was recovered in phase 1 (0–12 h) and 98.9% in phase 2 (12–24 h). In phase 1, 87.9% of glucose carbon was converted as lactate, while 6.2% of carbon was directed towards biomass formation. However, no carbon was utilized for biomass in the second phase, while 0.6% of carbon was utilized for acetate and 0.3% for formate, thus yielding 98% of lactate in the second phase. The formate and acetate co-metabolites were produced in very low quantities of less than 1 g l^{-1}. The electron balance, calculated for the substrates and products, indicated 0.99–1.0 (Table 4). This means that the electrons, released during glucose oxidation, were completely recovered in the form of products. Based on the carbon distribution of glucose into metabolites (no metabolites were omitted during analysis), a carbon metabolic pathway for *E. faecalis* CBRD01 is proposed in Fig. 2, suggesting that this strain utilizes a homolactic fermentative pathway to produce lactic acid.

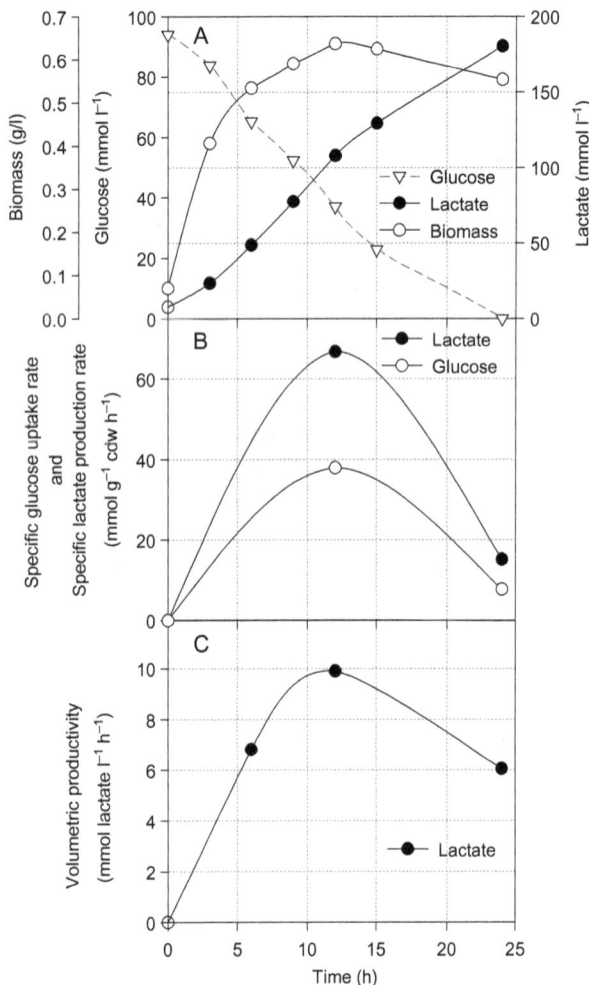

Fig. 1. Time-course profile of growth, glucose consumption and lactate production by *E. faecalis* CBRD01 under anaerobic batch fermentation.
A. Glucose consumption, biomass and lactate production.
B. Specific rates of glucose uptake and lactate production.
C. Volumetric productivity of lactate.

Table 3. *Enterococcus faecalis* CBRD01 fermentative characteristics of lactate production from glucose under anaerobic batch fermentation.

Biomass and lactate production	Cultivation time		
	0–12 h	12–24 h	0–24 h
Biomass (g l^{-1})	0.57	0.00	0.57
Glucose utilized (mmol l^{-1})	57.04	37.02	94.06
Lactate produced (mmol l^{-1}) [a]	100.25	72.55	172.8
Maximum specific growth rate (h^{-1})[b]	0.59	0.00	–
Biomass yield (g cdw[c] g^{-1} glucose)	0.06	0.00	0.03
Specific glucose uptake rate (mmol g^{-1} cdw h^{-1})	37.96	7.75	32.53
Specific lactate production rate (mmol g^{-1} cdw h^{-1})	66.72	15.20	59.76
Lactate yield (mol mol^{-1} glucose)	1.75	1.96	1.84
Lactate yield (g g^{-1} glucose)	0.88	0.98	0.92
Lactate final yield (% of theoretical)[d]			91.86

a. Optically pure L(+)-lactic acid.
b. Calculated between 0 and 3 h.
c. cdw, cell dry weight.
d. Accounted for 0–24 h.

Production of lactic acid by E. faecalis *CBRD01* in fed-batch process at bioreactor scale

To further evaluate the potential of *E. faecalis* CBRD01 for lactic acid production, a fed-batch process was carried out in a 1 l glass jar bioreactor with a 0.5 l working volume of LA5 medium at pH 7.0 and 37°C for 87 h under anaerobic conditions. The initial glucose concentration was 86.4 mM. During the fermentation period, a total of 825 mM of glucose was added in 14 batches (Fig. 3). The fed-batch process was initiated by adding 3.0 g l^{-1} of *E. faecalis* CBRD01 cells, grown under anaerobic conditions, to the culture medium. The total glucose consumption at the end of the fermentation process was

823 mmol l^{-1}. The strain produced 973 mmol l^{-1} lactate in 87 h. This translates into a lactate yield of 59% from glucose, with an average productivity of 1.0 g l^{-1} h^{-1} (Fig. 3). However, the lactate yield in the first phase (0–12 h) was above 91%. When compared with the batch production of lactic acid after 24 h (Fig. 1), the fed-batch process increased lactic acid production approximately threefold (532 mM).

Fed-batch bioreactor production of lactic acid by E. faecalis *CBRD01*at high cell density culture

In an attempt to increase the lactic acid titres, fed-batch experiments were performed using a high cell density

Table 4. Carbon material balance of glucose metabolism by *E. faecalis* CBRD01 under anaerobic batch fermentation.

Substrate and metabolites[a]	Molecular formula	Cultivation time										
		Phase I (0–12 h)					Phase II (12–24 h)				
		mM	C-mM	C (%)	γ_i[b]	γ_i balance (mol l^{-1})[c]	mM	C-mM	C (%)	γ_i[b]	γ_i balance (mol l^{-1})[c]
Substrate											
Glucose	$C_6H_{12}O_6$	57.04	342.24	100.00	4	1.37	37.02	222.12	100.00	4	0.89
Biomass	$CH_{1.8}O_{0.5}N_{0.2}$	0.52[d]	21.22[e]	6.20	4.8	0.10	0.00[d]	0.00	0.00[e]	4.8	0.00
Metabolites											
Lactate[f]	$C_3H_6O_3$	100.25	300.75	87.88	4	1.20	72.55	217.65	97.99	4	0.87
Acetate	$C_2H_4O_2$	6.51	13.01	3.80	4	0.05	0.65	1.30	0.59	4	0.01
Formate	CH_2O_2	11.22	11.22	3.28	2	0.02	0.67	0.67	0.30	2	0.00
Carbon dioxide	CO_2	0.00	0.00	0.00		0.00	0.00	0.00	0.00		0.00
Total products			346.20			1.38		1350.13			0.88
Reduction deg. balance						1.00					0.99
Carbon recovery %[g]				101.16					98.88		

a. The standard deviation of the measurements was less than 3% for the substrate and the metabolites.
b. Degree of reduction (γ_i), which was calculated using the formula: $\gamma_i = (4a + b - 2c + 6e + 5f)/a$, where *a* denotes carbon, *b* denotes hydrogen, *c* denotes oxygen, *e* denotes sulfur, and *f* denotes phosphorus (Gustafsson *et al.* 1993).
c. Degree of reduction (γ_i) balance, which was calculated as γ_i for total products/γ_i for substrate (expressed in mol l^{-1}).
d. g l^{-1}.
e. An average molecular weight of 24.6, which corresponds to an average cell with a molecular formula of $CH_{1.8}O_{0.5}N_{0.2}$. The average ash content of 8% was deduced from the actual cell dry mass (Stephanopoulos *et al.*, 1998).
f. Optically pure L(+)-lactic acid.
g. Total carbon in biomass and metabolites, Total carbon in glucose^{-1} × 100.

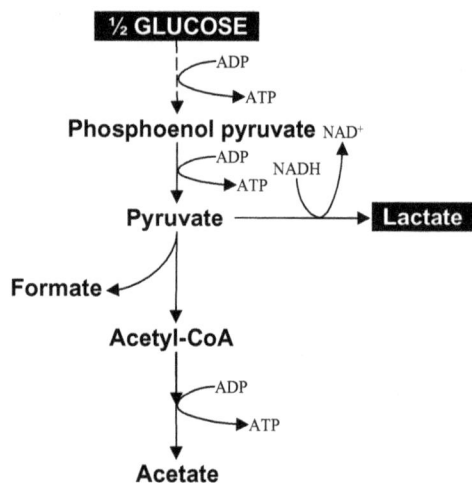

Fig. 2. Proposed metabolic pathway for lactate production from glucose in *E. faecalis* CBRD01.

culture (22 g l⁻¹) and LA6 culture medium in a 1 l glass jar bioreactor (0.3 l working volume). Lactic acid production was carried out under anaerobic conditions at pH 7.0 and 37°C for 38 h (Fig. 4). At the end of the fermentation process, *E. faecalis* CBRD01, grown in LA6 medium, was able to produce 2022 mmol lactate l⁻¹, which is equivalent to 182.1 g l⁻¹. The total glucose consumption was 210.01 g with an overall lactate yield of 86.7% from the theoretical maximum on glucose. The microbial biomass, retained at the end of fermentation, was 16.86 g cdw l⁻¹. About 811 mM of glucose was consumed in 13 h (before glucose addition) to produce 1484 mmol lactic acid l⁻¹, which represents a 91.5% yield of the theoretical maximum on glucose or 1.83 mol lactate mol⁻¹ glucose. The volumetric productivity accounted between 0

and 13 h was 114 mmol⁻¹ l⁻¹ h⁻¹ (10.3 g l⁻¹ h⁻¹), which is the highest volumetric productivity ever reported for microbial production of lactic acid. As the energy requirement for maintaining high cell density is very high, there was no net cell growth. The biomass production remained unchanged for 13 h and thereafter declined due to the dilution effect that occurred with the glucose spiking.

Discussion

The production of lactic acid for commercial applications requires the use of less fastidious, but robust microorganisms that can meet the industrial needs for high titre, high productivity and high yield of lactate of at least 100 g l⁻¹, 2.5 g l⁻¹ h⁻¹ and 80% respectively (Werpy and Petersen, 2004). As purification costs comprise a significant portion of the overall production costs, the purity of food grade (80–90% purity) and pharmaceutical grade (> 90%), lactic acid determines its price (John *et al.*, 2007). Therefore, to minimize the production and purification costs, there is a need of new microbial isolates with enhanced capabilities for high-purity lactic acid production.

Here, we describe a wild-type bacterial strain, recently isolated and identified as *E. faecalis* CBRD01. This strain is different from other known lactic-producing *Enterococcus* sp., such as *E. asini*, *E. durans*, *E. hirae*, *dispar*, *E. facium*, *E. solitaries*, *E. malodoratus*, *E. pseudoavium* and *E. saccharolyticus* in that it utilizes mannitol, sorbitol, ribose and melezitose as carbon sources for lactic acid production. However, while *E. faecalis* CBRD01 does not produce acid from arabinose, xylose, melibiose and raffinose, most of the other *Enterococcus* sp. such as *E. avium*, *E. casseliflavus*, *E. gallinarum*, *E. mundtii*, *E. cecorum*, *E. raffinosus* and *E. saccharolyticus* can

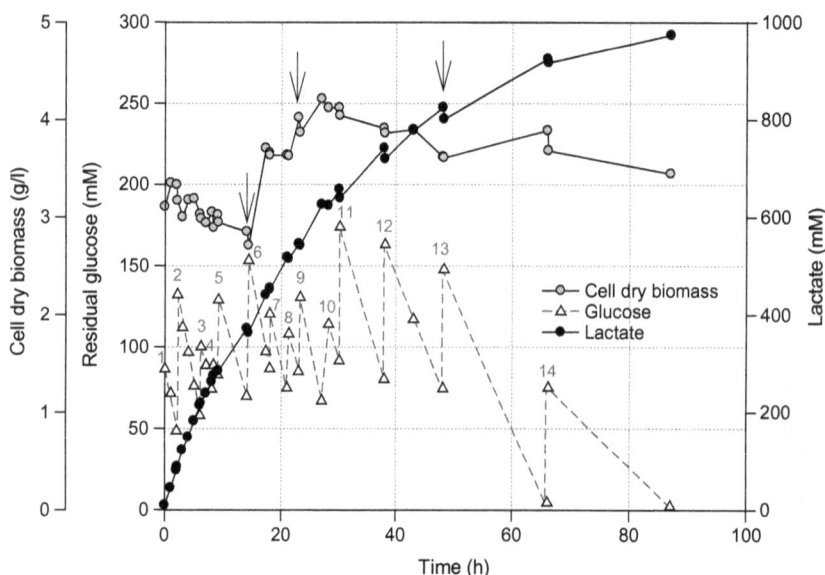

Fig. 3. Time-course profile of growth, glucose consumption and lactate production by *E. faecalis* CBRD01 under anaerobic fed-batch fermentation. Arrows indicate addition of yeast extract at 5 g l⁻¹. The numbers 1–14 in grey indicate glucose addition at different time intervals and different concentrations (1–86.4 mM; 2–83.74 mM; 3–42.07 mM; 4–14.63 mM; 5–46.04 mM; 6–83.59 mM; 7–33.82 mM; 8–33.17 mM; 9–45.71 mM; 10–46.98 mM; 11–82.34 mM; 12–82.80 mM; 13–73.24 mM; and 14–70.75 mM).

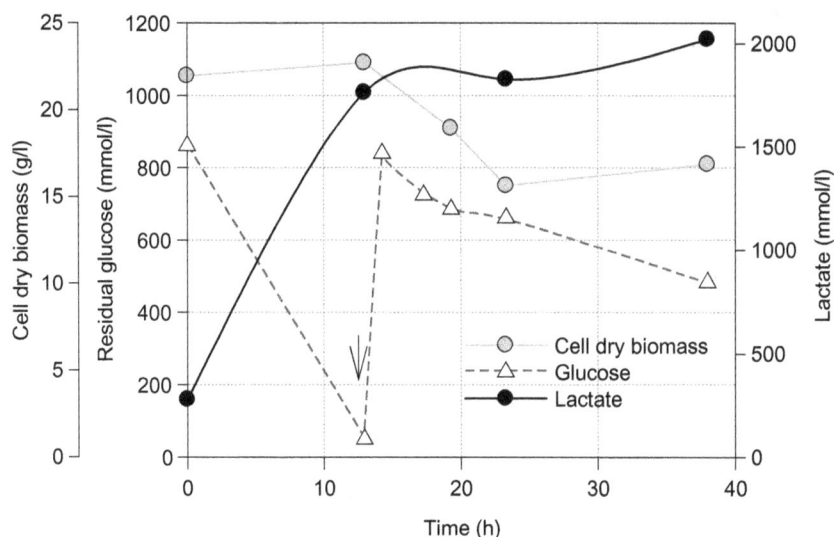

Fig. 4. Time-course profile of growth, glucose consumption and lactate production by *E. faecalis* CBRD01under anaerobic fed-batch fermentation with initial cell density of 22 g l^{-1}. Arrow indicates glucose addition.

ferment the previous carbon sources to lactic acid (Manero and Blanch, 1999).

Enterococcus faecalis CBRD01was able to produce high yields and titres of lactic acid with minimal impurities on glucose. The response of CBRD01 to different glucose concentrations is of particular interest because of the direct impact of parameters such as substrate inhibition, viscosity of culture medium, by-product formation on the cellular metabolisms and product formation (Booth, 1985; Hall *et al.*, 1995). Initial glucose concentrations of up to 110 mM had no adverse effect on cell growth of CBRD01 as μ_{max} (estimated between 0 and 3 h) remained constant (0.59 h^{-1}) at 29 and 56 mM glucose, and slightly increased (0.64 h^{-1}) at 110 mM glucose. Acetate and formate were the only co-metabolites detected at concentrations of less than 1 mM, suggesting that the formation of lactate by-products and their inhibitory role on glucose fermentation was negligible. However, the increase in the initial glucose concentrations from 29 to 110 mM did not improve glucose consumption. In fact, at 110 mM glucose, the total glucose consumption was lower than at 56 mM glucose. Furthermore, the rate of glucose uptake did not change substantially and remained in the range of 19.9–23.9 mmol g^{-1} cdw h^{-1} at all three glucose concentrations. This suggests that (i) the CBRD01 cells were not under stress at high glucose concentrations (110 mM), and (ii) the cellular metabolism of lactate production was not influenced by the glucose concentration. The latter was evident from the lactate yield that was in the range of 90–99% at all three glucose concentrations.

While examining the pH of the culture broth after 12 h of fermentation, it was noticed that pH decreased to 5.3 ± 0.3 from the initial pH 7.0, regardless of the initial glucose concentrations. The pH drop, which was due to the formation of lactic acid, could be responsible for

the cease of further biomass and lactate production in *E. faecalis* CBRD01. In order to understand the effect of pH on improving lactate production, a batch process was conducted under pH-controlled conditions in a 1 l glass jar bioreactor. When the pH of the culture medium was maintained at 7.0 ± 0.2 throughout fermentation, 173 mmol lactate l^{-1} was produced from 94 mmol glucose l^{-1}, which is equivalent to a 92% lactate yield on glucose carbon. Hence, the culture pH had a major impact on glucose metabolism and lactate production.

Although the batch process showed a 2.9-fold improvement in lactate production compared with shake flask experimental results, the supply of initial glucose was seen as a limiting factor for increased lactate production. Therefore, a fed-batch process was designed to better understand the lactate production potential of *E. faecalis* CBRD01. In a separate study, the effect of inoculum load was examined (data not shown). It was observed that at a given lactate titre, the use of higher inoculum densities (on dry weight) reduced the fermentation time to reach that titre. Alternatively, a higher lactate titre can be attained at the same fermentation time. For example, a fed-batch process with 3 g l^{-1} inoculum increased lactate production to 973 mmol l^{-1} from 823 mM of glucose in 87 h. When accounted at 24 h, the lactate titre was 541 mmol l^{-1}, and the total glucose consumed was estimated at 338 mmol l^{-1}. This corresponds to a lactate yield of 80% on glucose, which is threefold higher than the batch production of lactate after 24 h. However, at 13 h of fed-batch fermentation, the biomass concentration declined to 2.85 g l^{-1} from 3.0 g l^{-1}. Although the culture medium was maintained at glucose concentrations in excess of 60 mM, the carbon to cell growth and biomass maintenance with CBRD01 were low. This necessitated addition of complex nitrogen sources such as yeast extract to maintain biomass concentration.

To further improve the lactate production in *E. faecalis* CBRD01, fermentation parameters including medium composition, initial glucose concentrations and inoculum density were varied and modified. CBRD01 growth and lactate production were examined as a function of these parameters (data not shown). Results suggested the possibility of improving lactate production by using a high cell density culture of 22 g l^{-1} cells and a high glucose concentration of up to 1 M on an altered medium LA6. Under those conditions, CBRD01 produced 2.02 mol lactate l^{-1} (equivalent to 182 g lactate l^{-1}) at an overall lactate yield of 87% on glucose carbon. The overall volumetric rate of lactate production was estimated to be 56 mmol l^{-1} h^{-1} (equivalent to 5 g lactate l^{-1} h^{-1}). A similar strain, *E. faecalis* RKY1 was previously reported to ferment glucose at an optimal concentration of 150 g l^{-1} to lactic acid with a final titre of 144 g l^{-1}, at an optimal productivity of 5.1 g l^{-1} h^{-1} (Yun *et al.*, 2003; Wee *et al.*, 2004; 2006a,b). In comparison, the present investigation demonstrates a higher lactate titre at a similar volumetric productivity.

Recently, two *Bacillus* strains were used to produce lactic acid (Ou *et al.*, 2011; Meng *et al.*, 2012). High lactic acid titres were obtained on glucose and xylose. For example, a thermotolerant *B. coagulans* 36D1 strain produced 182 g l^{-1} of lactic acid with a carbon yield of 92.3% on glucose following a fed-batch fermentation for 261 h. The same strain could also produce lactic acid at 163 g l^{-1} from 86.3 g l^{-1} of xylose, thus yielding 87.3% on xylose carbon equivalents. However, the volumetric productivities of lactic acid production estimated from the fed-batch process with both glucose and xylose were less than 0.84 g l^{-1} h^{-1} (Ou *et al.*, 2011). In another report (Meng *et al.*, 2012), an alkaliphilic *Bacillus* sp. WL-S20 was shown to produce a maximum of 225 g l^{-1} of lactic acid in a fed-batch process at a productivity rate of 1.04 g l^{-1} h^{-1}, while a single-pulse feeding of glucose yielded a maximum of 180 g l^{-1} of lactic acid at 1.61 g l^{-1} h^{-1} after 112 h of fermentation. Both *Bacillus* strains were very efficient in producing lactic acid at high concentrations, comparable or exceeding those of *E. faecalis* CBRD01, but at threefold to fivefold lower volumetric productivity rates.

Enterococcus faecalis CBRD01, believed to use a homolactic fermentative pathway to produce L(+)-lactic acid, utilized 2–13% of carbon for its growth and energy metabolism, with the rest of the carbon converted to lactic acid at an overall volumetric productivity of 5 g l^{-1} h^{-1}. Formation of other metabolites such as acetate and formate was less than 0.1%. The fermentation characteristics and lactic acid-producing capabilities of CBRD01, as presented in this work, can certainly compete with existing published literature. Following further optimization, *E. faecalis* CBRD01 may become a strong candidate for industrial production of lactic acid.

Experimental procedures

Materials

Yeast extract (Cat. 212750) and tryptone (Cat. 211705) were purchased from Difco (Becton Dickinson, Franklin Lakes, NJ, USA). Glucose, lactic acid and all other chemicals and reagents, unless otherwise indicated, were purchased from Sigma-Aldrich (St. Louis, MO, USA).

Culture enrichment, isolation and identification of isolate CBRD01

Soil samples, collected from the Material Recovery Facility in Rapid City, SD, USA, were placed in 100 ml sterile serum bottles containing 50 ml culture enrichment medium and tightly sealed using butyl rubber stoppers with aluminium caps. Before inoculating the samples, the serum bottles were flushed with nitrogen gas (99.99%) for 15 min to ensure that the bottles were completely deprived of oxygen. The samples were then processed for culture enrichment for lactic acid production. The culture enrichment medium for lactic acid production had the following components per litre of deionized water: NH_4Cl, 1.6 g; $MgSO_4 \cdot 7H_2O$, 0.25 g; NaCl, 1.0 g; $FeCl_3 \cdot 6 H_2O$, 2.5 mg; $Na_2S_2O_3 \cdot 5 H_2O$, 1.0 g; yeast extract, 5.0 g; tryptone, 5.0 g; glucose, 100 mmol (18.0 g); Pfennig and Lippert's trace element solution (PL 9), 1.0 ml. The medium was supplemented with 100 mM potassium phosphate buffer at pH 7.0. The Pfennig and Lippert's trace element solution contained the following components per litre of 77 mM HCl: $FeSO_4 \cdot 7H_2O$, 0.21 g; $ZnSO_4 \cdot 7H_2O$, 50 mg; $MnCl_2 \cdot 4H_2O$, 50 mg; H_3BO_3, 15 mg; $CoCl_2 \cdot 6H_2O$, 100 mg; $CuSO_4 \cdot 5H_2O$, 6.7 mg; $NiCl_2 \cdot 6H_2O$, 10 mg; Na_2SeO_3, 5.0 mg; $Na_2MoO_4 \cdot 2H_2O$, 15 mg. Enrichment cultures, displaying positive growth with high lactic acid titre after three consecutive transfers in the selective medium, were chosen for isolation.

A 0.1 ml of lactate-producing microbial culture was placed onto agar plates, containing selective medium, and incubated under anaerobic condition. In total, 55 different microbial strains were isolated from the soil samples (data not shown). Cultures that displayed rapid growth on glucose with lactic acid as the primary metabolite were resorted for further isolation and verified for purity by streaking on medium solidified with agar. The most promising isolate with high lactate-producing capability on glucose was referred to as CBRD01 and used in this study. The strain CBRD01 was biochemically characterized (Table 1) and identified as *E. faecalis* by DSMZ-Germany (http://www.dsmz.de/). This strain was referred to as *E. faecalis* CBRD01.

Flask cultivation of E. faecalis CBRD01

To examine the production of lactic acid, the strain *E. faecalis* CBRD01 was cultured in 100 ml serum bottles with 50 ml working volume in an orbital incubator shaker (Innova R42, Eppendorf, USA) at 150 r.p.m. under anaerobic condition. The lactic acid medium (LA5 medium) contained the following components per litre of deionized water: NH_4Cl, 1.6 g; $MgSO_4 \cdot 7H_2O$, 0.25 g; NaCl, 1.0 g; $FeCl_3 \cdot 6 H_2O$, 2.5 mg; $Na_2S_2O_3 \cdot 5 H_2O$, 1.0 g; yeast extract, 3.0 g; Pfennig and Lippert's trace element solution (PL 9), 1.0 ml. The pH of the culture medium was adjusted to pH 7.0 with 100 mM potassium phosphate buffer. Three different concentrations of glucose, 25 mM (4.5 g l^{-1}), 50 mM (9 g l^{-1}) and 100 mM (18 g l^{-1}), were used in the fermentation studies. The cells were inoculated at 0.1 ± 0.01 OD$_{600}$, and fermentation was carried out at 37°C for 12 h. Samples were withdrawn periodically to determine cell mass, residual glucose and metabolites.

Bioreactor cultivation of E. faecalis CBRD01

A 1 l glass jar bioreactor (DASGIP-FB04CS, Eppendorf, Germany) with a working volume of 500 ml of LA5 medium containing 100 mM potassium phosphate buffer (pH 7.0) was used. Glucose was added at 86.4 mM. The dissolved oxygen was expelled by flushing the reactor with nitrogen gas (99.99%) for 30 min at 1.0 vol vol^{-1} min^{-1}. The strain was grown overnight and then inoculated at an OD$_{600}$ of 0.2 into the medium. The temperature and agitation rate were maintained at 37°C and 250 r.p.m. respectively. The pH was maintained at 7.0 ± 0.1 with 2.5 N NaOH and 2.5 N HCl. Samples were withdrawn periodically to determine the biomass, residual substrate and metabolite concentration. Fermentation was continued for 72 h. For fed-batch fermentation, concentrated glucose (4.0 M) was added intermittently. The glucose content in the medium was monitored periodically using high-performance liquid chromatography (HPLC), and its concentration was maintained above 60 mM until 60 h.

An improved fed-batch process was carried out at 37°C in a 1 l glass jar bioreactor with a 0.3 l working volume at high cell density using LA6 medium, which contained the following components per litre of deionized water: yeast extract, 18 g; tryptone, 9.0 g; dipotassium phosphate, 9.38 g; monopotassium phosphate, 1.73 g; Lippert's trace element solution, 1.0 ml. After sterilization, the fermenters were flushed with N$_2$ gas for 1 h to ensure oxygen-free environment. Fermentation was initiated by adding E. faecalis CBRD01 cells, grown under anaerobic conditions, at 22 g l^{-1}. Glucose was added at 0.86 M to the fermenter. The pH of LA6 was maintained at pH 7.0 ± 0.2 using 10 N NaOH and 5.0 N HCl solution throughout fermentation. Experiments were conducted for 38 h.

Analytical methods

Cell concentrations were measured in a 10 mm path length cuvette using a UV-2450 double-beam spectrophotometer (Shimadzu, Kyoto, Japan) at 600 nm. One unit of absorbance at 600 nm corresponded to 0.3 g cell dry weight (cdw) l^{-1}. Concentrations of glucose and fermentation metabolites (lactate, acetate, formate) were determined by the method described previously using Shimadzu LC20 HPLC (Shimadzu) equipped with a refractive index detector (model–RID-10A, Shimadzu) (Talluri et al., 2013). Briefly, the supernatants, obtained by centrifugation of the culture samples at 10 000 × g for 10 min, were filtered through the Nylon-membrane (Cole-Parmer, Vernon Hills, IL, USA) and eluted through a 300 × 7.8 mm Aminex HPX-87H (Bio-Rad, Hercules, CA, USA) column at 60°C using 5.0 mM H$_2$SO$_4$.

Calculation of fermentation parameters

The maximum specific growth rate (μ_{max}; h^{-1}) was estimated by plotting the total cell concentration against time in a log-linear plot. The slope of the straight line during exponential growth was used to find the average specific growth rate. Specific rates (glucose uptake rate or lactate production rate) were calculated by dividing the volumetric rates by the time average concentration of cells (mmol g cdw^{-1} h^{-1}). Consumption of glucose and production of lactate were used to calculate the volumetric rates or productivities (mmol l^{-1} h^{-1}). Lactate yield on glucose (mol mol^{-1}) was estimated by dividing lactate concentration (mol) by glucose consumption (mol). Biomass yield (g cdw g glucose^{-1}) was estimated from the cell mass produced (on cell dry weight) per gram of glucose consumed.

Conflict of interest

None declared.

References

Bajpai, P. (2013) Biorefinery in the Pulp and Paper Industry. London, UK: Academic Press.

Booth, I.R. (1985) Regulation of cytoplasmic pH in bacteria. Microbiol Rev 49: 359–378.

Bozell, J.J., and Petersen, G.R. (2010) Technology development for the production of biobased products from biorefinery carbohydrates – the US Department of Energy's 'Top 10' revisited. Green Chem 12: 539–554.

Garlotta, D. (2002) A literature review of poly(lactic acid). J Polymer Environ 9: 63–84.

Global Industry Analysts Inc. (2011) Global market opening for lactic acid [WWW document]. URL http://www.prweb.com/releases/2011/1/prweb8043649.htm/.

Gustafsson, L.R., Larsson, O.K., Larsson, C., and Adler, L. (1993) Energy balance calculations as a tool to determine maintenance energy requirements under stress conditions. Pure Appl Chem 65: 1893–1898.

Hall, H.K., Krem, K.L., and Foster, J.E. (1995) Molecular responses of microbes to environmental pH stress. Adv Microb Physiol 37: 229–272.

Hofvendahl, K., and Hahn-Hagerdal, B. (2000) Factors affecting the fermentative lactic acid production from renewable resources. Enzyme Microb Technol 26: 87–107.

Ilmen, M., Koivuranta, K., Ruohonen, L., Rajgarhia, V., Suominen, P., and Penttila, M. (2013) Production of L-lactic acid by the yeast Candida sonorensis expressing heterologous bacterial and fungal lactate dehydrogenases. Microb Cell Fact 12: 1–15.

Jarvis, L. (2001) Lactic acid outlook up as polylactide nears market. Chem Market Rep 259: 5–14.

John, R.P., Nampoothiri, K.M., and Pandey, A. (2007) Fermentative production of lactic acid from biomass: an overview on process developments and future perspectives. Appl Microbiol Biotechnol 74: 524–534.

Maas, R.H., Bakker, R.R., Eggink, G., and Weusthuis, R.A. (2006) Lactic acid production from xylose by the fungus Rhizopus oryzae. Appl Microbiol Biotechnol 72: 861–868.

Manero, A., and Blanch, A.R. (1999) Identification of *Enterococcus* spp. with a biochemical key. *Appl Environ Microbiol* **65:** 4425–4430.

Meng, Y., Xue, Y., Yu, B., Gao, C., and Ma, Y. (2012) Efficient production of ʟ-lactic acid with high optical purity by alkaliphilic *Bacillus* sp. WL-S20. *Bioresour Technol* **116:** 334–339.

Narayanan, N., Roychoudhury, P.K., and Srivastava, A. (2004) L (+) lactic acid fermentation and its product polymerization. *Electronic J Biotechnol* **7:** 167–179.

Ou, M.S., Ingram, L.O., and Shanmugam, K.T. (2011) L(+)-Lactic acid production from non-food carbohydrates by thermotolerant *Bacillus coagulans*. *J Ind Microbiol Biotechnol* **38:** 599–605.

Park, Y.H., Cho, K.M., Kim, H.W., and Kim, C. (2010) Method for producing lactic acid with high concentration and high yield using lactic acid bacteria. CJ Cheiljedang Corp. U.S. Patent 7682814 B2.

Porro, D., Bianchi, M., Ranzi, B.M., Frontali, L., Vai, M., Winkler, A.A., and Alberghina, L. (2008) Yeast strains for the production of lactic acid. Tate & Lyle Ingredients Americas, Inc. U.S. Patent 7326550 B2.

Södergård, A., and Stolt, M. (2010) Industrial production of high molecular weight Poly(Lactic Acid). In *Poly(Lactic Acid): Synthesis, Structures, Properties, Processing, and Applications*. Auras, R., Lim, L.T., Selke, S.E.M., and Tsuji, H. (eds). Hoboken, NJ, USA: John Wiley & Sons, Inc., pp. 27–41. doi: 10.1002/9780470649848.ch3.

Stephanopoulos, G.N., Aristidou, A.A., and Nielsen, J. (1998) *Metabolic Engineering: Principles and Methodologies*. San Diego, CA, USA: Academic Press.

Talluri, S., Raj, S.M., and Christopher, L.P. (2013) Consolidated bioprocessing of untreated switchgrass to hydrogen by the extreme thermophile *Caldicellulosiruptor saccharolyticus* DSM 8903. *Bioresour Technol* **139:** 272–279.

Wee, Y.J., Yun, J.S., Park, D.H., and Ryu, H.W. (2004) Biotechnological production of L(+)-lactic acid from wood hydrolyzate by batch fermentation of *Enterococcus faecalis*. *Biotechnol Lett* **26:** 71–74.

Wee, Y.J., Kim, J.N., and Ryu, H.W. (2006a) Biotechnological production of lactic acid and its recent applications. *Food Technol Biotechnol* **44:** 163–172.

Wee, Y.J., Yun, J.S., Kim, D., and Ryu, H.W. (2006b) Batch and repeated batch production of L (+)-lactic acid by *Enterococcus faecalis* RKY1 using wood hydrolyzate and corn steep liquor. *J Ind Microbiol Biotechnol* **33:** 431–435.

Werpy, T., and Petersen, G. (2004) *Top value added chemicals from biomass: volume i – results of screening for potential candidates from sugars and synthesis gas* [WWW document]. PNRL and NREL, US DoE URL http://www.nrel.gov/docs/fy04osti/35523.pdf.

Yun, J.S., Wee, Y.J., and Ryu, H.W. (2003) Production of optically pure L(+)-lactic acid from various carbohydrates by batch fermentation of *Enterococcus faecalis* RKY1. *Enzyme Microb Technol* **33:** 416–423.

Zhang, Z.Y., Jina, B., and Kelly, J.M. (2007) Production of lactic acid from renewable materials by *Rhizopus* fungi. *Biochem Engg J* **35:** 251–263.

Zhou, S., Causey, T.B., Hasona, A., Shanmugam, K.T., and Ingram, L.O. (2003) Production of optically pure D-lactic acid in mineral salts medium by metabolically engineered *Escherichia coli* W3110. *Appl Environ Microbiol* **69:** 399–407.

16

Molecular annotation of ketol-acid reductoisomerases from *Streptomyces* reveals a novel amino acid biosynthesis interlock mediated by enzyme promiscuity

Karina Verdel-Aranda,[1] Susana T. López-Cortina,[2] David A. Hodgson[3] and Francisco Barona-Gómez[1]*

[1]*Evolution of Metabolic Diversity Laboratory, Unidad de Genómica Avanzada (Langebio), Cinvestav-IPN, Km 9.6 Libramiento Norte, Irapuato, Guanajuato CP36822, México.*

[2]*Facultad de Ciencias Químicas, Universidad Autónoma de Nuevo León, San Nicolás de los Garza, Nuevo León, México.*

[3]*School of Life Sciences, University Warwick, Coventry, UK.*

Summary

The 6-phosphogluconate dehydrogenase superfamily oxidize and reduce a wide range of substrates, making their functional annotation challenging. Ketol-acid reductoisomerase (KARI), encoded by the *ilvC* gene in branched-chain amino acids biosynthesis, is a promiscuous reductase enzyme within this superfamily. Here, we obtain steady-state enzyme kinetic parameters for 10 IlvC homologues from the genera *Streptomyces* and *Corynebacterium*, upon eight selected chemically diverse substrates, including some not normally recognized by enzymes of this superfamily. This biochemical data suggested a *Streptomyces* biosynthetic interlock between proline and the branched-chain amino acids, mediated by enzyme substrate promiscuity, which was confirmed via mutagenesis and complementation analyses of the *proC*, *ilvC1* and *ilvC2* genes in *Streptomyces coelicolor*. Moreover, both *ilvC* orthologues and paralogues were analysed, such that the relationship between gene duplication and functional diversification could be explored. The KARI paralogues present in *S. coelicolor* and *Streptomyces lividans*, despite their conserved high sequence identity (97%), were shown to be more promiscuous, suggesting a recent functional diversification. In contrast, the KARI paralogue from *Streptomyces viridifaciens* showed selectivity towards the synthesis of valine precursors, explaining its recruitment within the biosynthetic gene cluster of valanimycin. These results allowed us to assess substrate promiscuity indices as a tool to annotate new molecular functions with metabolic implications.

Introduction

It is well acknowledged that enzymes can be promiscuous or multifunctional, catalysing different chemical transformations upon one or more substrates, or catalysing identical reactions using several related or unrelated substrates (O'Brien and Herschlag, 1999; Khersonsky *et al.*, 2006; Khersonsky and Tawfik, 2010). Since the recognition of this phenomenon, enzyme promiscuity has been defined as the ability of enzymes to exert other activities beyond those for which they have evolved, implying that such activities are overall not relevant for the physiology of the organism (Copley, 2003; Khersonsky *et al.*, 2006; Khersonsky and Tawfik, 2010). It has also been hypothesized that promiscuous enzymatic activities serve as evolutionary starting points for the appearance of new functions (Jensen, 1976; Khersonsky *et al.*, 2006; Piatigorsky, 2007; Khersonsky and Tawfik, 2010). After duplication, for instance, of a metabolic gene encoding for a promiscuous enzyme, subsequent mutations in one of the paralogue could lead to a novel function (Ohno, 1970). Indeed, enzyme promiscuity followed by enzyme recruitment, seems to have given rise to many peripheral metabolic pathways, such as natural products biosynthetic pathways (Vining, 1992).

It could be argued that one of the main challenges in molecular biology is the correct functional annotation of proteins, a situation that is accentuated in the context of promiscuous enzymes. This challenge has been addressed either by reductionist studies focusing in a single protein, or after high-throughput systems-level analyses involving many proteins (Laskowski *et al.*, 2005; Redfern *et al.*, 2009; Schnoes *et al.*, 2013). The trade-off between these approaches represents a conundrum for annotation of enzyme superfamilies, defined as

*For correspondence. E-mail fbarona@langebio.cinvestav.mx

Funding Information This work was supported by Conacyt grant to FBG (No. 179290) and PhD scholarship to KVA (No. 202715).

structurally and functionally related enzymes that can catalyse similar reactions upon quite different substrates (Gerlt and Babbitt, 2001; Gerlt et al., 2012). Unfortunately, a deep understanding of the relationships among sequence, structure and function of enzyme superfamilies is limited to few cases, such as the enolase superfamily (Gerlt et al., 2012).

Annotation of the 6-phosphogluconate dehydrogenase (6PGDH) superfamily, despite including 12 enzyme families in Structural Classification of Proteins (SCOP) database (Andreeva et al., 2008) involved in many fundamental metabolic pathways (Fig. 1), represents a complicated challenge. At the sequence and structural levels, the superfamily is characterized by a broadly occurring G-X-G-X-X-G sequence motif, which is actually a feature of many dehydrogenases that use NAD(H)$^+$/NADP(H)$^+$ as cofactors, and a domain comprising the universal Rossmann fold. Annotation of enzymes belonging to this superfamily is further confounded by the broad range of chemically diverse substrates that this superfamily can convert (Mondal et al., 2010). For example, the NADH-dependent D-2-hydroxyacid dehydrogenases from the bacteria Enterococcus faecalis and Lactococcus lactis have been incorrectly annotated as ketopantoate reductases [KPR, Enzyme Commission (EC) 1.1.1.169, panE gene] (Wada et al., 2008; Chambellon et al., 2009).

Family 3 of this superfamily solely includes ketol-acid reductoisomerases (KARI, EC 1.1.1.86, ilvC gene). These are particularly interesting enzymes, as they perform two different reactions, and their substrate promiscuity is essential for the biosynthesis of the branched-chain amino acids, namely, valine, isoleucine and leucine. Previous enzyme mechanistic studies in Escherichia coli, Pseudomonas aeruginosa and plants (Dumas et al., 1994a,b; Biou et al., 1997; Ahn et al., 2003; Tyagi et al., 2005) have shown that the isomerization and the reduction activities of KARI can be separated and measured in vitro. Thus, KARI catalyses an alkyl migration followed by an NADP(H)$^+$-dependent reduction. An implication of this is that the physiologically relevant substrates of KARI are 2-hydroxy-2-methyl ketobutyrate (2H2M3KB) in valine biosynthesis and 2-hydroxy-2-ethyl ketobutyrate (2H2E3KB) in isoleucine biosynthesis. Moreover, the reductase activity of KARI is exerted upon the intermediates 3-hydroxy-3-methyl-2-ketobutyrate (3H3M2KB) and 3-hydroxy-3-ethyl-2-ketobutyrate (3H3E2KB) respectively (Fig. 1). Therefore, KARI enzymes have at least four different substrates; however, only one EC number has been used to classify this enzyme.

Despite the large number of qualitative studies on promiscuous enzymes from different perspectives (O'Brien and Herschlag, 1999; Copley, 2003; Bornscheuer and Kazlauskas, 2004; Khersonsky et al., 2006; Hult and Berglund, 2007; Kim and Copley, 2007; Patrick et al.,

2007; Khersonsky and Tawfik, 2010), there have been few efforts to systematically quantify this important property (Chakraborty and Rao, 2012). In 2008, Nath and colleagues published the index of substrate promiscuity, which is an entropy-based metric used to compare the promiscuous behaviour of enzymes. The promiscuity index uses information theory to describe catalytic efficiency of a set of enzymes towards various substrates. This metric is the probability that any given substrate will be the first one to be converted when an enzyme is simultaneously exposed to equal, low concentrations of all available substrates (Nath and Atkins, 2008; Nath et al., 2010). Moreover, this index, which ranges from 0 to 1, has two forms: a standard (I) and a weighted index (J). While I only takes into account the catalytic efficiency, J incorporates the chemical similarity of substrates, as given by a Tanimoto coefficient (Willet et al., 1998).

Here, we report the biochemical characterization of 10 enzymes annotated as KARIs from species belonging to the order Actinomycetales, renowned to include prolific producers of natural products, such as Streptomyces. In vitro results, using a total of eight chemically diverse substrates revealed a differential promiscuous behaviour, including: (i) reduction of pyrroline-5-carboxylate (P5C), a substrate chemically unrelated to the keto acids typically converted by KARIs and (ii) diversification or specialization of ilvC paralogues, which seemed to depend on their genomic and metabolic contexts. These observations guided in vivo genetic-based experiments, allowing for the discovery of a biosynthetic interlock between proline and the branched-chain amino acids in Streptomyces, as well as an assessment of the trade-off between in vitro and in vivo data for molecular functional annotation, by means of using substrate promiscuity indices.

Results and discussion

Chemical universe of the 6PGDH enzyme superfamily

Given that KARI, the central subject of this study, is a reductase, we arbitrary limited our analyses to the sub-set of the 6PGDH enzyme superfamily that are reductases (Fig. 1). This includes, in addition to family 3 (KARI), family 9 (KPR) and 10 (P5CR), and the enzyme 2-hydroxy-3-oxopropionate reductase (2H3OPR) of family 1 [hydroxyisobutyrate dehydrogenase (HBDH)-6PGDH]. KPR of family 9, an enzyme involved in pantothenate biosynthesis, reduces a hydroxyl keto acid into a dihydroxy acid, but without being preceded by an isomerization step as in KARI (Zheng and Blanchard, 2000). In contrast, the reductase activity of P5CR of family 10, involved in proline biosynthesis, catalyses a different reduction as reflected by the second digit of its EC number (i.e. 1.5.1.2). The bond being reduced by P5CR involves the nitrogen atom of pyrroline, rather than the oxygen

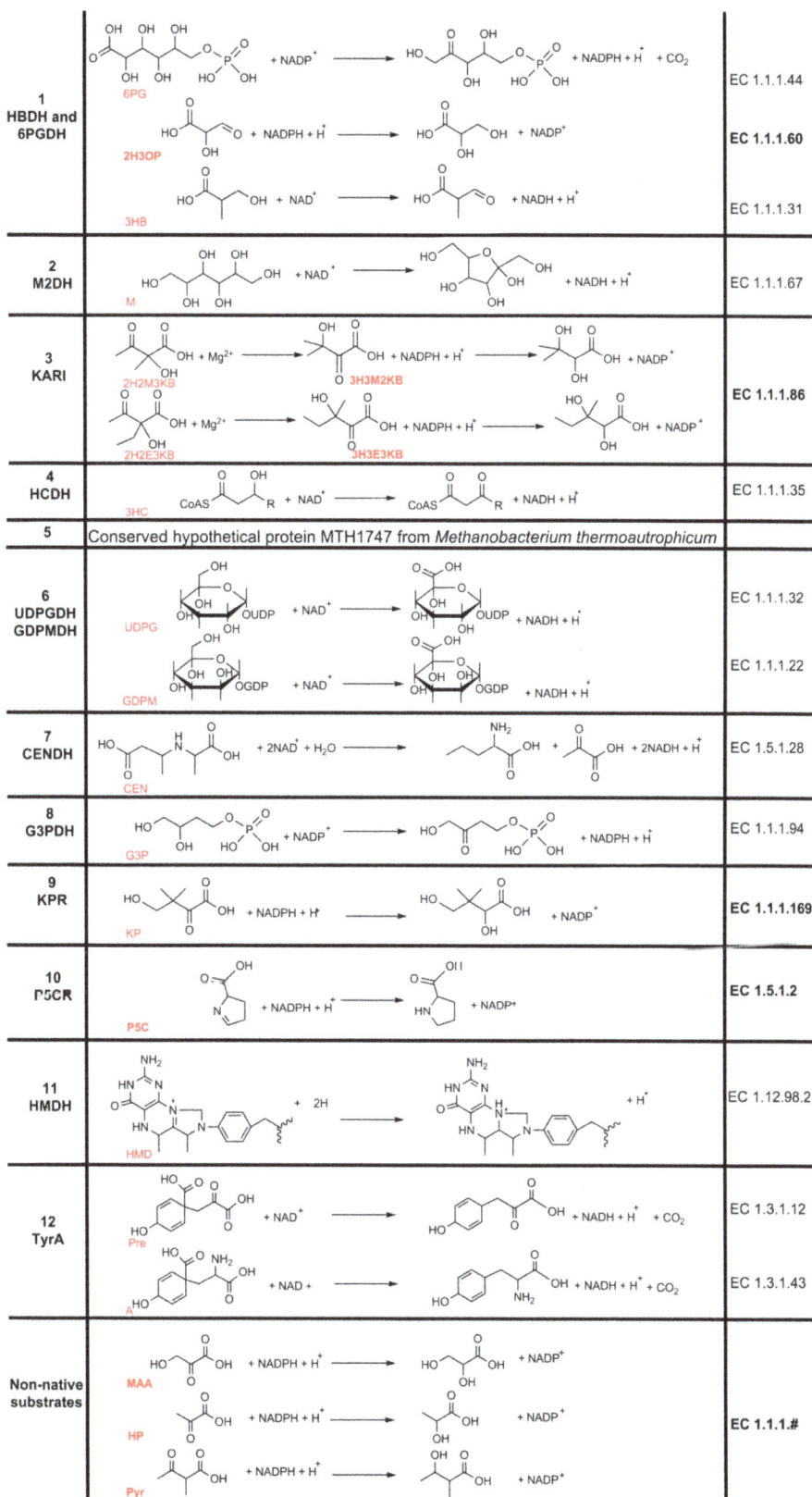

1 **HBDH and** **6PGDH**		EC 1.1.1.44 **EC 1.1.1.60** EC 1.1.1.31
2 **M2DH**		EC 1.1.1.67
3 **KARI**		**EC 1.1.1.86**
4 **HCDH**		EC 1.1.1.35
5	Conserved hypothetical protein MTH1747 from *Methanobacterium thermoautrophicum*	
6 **UDPGDH** **GDPMDH**		EC 1.1.1.32 EC 1.1.1.22
7 **CENDH**		EC 1.5.1.28
8 **G3PDH**		EC 1.1.1.94
9 **KPR**		**EC 1.1.1.169**
10 **P5CR**		**EC 1.5.1.2**
11 **HMDH**		EC 1.12.98.2
12 **TyrA**		EC 1.3.1.12 EC 1.3.1.43
Non-native **substrates**		**EC 1.1.1.#**

Fig. 1. Enzyme members of the 6PGDH superfamily in SCOP. 1: Hydroxyisobutyrate-6-phosphogluconate dehydrogenase (HBDH-6PGDH); 6-phosphogluconate dehydrogenase (6PGDH, EC 1.1.1.44), 2-hydroxy-3-oxopropionate reductase (2H3OPR, EC 1.1.1.60), 3-hydroxyisobutyrate dehydrogenase (3HBDH, EC 1.1.1.31). 2: Mannitol 2-dehydrogenase (M2DH, EC 1.1.1.67). 3: Ketol-acid reductoisomerase (KARI, EC 1.1.1.86). 4: 3-hydroxyacyl-CoA dehydrogenase (3HCDH, EC 1.1.1.35). 5: Conserved hypothetical protein MTH1747. 6: uridine diphosphate (UDP)-glucose/GDP-mannose dehydrogenase (UDPGDH/GDPMDH, EC 1.1.1.22/EC 1.1.1.132). 7: N-(1-D-carboxylethyl)-L-norvaline dehydrogenase (CENDH, EC 1.5.1.28). 8: Glycerol-3-phosphate dehydrogenase (G3PDH, EC 1.1.1.94). 9: Ketopantoate reductase (KPR, EC1.1.1.169). 10: Pyrroline-5-carboxylate reductase (P5CR, EC 1.5.1.2.). 11: 10-methenyltetrahydromethanopterin hydrogenase (HMD, EC 1.12.98.2). 12: TyrA, prephenate dehydrogenase/arogenate dehydrogenase (PreDH/ADH, EC 1.3.1.12/EC 1.3.1.43). The EC numbers of reductases are shown in bold. The last three reactions belong to the non-native substrates.

◄───

of more frequently encountered carbonyl (Nocek *et al.*, 2005). 2H3OPR of family 1, which also includes the dehydrogenases of 6-phosphogluconate and 3-hydroxyisobutyrate, catalyses the reduction of 2-hydroxy-3-oxopropionate to glycerate (Osipiuk *et al.*, 2009).

To better understand the relationships between these reductases, we aimed to explore the chemical distribution of the native substrates of all enzymes of 6PGDH superfamily. We also analysed three keto acids that share chemical similarity with substrates of this superfamily, but which are not known to be the main substrate of members of this superfamily. This latter sub-set, which includes pyruvate (Pyr), hydroxypyruvate (HP) and methylacetoacetate (MAA), is referred to as non-native substrates. A modified list of descriptors published previously by Nath and Atkins (2008), including novel *ad hoc* descriptors for keto acids, was then obtained (Table S1). This took into account: (i) the position of the hydroxyl group (primary, secondary, tertiary alcohol), (ii) the nature of the alkyl groups (methyl, ethyl) and (iii) the position adopted by these alkyl groups.

The result obtained after this analysis is a symmetrical distance matrix and can be displayed two-dimensionally by using multidimensional scaling. The resulting perceptual map is a graphic representation within a coordinate system of different features or distances of objects, in this case, substrates. In such a Cartesian view, distances between each substrate are represented as normalized chemical dissimilarity scores, which therefore lack units. Overall, the results shown in Fig. 2 highlight the sensitivity of the keyset of descriptors used to calculate the Tanimoto coefficient (Table S4). The result of this analysis shows that substrates are evenly distributed throughout all four quadrants and they seem to cluster in accordance to conserved functional groups.

Given that most enzyme families act upon chemically closely related substrates, it was unexpected to find that members of family 1 recognize substrates distributed throughout two quadrants. Such wide distribution may be due to differences in functional groups between these three substrates: 6-phosphogluconate has a phosphate group that is absent from the other two substrates; whereas 2-hydroxy-3-oxopropionate is a hydroxyacid

semialdehyde and 3-hydroxyisobutyrate is a hydroxyacid that lacks a carbonyl group. Figure 2 also shows the distribution of the three non-native keto acids. MAA has a keto group prone to be reduced, as well as an alkyl moiety prone to migration as in some native substrates. This substrate clusters together with Pyr, yet slightly away from HP, which is the only non-native substrate with a hydroxyl group.

Selection and chemical synthesis of substrates

In addition to the substrates of KARI (2H2M3KB/3H3M2KB for valine and 2H2E3KB/3H3E2KB for isoleucine), the substrate of family 10, P5C, which is the direct precursor of proline, was selected. P5C is the most dissimilar substrate that is subject to reduction, and to our knowledge, it has never been tested as a KARI substrate, although this enzyme has been extensively biochemically characterized (Primerano and Burns, 1983; Dumas *et al.*, 1994a,b; 1995; 2001; Biou *et al.*, 1997; Ahn *et al.*, 2003; Tyagi *et al.*, 2005). Following adaptations from previously reported protocols (Williams and Frank, 1975; Chunduru *et al.*, 1989; Tyagi *et al.*, 2005), these substrates, as well as the non-native substrate MAA, were chemically synthesized as described in Fig. 3.

The substrate ketopantoate (KP), unfortunately, could not be included because its precursor for chemical synthesis, dihydro-4,4-dimethyl-2,3-furandione, is subject to strict banning policies where this work was performed. Furthermore, the substrate 2H3OP of 2H3OPR, as the sole reductase of the multi-enzyme family 1, could not be purified after synthesis. Given that 3H3OP and HP are tautomers, their separation is difficult. The previously catalytic parameters of the enzyme 3H3OPR from *Pseudomonas putida* and *Pseudomonas acidovorans*, where D-glycerate metabolism involving this enzyme was first described (Kohn and Jakoby, 1968), were actually measured in the presence of HP. Thus, use of HP was used as a substitute for 3H3OP. In total, eight ligands, including five native and three non-native substrates, belonging to two of the four quadrants of the Cartesian plane (Fig. 2), were experimentally characterized. The set of enzymes used for these experiments were selected as described in the following section.

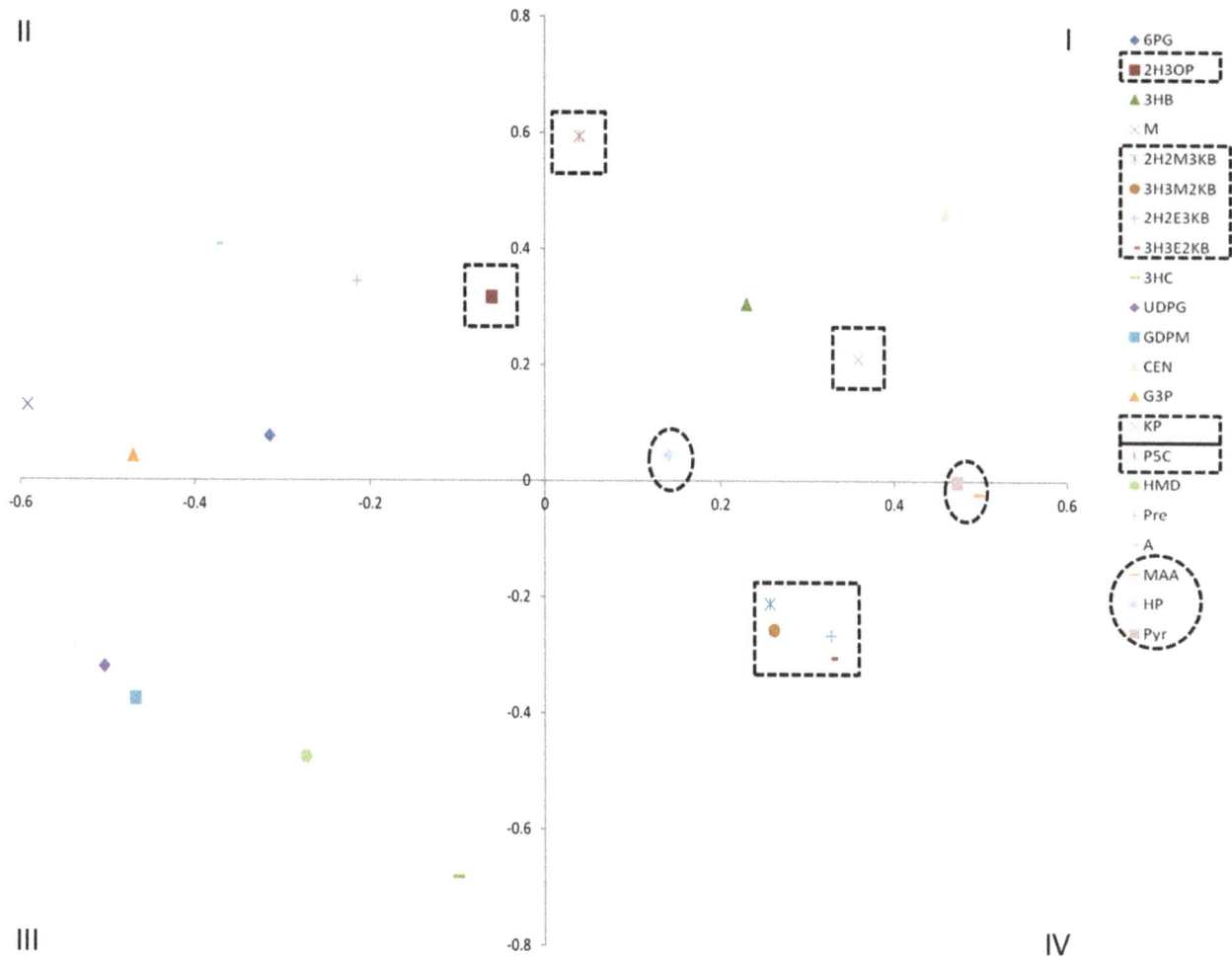

Fig. 2. Chemical diversity of the 6PGDH superfamily. Two-dimentional representation of the chemical distance matrix using multidimensional scaling. Distribution represents normalized Tanimoto dissimilarity scores between each substrate. The *quadrants* are labelled with *Roman numerals*. Native substrates of reductases from 6PGDH superfamily are shown in dotted boxes. 2H2M3KB, 3H3M2KB, 2H2E3KB and 3H3E2KB cluster together in the fourth quadrant with non-native substrates (dashed circles) MAA and Pyr. Most distant P5C and HP are distributed in quadrant one, while 2H3OP is in quadrant two.

Selection and biochemical analysis of actinobacterial KARI homologues

After reconstruction of a phylogenetic tree using actinobacterial sequences of KARI homologues obtained from public databases, as well as sequences of the taxonomic marker RpoB (Fig. S1), around two dozen of KARI homologues were selected for functional analysis. Based on the phylogenetic distribution of the selected KARIs, covering different genera within the *Actinobacteria*, we decided to focus in the genus *Streptomyces*. The genomic coverage of this genus has witnessed a substantial increase, allowing for the identification of lineages where gene duplication events have occurred. A diagram that aims to recapitulate the evolutionary history of the selected *ilvC* homologues, including two independent paralogous events, is shown in Fig. 4A. As an out-group,

the KARI enzyme from the actinobacterium *Corynebacterium glutamicum*, Cgl, was adopted. We selected Cgl as an out-group because this enzyme has been shown to compensate for the lack of a *panE* gene in this organism (Merkamm *et al.*, 2003), and thus a physiologically relevant dual-substrate specificity (KPR and KARI activities) can be safely assumed.

The KARI homologues that could be successfully cloned for heterologous expression and purification purposes (Table S2) led to a final list of 10 enzymes. Nine of these enzymes belong to the genus *Streptomyces* and one to *C. glutamicum*. The only organism that was analysed that lacks a complete genome sequence is *Streptomyces viridifaciens*, although it has been proposed that its KARI homologue, Svi2, is the result of a gene duplication event (Garg *et al.*, 2002). This is supported by the fact that this gene is part of an *ilvBNCE* operon that is

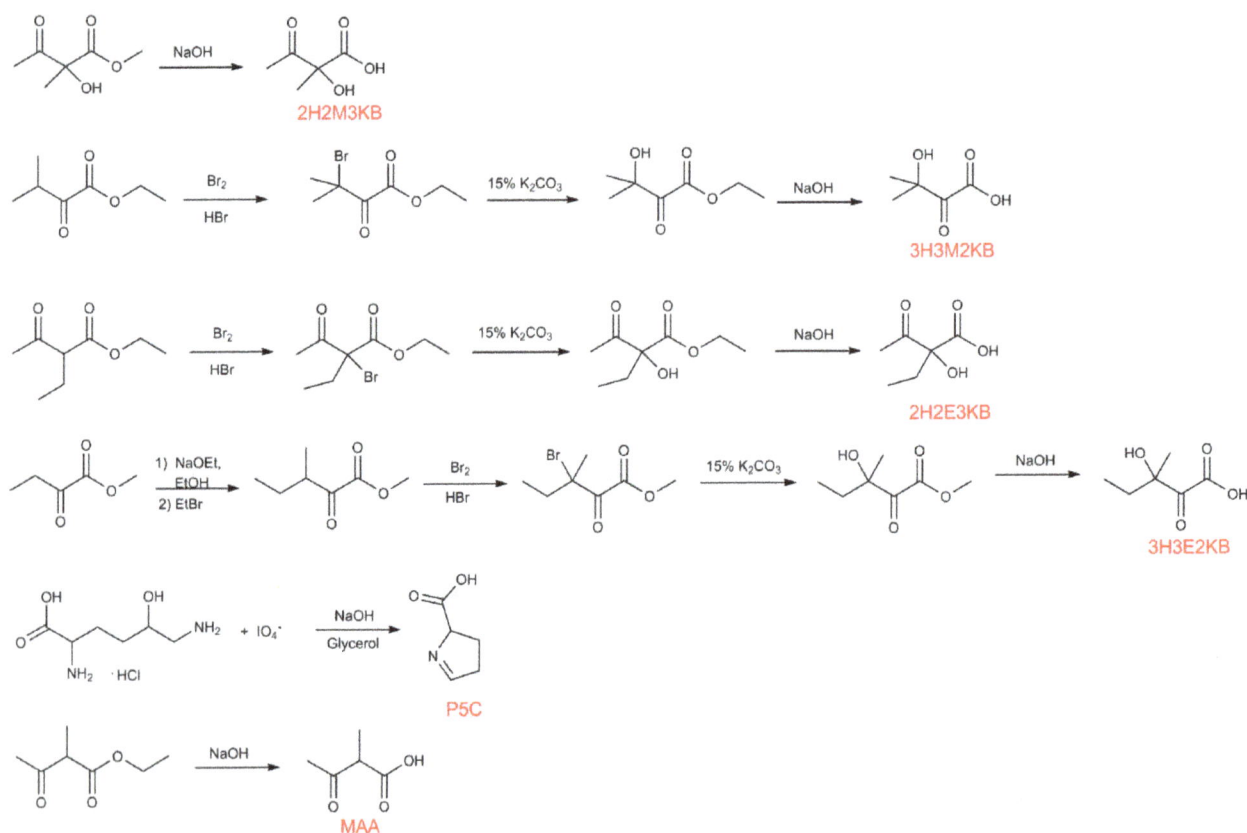

Fig. 3. Chemical synthesis of substrates of KARI. Synthesis of 2-hydroxy-2-methyl-3-ketobutyrate (2H2M3KB); 3-hydroxy-3-methyl-2-ketobutyrate (3H3M2KB); 2-hydroxy-2-ethyl-3-ketobutyrate (2H2E3KB); 3-hydroxy-3-ethyl-2-ketobutyrate (3H3E2KB); Pyrroline-5-carboxylate (P5C); and 2-methylacetoacetate (MAA).

next to the valanimycin biosynthetic gene cluster, which differs from the conserved *ilvBNC* operon – without *ilvE* – present in all of the available *Streptomyces* genomes (Fig. S2). The other paralogous relationship that was included is that of Sco1/Sco2 and Sli1/Sli2 from *S. coelicolor* and *S. lividans* respectively. Given that Sco1 and Sli1 are identical, and Sco2 and Sli2 are also identical, it is reasonable to conclude that this duplication must have occurred very recently but before speciation of these organisms. Thus, a single paralogous relationship for these four enzymes was considered.

Steady-state enzyme kinetic parameters were obtained for the 10 soluble KARI homologues, using both native (Table 1) and non-native (Table 2) substrates, and IlvC from *E. coli* (Eco) as a negative control for all substrates other than valine precursors (ValP = 2H2MKB and 3H3MKB) and isoleucine precursors (IleP = 2H2E3KB and 3H3E2KB), which served as positive controls. Overall, the K_M and k_{cat} parameters were found to be better for ValP than IleP. Highlighting this trend is Svi2 from *S. viridifaciens*, which has the highest k_{cat}/K_M of all enzymes, due to slightly better K_M and k_{cat} parameters for ValP, but yet the worst K_M parameter for

IleP. As a result, a 100-fold better catalytic efficiency (k_{cat}/K_M) for ValP can be seen. Thus, it is reasonable to conclude that Svi2 has a specialized function for the synthesis of valine, which serves as precursor for valanimycin, a natural product that has never been shown to incorporate isoleucine or leucine into its structure (Fig. S2). In fact, an amino acid decarboxylase encoded by the *vlmD* gene of the valanimycin biosynthetic gene cluster has been shown to be highly specific towards valine (Garg *et al.*, 2002).

The paralogues Sco2 and Sli2 show a 15-fold lower k_{cat}/K_M catalytic efficiency for ValP compared with their corresponding Sco1 and Sli1 paralogous partners. The trade-off of the kinetic parameters of Sco2 and Sli2 could be a sign of functional divergence, implying a yet-to-be discovered enzyme function. As in the case of other promiscuous reductases involved in the chlorophyll cycle (Ito and Tanaka, 2014), it could be that the dehydrogenation and not the reduction reaction is favoured. This would imply that Sco2 and Sli2 are better suited for catalysing the reverse reaction. This may be related to the fact that Sco1 and Sli1 are unable to reduce HP, which contradicts all other KARIs, including Sco2 and Sli2 (Table 2). These

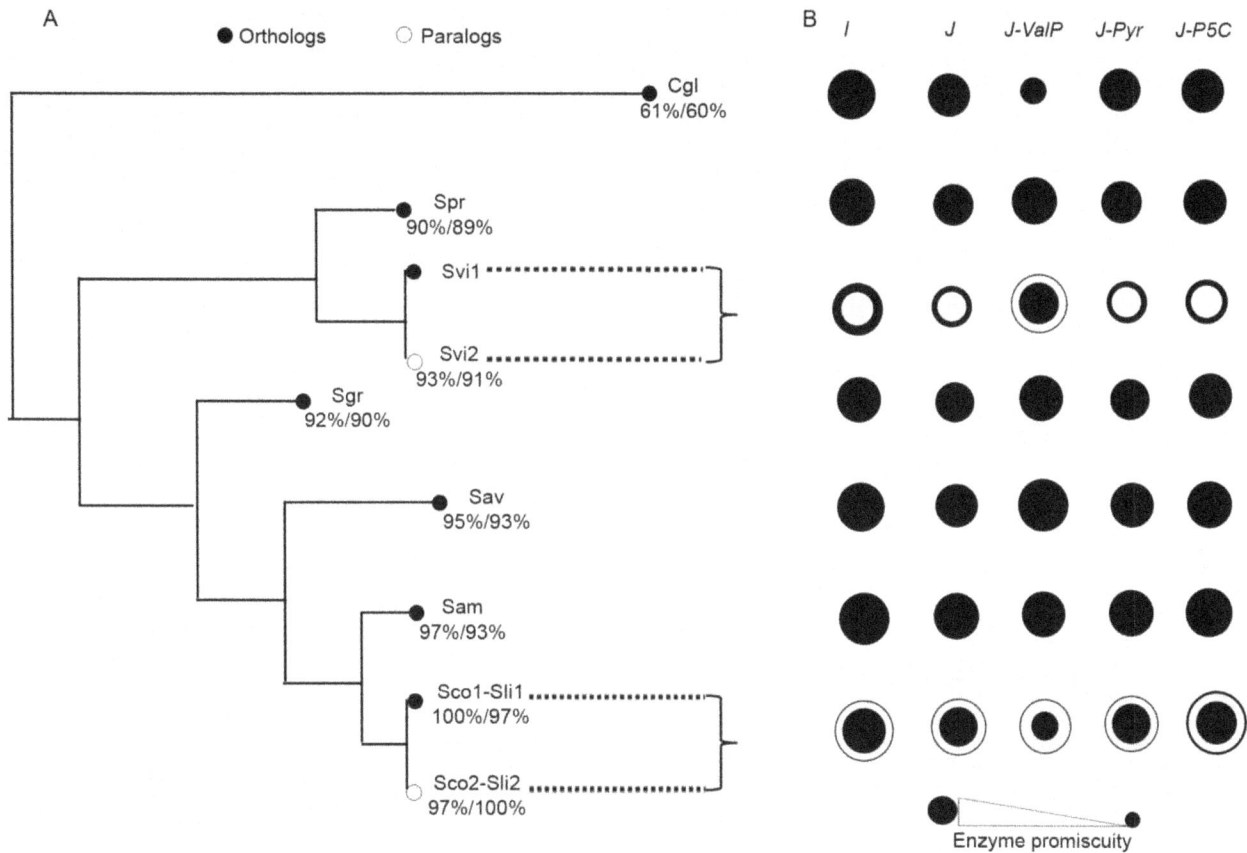

Fig. 4. Promiscuous behaviour of phylogenetically related KARI homologues.
A. Evolutionary relationships of KARI homologues based on RpoB and KARI phylogenies. Percentages provided are sequence identity calculated with respect to Sco1/Sco2.
B. Substrate promiscuity indices *I* and *J* for orthologues are shown as black circles, other than paralogues that are shown as white circles. The two paralogous independent events investigated in this study are highlighted with a key. Names of enzymes are as in Table 1. Refer to text for further details.

subtle functional differences become relevant in the context of the high sequence identity (97%) shared between Sco1 with Sco2, and Sli1 with Sli2, warranting further investigation.

All KARI homologues, moreover, could be saturated with the different substrates tested, with the exception of P5C (Table 1). The activity upon this substrate was found to be very low but present in all *Streptomyces* KARIs. The positive control for these experiments was P5CR or ProC from *S. coelicolor*. For the *Streptomyces* KARIs, therefore, only the catalytic efficiencies and not K_M and k_{cat} parameters determined independently could be obtained. Interestingly, we could not detect signs of conversion of P5C by Cgl, not even using highly sensitive *in vivo* complementation assays based in high copy number plasmids and an *E. coli proC* minus proline auxotroph (see below). This could be attributable to the fact that Cgl has a physiologically relevant KPR activity (Merkamm *et al.*, 2003), implying a P5CR and KPR activity trade-off during the course of evolution.

In vivo genetic-based analysis of functional predictions

P5CR has been postulated to be a remote homologue of KARI, as these enzymes share 19% at the sequence level and they show strong structural similarities (Nocek *et al.*, 2005). As a quick method to look for P5CR activity, we used an *E. coli* proline auxotrophic mutant that lacks the *proC* gene (Baba *et al.*, 2006). The inability of this strain to grow in M9 minimal media without proline could be rescued with all of the different KARI homologues, previously cloned into a pASK plasmid derivative suitable for complementation assays (Table S2), other than Cgl from *C. glutamicum*. Moreover, since all of these KARI homologues do complement an *ilvC* mutant from the same collection, this result supports the proposed trade-off between the P5CR and KPR activities of Cgl during the course of evolution.

We then aimed to construct a set of mutants in *S. coelicolor*, useful for *in vivo* complementation assays. Previously, we have shown that a *proC* minus mutant of

Table 1. Kinetic parameters of selected KARI homologues against native substrates.

Substrate	Kinetic parameters[a]	Eco	Sco1	Sco2	Sli1	Sli2	Sam	Enzyme Sav	Sgr	Spr	Svi2	Cgl	P5CR
2H2M3KB	k_{cat} (s^{-1})	2 ± 0.097	1.1 ± 0.2	0.24 ± 0.01	0.7 ± 0.59	0.28 ± 0.02	0.63 ± 0.5	0.24 ± 0.2	0.42 ± 0.04	3.6 ± 0.3	2 ± 0.07	3.2 ± 0.4	ND[b]
	K_M (mM)	0.28 ± 0.03	1.6 ± 0.2	12 ± 0.8	1.2 ± 0.07	9.3 ± 0.8	2 ± 0.15	0.8 ± 0.09	1.2 ± 0.7	9 ± 0.7	1.8 ± 0.9	8 ± 0.9	ND[b]
	k_{cat}/K_M (M^{-1} s^{-1})	7142 ± 689	$687. \pm 43$	20 ± 3.2	583 ± 64	30 ± 3.8	315 ± 33	300 ± 39	350 ± 37	400 ± 38	1111 ± 115	400 ± 15	ND[b]
3H3M2KB	k_{cat} (s^{-1})	3.1 ± 0.25	2.2 ± 0.15	0.3 ± 0.04	2.8 ± 0.3	0.4 ± 0.05	0.8 ± 0.1	1.2 ± 0.17	0.5 ± 0.06	5.2 ± 0.6	2.5 ± 0.3	0.9 ± 0.08	ND[b]
	K_M (mM)	0.26 ± 0.03	3 ± 0.2	10 ± 0.11	4 ± 0.5	8.6 ± 0.9	4 ± 0.3	13 ± 0.9	8 ± 0.9	8.6 ± 0.9	5 ± 4.8	2.3 ± 0.25	ND[b]
	k_{cat}/K_M (M^{-1} s^{-1})	11923 ± 980	733 ± 7	30 ± 3.3	700 ± 71	45 ± 4	200 ± 54	520 ± 58	850 ± 90	530 ± 50	500 ± 48	391 ± 38	ND[b]
2H2E3KB	k_{cat} (s^{-1})	3.5 ± 0.031	3.2 ± 0.2	0.39 ± 0.04	3.6 ± 0.4	0.12 ± 0.02	1.3 ± 0.09	0.4 ± 0.04	0.44 ± 0.03	0.18 ± 0.03	0.3 ± 0.02	0.9 ± 0.08	ND[b]
	K_M (mM)	0.3 ± 0.028	67 ± 7	12 ± 1	71 ± 5	5 ± 0.3	6.7 ± 0.7	2.6 ± 0.3	2.5 ± 0.3	10 ± 0.9	25 ± 6	75 ± 5	ND[b]
	k_{cat}/K_M (M^{-1} s^{-1})	11666 ± 1000	47 ± 4	32 ± 2	50 ± 0.4	48 ± 5	194 ± 20	153 ± 10	176 ± 12	18 ± 1	12 ± 1	12 ± 2	ND[b]
3H3E2KB	k_{cat} (s^{-1})	3.8 ± 0.039	1.8 ± 0.2	0.8 ± 0.02	1.3 ± 0.04	0.9 ± 0.03	0.9 ± 0.1	0.8 ± 0.08	1 ± 0.05	1.4 ± 0.9	0.5 ± 0.03	0.18 ± 0.01	ND[b]
	K_M (mM)	0.29 ± 0.03	45 ± 3	20 ± 3	50 ± 6	20 ± 1.9	20 ± 2	40 ± 4	35 ± 3	50 ± 6	50 ± 3	5 ± 1	ND[b]
	k_{cat}/K_M (M^{-1} s^{-1})	13103 ± 1000	40 ± 3	40 ± 5	46 ± 3	46 ± 4	45 ± 4	20 ± 3	28 ± 2	28 ± 3	10 ± 0.9	36 ± 4	ND[b]
P5C	k_{cat}/K_M (M^{-1} s^{-1})	ND[b]	2.4 ± 0.3	3 ± 0.4	3.7 ± 0.4	6.1 ± 0.6	5.5 ± 0.6	5.6 ± 0.5	2.4 ± 0.2	2.8 ± 0.3	3.8 ± 0.3	ND[b]	601 ± 70

a. Kinetic parameters shown are means and standard errors of three enzymatic reaction.
b. ND, activity not determined because it is below the limit of detection of the enzyme assay, which is $k_{cat}/K_M = 0.000013$ M^{-1} s^{-1}.
Enzymes nomenclature: Eco: KARI, ilvC, Escherichia coli; Sco1: KARI1, ilvC1, Streptomyces coelicolor; Sco2: KARI2, ilvC2, Streptomyces coelicolor; Sli1: KARI1, ilvC1, Streptomyces lividans; Sli2: KARI2, ilvC2, Streptomyces lividans; Sam: KARI, ilvC, Streptomyces ambofaciens; Sav: KARI, ilvC, Streptomyces avermitilis; Sgr: KARI, ilvC, Streptomyces griseus; Spr: KARI, ilvC, Streptomyces pristinaespiralis; Svi2: KARI2, ilvC2, Streptomyces viridifaciens; Cgl: KARI, ilvC-panE, Corynebacterium glutamicum; P5CR: P5CR, proC, Streptomyces coelicolor.

Table 2. Kinetic parameters of selected KARI homologues against non-native substrates.

Substrate	Kinetic parameters[a]	Sco1	Sco2	Sli1	Sli2	Sam	Enzyme[c] Sav	Sgr	Spr	Svi2	Cgl	P5CR
MAA	k_{cat} (s^{-1})	1 ± 0.2	0.6 ± 0.05	0.675 ± 0.06	0.5 ± 0.05	1.2 ± 0.09	1.8 ± 0.1	1.05 ± 0.1	2 ± 0.2	2 ± 0.2	0.4 ± 0.03	ND[b]
	K_M (mM)	50 ± 4	60 ± 5	45 ± 4	50 ± 4	40 ± 4	30 ± 2	35 ± 4	40 ± 5	50 ± 5	40 ± 1.5	ND[b]
	k_{cat}/K_M (M^{-1} s^{-1})	20 ± 1.8	10 ± 1	15 ± 1.2	10 ± 0.9	30 ± 3	60 ± 7	30 ± 2.5	50 ± 6	40 ± 5	10 ± 1.3	ND[b]
HP	k_{cat} (s^{-1})	ND[b]	0.5 ± 0.05	ND[b]	0.7 ± 0.06	1 ± 0.09	2 ± 0.2	1.5 ± 0.1	2 ± 0.2	2 ± 0.2	0.5 ± 0.04	ND[b]
	K_M (mM)	ND[b]	40 ± 3	ND[b]	50 ± 5	50 ± 4	60 ± 5	30 ± 4	60 ± 5	50 ± 5	40 ± 4	ND[b]
	k_{cat}/K_M (M^{-1} s^{-1})	ND[b]	60 ± 7	ND[b]	40 ± 3.5	20 ± 2.2	50 ± 5	40 ± 3.8	30 ± 3	50 ± 4	60 ± 3.9	ND[b]
Pyr	k_{cat} (s^{-1})	2.8 ± 0.05	5.6 ± 0.5	6.7 ± 0.7	3 ± 0.2	6.3 ± 0.5	3.6 ± 0.4	1 ± 0.1	3 ± 0.2	5.7 ± 0.5	8.6 ± 0.9	ND[b]
	K_M (mM)	950 ± 80	700 ± 70	1120 ± 110	1010 ± 100	1050 ± 99	1200 ± 110	1060 ± 120	1080 ± 100	1140 ± 112	1070 ± 118	ND[b]
	k_{cat}/K_M (M^{-1} s^{-1})	3 ± 0.2	8 ± 0.7	6 ± 0.5	3 ± 0.2	6 ± 0.5	3 ± 0.3	1 ± 0.2	3 ± 0.3	5 ± 0.4	8 ± 0.7	ND[b]

a. Kinetic parameters shown are means and standard errors of three enzymatic reaction.
b. ND, activity not determined because it is below the limit of detection of the enzyme assay, which is $k_{cat}/K_M = 0.000013$ M^{-1} s^{-1}.
c. Enzymes nomenclature: Eco: KARI, ilvC, Escherichia coli; Sco1: KARI1, ilvC1, Streptomyces coelicolor; Sco2: KARI2, ilvC2, Streptomyces coelicolor; Sli1: KARI1, ilvC1, Streptomyces lividans; Sli2: KARI2, ilvC2, Streptomyces lividans; Sam: KARI, ilvC, Streptomyces ambofaciens; Sav: KARI, ilvC, Streptomyces avermitilis; Sgr: KARI, ilvC, Streptomyces griseus; Spr: KARI, ilvC, Streptomyces pristinaespiralis; Svi2: KARI2, ilvC2, Streptomyces viridifaciens; Cgl: KARI, ilvC-panE, Corynebacterium glutamicum; P5CR: P5CR, proC, Streptomyces coelicolor.

S. coelicolor, termed WP101, is prototrophic (Barona-Gomez and Hodgson, 2010). Based on our biochemical analysis, it is tempting to speculate that the remaining P5CR activity of strain WP101 could be related to the promiscuous P5CR activity of Sco1 and/or Sco2. Construction of a triple *proC*, *ilvC1* and *ilvC2* mutant, ideally unmarked to avoid polar effects (*proC* and *ilvC1* seem to be part of operons), was therefore attempted. After an unsuccessful comprehensive screening to isolate double cross-over recombination events, or marker excision using a phage display method (Khodakaramian et al., 2006), we concluded that strong counter-selection for mutation of *ilvC1* in the absence of *proC* and *ilvC2* is operating. This counter-selection was overcome, however, by the use of two different resistant markers. Using as genetic background, a *proC::scar* unmarked mutant, the streptomycin (*aadA*) and apramycin [*aac(3)IV*] resistance cassettes were used to replace the *ilvC1* and *ilvC2* genes respectively.

The *S. coelicolor* mutants constructed in this study, as well as their growth requirements, are shown in Table S3. The *S. coelicolor ilvC1* and *ilvC2* double mutant, as expected, is auxotrophic for valine, leucine and isoleucine, while the triple mutant, in addition to being auxotrophic to these amino acids, is also auxotrophic for proline. To further analyse these mutants, five *ilvC* genes representative of different genetic backgrounds and catalytic efficiencies were cloned into a modified version of the pAV11B plasmid called pAV11B_FBG (Jyothikumar et al., 2012) (Table S2). This plasmid drives gene expression from a tetracycline-inducible *tcp830* promoter that can be strongly induced with anhydrotetracycline, and thus poor enzyme activities can be revealed. The selected *ilvC* genes include those coding for the proteins Sco1 and Sco2, as well as Svi2, as representatives of paralogous enzymes with different promiscuous behaviour; the *ilvC* gene from *S. griseus* as an example of a gene encoding a non-duplicated KARI and the *ilvC* gene of *C. glutamicum* as the sole member analysed here that lacks P5CR activity.

All *ilvC* genes, including Cgl, rescued growth of the *S. coelicolor* double mutant, as expected. Induction of expression with different anhydrotetracycline concentrations only reduced growth time for the case of the complemented strain with the Cgl-containing construct. In contrast, exactly under the same conditions, Cgl failed to rescue growth of the triple mutant, which is in agreement with the fact that a *C. glutamicum proC* mutant is a proline auxotroph (Ankri et al., 1996). Moreover, all of the *Streptomyces* enzymes did complement equally well the triple mutant (Fig. 5). Interestingly, the differential promiscuous behaviour of the paralogues Sco1, Sco2 and Svi2 was not reflected by these complementation assays.

Thus, we conclude, first, that promiscuous activities *in vivo*, even if small, are usually enough to support growth, something that has been previously broadly reported (Kim and Copley, 2007; Patrick et al., 2007) and second, that evolutionary trade-offs could drive or restrain enzyme promiscuity, even if chemically speaking reactions are actually feasible. These two conclusions provide insights into the future development of enzyme promiscuity indices, as further discussed below.

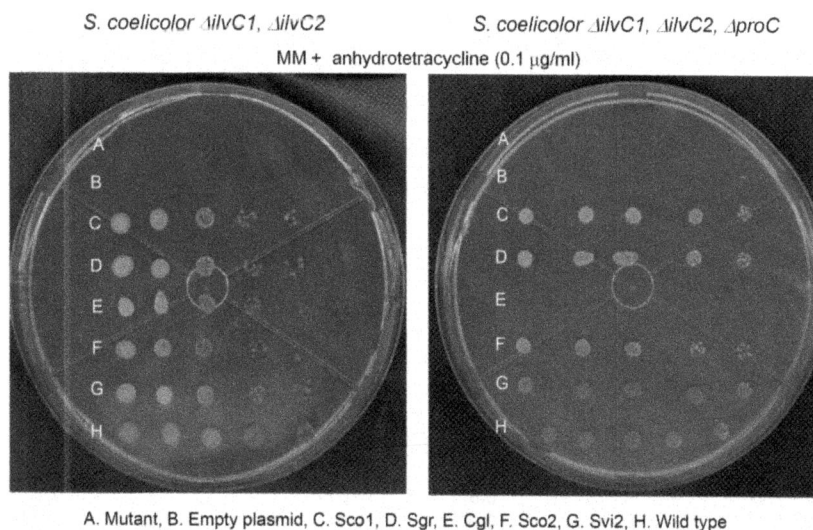

S. coelicolor ΔilvC1, ΔilvC2   *S. coelicolor ΔilvC1, ΔilvC2, ΔproC*

MM + anhydrotetracycline (0.1 µg/ml)

A. Mutant, B. Empty plasmid, C. Sco1, D. Sgr, E. Cgl, F. Sco2, G. Svi2, H. Wild type

Fig. 5. Functional *in vivo* analysis of selected KARIs. Complementation assays in *S. coelicolor* double (*ΔilvC1, ΔilvC2*) and triple (*ΔilvC1, ΔilvC2, ΔproC*) mutants in *Streptomyces* minimal medium (MM) supplemented with anhydrotetracycline (0.1 µg ml^{-1}). Picture was taken after 4 days of growth. All strains other than A and H, i.e. from B to G, are the corresponding *S. coelicolor* mutant [marked as A: double *ilvC* mutant, left panel; triple *ilvC* and *proC* mutant, right panel] complemented with (B) empty plasmid, (C) Sco1, (D) Sgr, (E) Cgl, (F) Sco2 and (G) Svi2 (Names of enzymes are as in Table 1). H refers to *S. coelicolor* wild-type strain.

Assessment of indices of substrate promiscuity

Standard *I* and weighted *J* indices of substrate promiscuity were calculated for the ten KARI homologues and the eight substrates that were experimentally tested (Table 3). In addition to the experimentally based calculations, for some cases where either the enzyme or the substrate could not be obtained, theoretical estimations were performed. These estimations were based on conservative assumptions considering physiological expectations. On one hand, as the substrate of KP could not be obtained and a physiologically relevant KPR activity has been demonstrated (Merkamm *et al.*, 2003), we assumed that Cgl has a catalytic efficiency for KP reduction in the same order of magnitude than that detected for ValP. On the other hand, for the remaining enzymes from *Streptomyces*, we estimated low KPR catalytic efficiencies (20-fold lower with respect to ValP), as these would be similar to the scenario previously reported for KARI in *Salmonella typhimurium*, which has a *panE* gene (Primerano and Burns, 1983). Also, we assumed that the central KARI of *S. viridifaciens*, i.e. the central metabolic paralogue of Svi2, has a catalytic efficiency for all substrates similar to the mean value of the parameters found in all KARI enzymes from *Streptomyces*.

The calculated *I* and *J* indices, including some perturbations where certain substrates were omitted as further discussed in the following paragraph, are shown in Table 3. Furthermore, these data are also shown in a graphical fashion in Fig. 4B, where circles are used to represent the degree of substrate promiscuity of the enzyme. Single copies are shown as black circles, whereas paralogues are shown as white circles. Overall,

we note that *I* and *J* indices are quite similar, as expected from the fact that all enzymes have high catalytic efficiencies for more than one substrate, e.g. ValP. High catalytic efficiency would therefore mask the fact that the substrates tested are chemically quite different. In other words, the contribution of poor activities upon chemically diverse substrates is not captured by either of these indices. Thus, we hypothesized that removal of ValP, which includes the substrates with the better enzyme kinetic parameters, might expose the contribution of low activities, as well as the contribution of chemical diversity. Likewise, removal of a similar substrate to ValP, e.g. Pyr, or a chemically dissimilar substrate, e.g. P5C, when poor enzyme kinetic parameters are recorded, would lead to marginal changes. For this purpose, we recalculate the Tanimoto distances (Table S6) as well as *I* and *J*, but omitting P5C, ValP and Pyr (Fig. 4B).

As KP and ValP are highly similar, when the latter is removed, Cgl appears to behave as a non-promiscuous enzyme, with specificity for KP, as expected. This perturbation also makes Sco1 and Sli1 appear less promiscuous, which would be equivalent to becoming specialized for IleP. Interestingly, this perturbation unmasked the specialized nature of Svi2 for ValP, as it seems to become a promiscuous enzyme with the remaining substrates, for which none has good enzyme kinetic parameters. The effect of removing ValP is such that the behaviour of the paralogues in *S. viridifaciens* is even inverted. In contrast, removal of Pyr or P5C only marginally affects the possible behaviour of all enzymes, the exception being the case of Sco2 and Sli2 for P5C. In this case, the apparent increase of the promiscuity of these enzymes relates to the fact that none of the remaining substrates, which are chemically

Table 3. Enzyme substrate promiscuity indices.

Enzyme[b]	Standard Index of substrate promiscuity (*I*)[a,c] $-\dfrac{1}{\log N}\sum_{i=1}^{N}\dfrac{e_i}{\sum_{j=1}^{N}e_j}\log\dfrac{e_i}{\sum_{j=1}^{N}e_j}$				Weighted Index of substrate promiscuity (*J*)[a,c] $\dfrac{-N}{\log N\sum_{i=1}^{N}\langle\delta_i\rangle}\sum_{i=1}^{N}\langle\delta_i\rangle\dfrac{e_i}{\sum_{j=1}^{N}e_j}\ln\dfrac{e_i}{\sum_{j=1}^{N}e_j}$			
	I	I-ValP	I-Pyr	I-P5C	J	J-ValP	J-Pyr	J-P5C
Cgl	0.71	0.57	0.74	0.75	0.61	0.39	0.61	0.69
Spr	0.67	0.71	0.70	0.70	0.59	0.67	0.60	0.64
Svi1	0.67	0.62	0.70	0.70	0.59	0.59	0.60	0.63
Svi2	0.47	0.81	0.49	0.49	0.42	0.79	0.43	0.46
Sgr	0.65	0.66	0.68	0.67	0.57	0.64	0.58	0.61
Sav	0.70	0.77	0.73	0.73	0.62	0.74	0.64	0.67
Sam	0.74	0.66	0.77	0.77	0.66	0.64	0.66	0.70
Sco2	0.83	0.75	0.83	0.83	0.77	0.73	0.76	0.83
Sli2	0.80	0.71	0.83	0.83	0.75	0.70	0.75	0.78
Sco1	0.64	0.42	0.67	0.67	0.56	0.40	0.57	0.61
Sli1	0.64	0.51	0.67	0.67	0.56	0.48	0.56	0.60
Sco_P5CR	0	0	0	0	0	0	0	0

a. Both indices range between 0 and 1.
b. Enzyme nomenclature is as in Table 1.
c. *N*, number of substrates; e, catalytic efficiency; $\delta_i = \delta_{ij}/\delta_{set}$, δ_{ij}, the mean Tanimoto distance from a member i to all the other members in the set; δ_{set}, the overall set dissimilarity.

quite dissimilar to P5C, is converted at a high catalytic efficiency. Indeed, the values of the standard I index for Sco2 and Sli2 (0.83 and 0.80 respectively) becomes virtually identical to the values obtained for the J-$P5C$ perturbation (0.83 and 0.78 respectively).

Our results disagree with previous analyses using highly efficient non-specific enzymes, such as gluthatione S-transferase, proteases and cytochrome P450, which were concluded to be highly promiscuous (Nath and Atkins, 2008; Nath et al., 2010). In our case, the I and J indices only partially paint the portrait of the promiscuous nature of KARI enzymes. Although the detoxifying enzymes have evolved broad substrate specificity, KARI enzymes seemed to have evolved low secondary promiscuous activities, as previously defined (O'Brien and Herschlag, 1999; Copley, 2003; Khersonsky et al., 2006; Khersonsky and Tawfik, 2010). Even when the fingerprint-based approaches used to calculate Tanimoto distances capture important elements of chemical similarity, it is clear that when low enzyme activities are considered, the model does not describe reality.

The fact that enzyme promiscuity has been shown to be essential for survival of the organism upon certain conditions or genetic backgrounds (Kim and Copley, 2007; Patrick et al., 2007) raises questions about the utility of I and J indices at their current state, as they fail to describe secondary and low enzyme promiscuous activities. This assessment therefore opens the possibility to further develop enzyme promiscuity metrics that would include other features of the substrates and enzymes, such as three-dimensional structural architecture, molecular mass, hydrophobicity and electrostatic charges (Ferro and Bredow, 2010), as well as the evolutionary history of enzymes.

Experimental procedures

Synthesis of KARI substrates

ValP and IleP substrates were obtained by alkaline hydrolysis of the corresponding esters by 1.1 equiv of potassium hydroxide followed by addition of 1 M Tris, pH 7.5. Ester of 2H2M3KB is commercially available, the rest of the esters were obtained as follows: synthesis of ethyl 3H3M2KB, the ester of 3H3M2KB, was prepared from bromination and hydroxylation of ethyl 3-methyl-2-ketobutyrate as described by (Chunduru et al., 1989). Ethyl 3H3E2KB, the ester of 3H3E2KB, was prepared using ethyl-2-ethylacetoacetate as precursor. Methyl 2-hydroxy-2-ethyl-3-ketobutyrate, the ester of 2H2E3KB, was prepared using methyl-2-oxobutanoate as the initial precursor, which required an alkylation step (Brändström, 1959), using sodium hydroxide in ethanol and ethyl bromide. Resulting compound methyl 2-ethyl-3-ketobutyrate was subjected to bromination and hydroxylation as described earlier. After purification, esters were characterized by ^{1}H nuclear magnetic resonance (NMR) and ^{13}C NMR. Synthesis, purification and quantification of P5C were made as described by Williams and Frank (1975). All precursors, as well as Pyr and HP, were purchased from Sigma-Aldrich®.

Calculation of Tanimoto distances

The modified list of descriptors used to help distinguish our substrates is shown in Table S1. We calculate the Tanimoto distance (Willet et al., 1998) by using a PERL script, available upon request, which creates a bit vector (0 for absent descriptor, 1 if it is present) for each substrate. The final output of this script is a distance matrix (Table S4), which was used for construction of a perceptual map showing the relationships between the substrates using PERMAT VERSION 11.8A.

Protein expression and purification

All ilvC genes, including that of E. coli as a negative control, were cloned into pET28a from Novagene (Table S2). Escherichia coli strain BL21 Star (DE3) (Invitrogen) was used as host for expression. Luria–Bertani media with 0.5 mM of isopropyl β-D-1-thiogalactopyranoside (IPTG), at an OD_{600nm} of 0.7 and 17°C for 12 h in agitation (200 r.p.m.), was used. Purification of enzymes was performed using Ni-NTA column VivaPure maxiprep MC (Santorius Stedim Biotech) at 4°C. As equilibration buffer, 20 mM Tris-HCl with 500 mM NaCl, pH 7.9, was used. Elution of proteins from the Ni-NTA column was done at 150 mM of imidazole, pH 7.9. Proteins were dialysed against several changes of Tris HCl buffer, pH 8, and concentrated using an Amicon Ultra centrifugal filter (Millipore). Final concentration was determined using Bio-Rad protein assay dye reagent.

Steady-state enzyme kinetics

Initial velocities (v_0) were determined at 25°C. The reaction buffer used contained 0.1 M Tris-HCl (pH 8), 10 mM $MgCl_2$, 0.22 mM NADP(H)$^+$, and different concentrations of the substrates (0–100 mM) up to saturation of the enzymes were used, in a final volume of 100 μl. The enzyme concentration for each assay was determined by measuring the rate of the reaction at a constant substrate concentration of 100 mM. The NADP(H)$^+$ oxidation was followed at 340 nm in a Cary spectrophotometer 100 Bio, after enzymes were added to the reaction mix. Kinetic constants K_M and k_{cat} were calculated by non-linear regression fit of the initial velocity data to the Michaelis–Menten equation, and the limit of detection was calculated to be $k_{cat}/K_M = 0.000013$ M^{-1} s^{-1}.

Bacterial molecular genetics

Oligonucleotides used in this study are shown in Table S5. Strains, plasmids and cosmids used in this study are listed in Table S2. All genes were amplified by polymerase chain reaction and cloned with standard methods, except for ilvC2 of S. viridifaciens (Garg et al., 2002), which was commercially synthesized (GenScript). Streptomyces strains were grown using mannitol soya medium and minimal medium media previously reported (Kieser et al., 2000). Antibiotics (Sigma-

Aldrich®) were added at the following concentrations: ampicillin (50 µg ml⁻¹), apramycin (50 µg ml⁻¹), chloramphenicol (25 µg ml⁻¹), kanamycin (50 µg ml⁻¹), nalidixic acid (25 µg ml⁻¹) and hygromycin (50 µg ml⁻¹). Gene disruptions in *S. coelicolor* were done as previously (Gust *et al.*, 2003). A *proC⁻* mutant from the Keio collection (Baba *et al.*, 2006) and the *ilvC* genes cloned into pASK plasmid were used for P5CR complementation in *E. coli*. Transformed cells were grown, washed and plated on M9 media. For complementation assays in *S. coelicolor*, the growth requirements of the double and triple mutants (*proC::scar ilvC1::aadA, ilvC2::aac(3)IV*) were obtained using MM supplemented with the relevant amino acids at a concentration of 7.5 µg ml⁻¹. Mutants were obtained after conjugation, using a methylation-deficient *E. coli* host ET12567 / pUZ8002 (Gust *et al.*, 2003) and pAV11B_FBG (a modified version of pAV11B; Jyothikumar *et al.*, 2012) bearing the *ilvC* genes. The same number of spores was plated for the complementation assays.

Substrate promiscuity indices

Standard *I* and weighted *J* indices (ranging between 0 and 1) were obtained using the equations 4 and 5 described in Nath and Atkins (2008). In order to recalculate the indices with different scenarios, we obtained the distance matrix with the Perl script omitting the valine precursors, P5C and Pyr (Table S6).

Acknowledgements

We would like to thank Hilda E. Ramos-Aboites and Christian E. Martínez-Guerrero for technical support, as well as Pablo Cruz-Morales, Noel Ferro, Marcus Moore, Rosario Muñoz-Clares and Dante Pertusi for critical reading of the paper.

Conflict of interest

The authors declare no conflict of interests.

References

Ahn, H.J., Eom, S.J., Yoon, H.J., Lee, B.I., Cho, H., and Suh, S.W. (2003) Crystal structure of class I acetohydroxy acid isomeroreductase from *Pseudomonas aeruginosa*. *J Mol Biol* **328**: 505–515.

Andreeva, A., Howorth, D., Chandonia, J.M., Brenner, S.E., Hubbard, T.J., Chothia, C., and Murzin, A.G. (2008) Data growth and its impact on the SCOP database: new developments. *Nucleic Acids Res* **36**: D419–D425.

Ankri, S., Serebrijski, I., Reyes, O., and Leblon, G. (1996) Mutations in the *Corynebacterium glutamicum* proline biosynthetic pathway: a natural bypass of th proA step. *J Bacteriol* **178**: 4412–4419.

Baba, T., Ara, T., Hasegawa, M., Takai, Y., Okumura, Y., Baba, M., *et al.* (2006) Construction of *Escherichia coli* K-12 in-frame, single-gene knockout mutants: the Keio collection. *Mol Syst Biol* **2**: 1–11. 2006 0008.

Barona-Gomez, F., and Hodgson, D.A. (2010) Multicopy proC in *Streptomyces coelicolor* A3(2) elicits a transient production of prodiginines, while proC deletion does not yield a proline auxotroph. *J Mol Microbiol Biotechnol* **19**: 152–158.

Biou, V., Dumas, R., Cohen-Addad, C., Douce, R., Job, D., and Pebay-Peyroula, E. (1997) The crystal structure of plant acetohydroxy acid isomeroreductase complexed with NADPH, two magnesium ions and a herbicidal transition state analog determined at 1.65 A resolution. *EMBO J* **16**: 3405–3415.

Bornscheuer, U.T., and Kazlauskas, R.J. (2004) Catalytic promiscuity in biocatalysis: using old enzymes to form new bonds and follow new pathways. *Angew Chem Int Ed Engl* **43**: 6032–6040.

Brändström, A. (1959) Sodium hydroxide in alcohol as a base for the alkylation of ethyl acetoacetate. *Acta Chem Scand* **13**: 607–608.

Chakraborty, S., and Rao, B.J. (2012) A measure of the promiscuity of proteins and characteristics of residues in the vicinity of the catalytic site that regulate promiscuity. *PLoS ONE* **7**: e32011.

Chambellon, E., Rijnen, L., Lorquet, F., Gitton, C., van Hylckama Vlieg, J.E., Wouters, J.A., and Yvon, M. (2009) The D-2-hydroxyacid dehydrogenase incorrectly annotated PanE is the sole reduction system for branched-chain 2-keto acids in *Lactococcus lactis*. *J Bacteriol* **191**: 873–881.

Chunduru, S.K., Mrachko, G.T., and Calvo, K.C. (1989) Mechanism of ketol acid reductoisomerase–steady-state analysis and metal ion requirement. *Biochemistry* **28**: 486–493.

Copley, S.D. (2003) Enzymes with extra talents: moonlighting functions and catalytic promiscuity. *Curr Opin Chem Biol* **7**: 265–272.

Dumas, R., Job, D., Douce, R., Pebay-Peyroula, E., and Cohen-Addad, C. (1994a) Crystallization and preliminary crystallographic data for acetohydroxy acid isomeroreductase from *Spinacia oleracea*. *J Mol Biol* **242**: 578–581.

Dumas, R., Cornillon-Bertrand, C., Guigue-Talet, P., Genix, P., Douce, R., and Job, D. (1994b) Interactions of plant acetohydroxy acid isomeroreductase with reaction intermediate analogues: correlation of the slow, competitive, inhibition kinetics of enzyme activity and herbicidal effects. *Biochem J* **301** (Part 3): 813–820.

Dumas, R., Butikofer, M.C., Job, D., and Douce, R. (1995) Evidence for two catalytically different magnesium-binding sites in acetohydroxy acid isomeroreductase by site-directed mutagenesis. *Biochemistry* **34**: 6026–6036.

Dumas, R., Biou, V., Halgand, F., Douce, R., and Duggleby, R.G. (2001) Enzymology, structure, and dynamics of acetohydroxy acid isomeroreductase. *Acc Chem Res* **34**: 399–408.

Ferro, N., and Bredow, T. (2010) Assessment of quantum-chemical methods for electronic properties and geometry of signaling biomolecules. *J Comput Chem* **31**: 1063–1079.

Garg, R.P., Ma, Y., Hoyt, J.C., and Parry, R.J. (2002) Molecular characterization and analysis of the biosynthetic gene cluster for the azoxy antibiotic valanimycin. *Mol Microbiol* **46**: 505–517.

Gerlt, J.A., and Babbitt, P.C. (2001) Divergent evolution of enzymatic function: mechanistically diverse superfamilies and functionally distinct suprafamilies. *Annu Rev Biochem* **70:** 209–246.

Gerlt, J.A., Babbitt, P.C., Jacobson, M.P., and Almo, S.C. (2012) Divergent evolution in enolase superfamily: strategies for assigning functions. *J Biol Chem* **287:** 29–34.

Gust, B., Challis, G.L., Fowler, K., Kieser, T., and Chater, K.F. (2003) PCR-targeted *Streptomyces* gene replacement identifies a protein domain needed for biosynthesis of the sesquiterpene soil odor geosmin. *Proc Natl Acad Sci USA* **100:** 1541–1546.

Hult, K., and Berglund, P. (2007) Enzyme promiscuity: mechanism and applications. *Trends Biotechnol* **25:** 231–238.

Ito, H., and Tanaka, A. (2014) Evolution of a new chlorophyll metabolic pathway driven by the dynamic changes in enzyme promiscuous activity. *Plant Cell Physiol* **55:** 593–603.

Jensen, R.A. (1976) Enzyme recruitment in evolution of new function. *Annu Rev Microbiol* **30:** 409–425.

Jyothikumar, V., Klanbut, K., Tiong, J., Roxburgh, J.S., Hunter, I.S., Smith, T.K., and Herron, P.R. (2012) Cardiolipin synthase is required for *Streptomyces coelicolor* morphogenesis. *Mol Microbiol* **84:** 181–197.

Khersonsky, O., and Tawfik, D.S. (2010) Enzyme promiscuity: a mechanistic and evolutionary perspective. *Annu Rev Biochem* **79:** 471–505.

Khersonsky, O., Roodveldt, C., and Tawfik, D.S. (2006) Enzyme promiscuity: evolutionary and mechanistic aspects. *Curr Opin Chem Biol* **10:** 498–508.

Khodakaramian, G., Lissenden, S., Gust, B., Moir, L., Hoskisson, P.A., Chater, K.F., and Smith, M.C. (2006) Expression of Cre recombinase during transient phage infection permits efficient marker removal in *Streptomyces*. *Nucleic Acids Res* **34:** e20.

Kieser, M.J.B., Buttner, M.J., Chater, K.F., and Hopwood, D.A. (2000) *Practical Streptomyces Genetics*. Norwich, UK: John Innes Foundation.

Kim, J., and Copley, S.D. (2007) Why metabolic enzymes are essential or nonessential for growth of *Escherichia coli* K12 on glucose. *Biochemistry* **46:** 12501–12511.

Kohn, L.D., and Jakoby, W.B. (1968) Tartaric acid metabolism. IV. Crystalline L-malic dehydrogenase from *Pseudomonas acidovorans*. *J Biol Chem* **243:** 2472–2478.

Laskowski, R.A., Watson, J.D., and Thornton, J.M. (2005) ProFunc: a server for predicting protein function from 3D structure. *Nucleic Acids Res* **33:** W89–W93.

Merkamm, M., Chassagnole, C., Lindley, N.D., and Guyonvarch, A. (2003) Ketopantoate reductase activity is only encoded by *ilvC* in *Corynebacterium glutamicum*. *J Biotechnol* **104:** 253–260.

Mondal, S., Nagao, C., and Mizuguchi, K. (2010) Detecting subtle functional differences in ketopantoate reductase and related enzymes using a rule-based approach with sequence-structure homology recognition scores. *Protein Eng Des Sel* **23:** 859–869.

Nath, A., and Atkins, W.M. (2008) A quantitative index of substrate promiscuity. *Biochemistry* **47:** 157–166.

Nath, A., Zientek, M.A., Burke, B.J., Jiang, Y., and Atkins, W.M. (2010) Quantifying and predicting the promiscuity

and isoform specificity of small-molecule cytochrome P450 inhibitors. *Drug Metab Dispos* **38:** 2195–2203.

Nocek, B., Chang, C., Li, H., Lezondra, L., Holzle, D., Collart, F., and Joachimiak, A. (2005) Crystal structures of delta1-pyrroline-5-carboxylate reductase from human pathogens *Neisseria meningitides* and *Streptococcus pyogenes*. *J Mol Biol* **354:** 91–106.

O'Brien, P.J., and Herschlag, D. (1999) Catalytic promiscuity and the evolution of new enzymatic activities. *Chem Biol* **6:** R91–R105.

Ohno, S. (1970) *Evolution by Gene Duplication*. New York, USA: Springer Verlag.

Osipiuk, J., Zhou, M., Moy, S., Collart, F., and Joachimiak, A. (2009) X-ray crystal structure of GarR-tartronate semialdehyde reductase from *Salmonella typhimurium*. *J Struct Funct Genomics* **10:** 249–253.

Patrick, W.M., Quandt, E.M., Swartzlander, D.B., and Matsumura, I. (2007) Multicopy suppression underpins metabolic evolvability. *Mol Biol Evol* **24:** 2716–2722.

Piatigorsky, J. (2007) *Gene Sharing and Evolution The Diversity of Protein Functions*. Cambridge, MA, USA: Harvard University Press, p. 336.

Primerano, D.A., and Burns, R.O. (1983) Role of acetohydroxy acid isomeroreductase in biosynthesis of pantothenic acid in *Salmonella typhimurium*. *J Bacteriol* **153:** 259–269.

Redfern, O.C., Dessailly, B.H., Dallman, T.J., Sillitoe, I., and Orengo, C.A. (2009) FLORA: a novel method to predict protein function from structure in diverse superfamilies. *PLoS Comput Biol* **5:** e1000485.

Schnoes, A.M., Ream, D.C., Thorman, A.W., Babbitt, P.C., and Friedberg, I. (2013) Biases in the experimental annotations of protein function and their effect on our understanding of protein function space. *PLoS Comput Biol* **9:** e1003063.

Tyagi, R., Lee, Y.T., Guddat, L.W., and Duggleby, R.G. (2005) Probing the mechanism of the bifunctional enzyme ketol-acid reductoisomerase by site-directed mutagenesis of the active site. *FEBS J* **272:** 593–602.

Vining, L.C. (1992) Secondary metabolism, inventive evolution and biochemical diversity – a review. *Gene* **115:** 135–140.

Wada, Y., Iwai, S., Tamura, Y., Ando, T., Shinoda, T., Arai, K., and Taguchi, H. (2008) A new family of D-2-hydroxyacid dehydrogenases that comprises D-mandelate dehydrogenases and 2-ketopantoate reductases. *Biosci Biotechnol Biochem* **72:** 1087–1094.

Willet, P., Barnard, J.M., and Downs, G.M. (1998) Chemical similarity searching. *J Chem Inf Comput Sci* **38:** 983–996.

Williams, I., and Frank, L. (1975) Improved chemical synthesis and enzymatic assay of delta-1-pyrroline-5-carboxylic acid. *Anal Biochem* **64:** 85–97.

Zheng, R., and Blanchard, J.S. (2000) Identification of active site residues in *E. coli* ketopantoate reductase by mutagenesis and chemical rescue. *Biochemistry* **39:** 16244–16251.

Supporting information

Additional Supporting Information may be found in the online version of this article at the publisher's web-site:

Fig. S1. Phylogenetic reconstruction using sequences of RpoB and KARI homologues from *Actinobacteria*.

Fig. S2. A. Genome context analysis of *ilvC* genes in *Streptomyces*.

C. Biosynthetic pathway for valanimycin.

Table S1. Descriptor keyset used for obtaining Tanimoto distance of KARI substrates.

Table S2. Plasmids generated in this study by sub-cloning.

Table S3. Growth requirements of *S. coelicolor* mutants.

Table S4. Tanimoto distance matrix calculated for all substrates of 6PGDH superfamily constructed using descriptors showed in Table S1.

Table S5. Oligonucleotides used in this study for PCR amplification and disruption of the genes *ilvC1*, *ilvC2*, *proC* in *S. coelicolor*.

Table S6. Recalculated matrix of Tanimoto distances matrix of substrates used to calculate promiscuity indices from KARI homologues including some perturbations.

17

Use of lectins to in situ visualize glycoconjugates of extracellular polymeric substances in acidophilic archaeal biofilms

R. Y. Zhang,[1] T. R. Neu,[2] S. Bellenberg,[1] U. Kuhlicke,[2] W. Sand[1] and M. Vera[1]*

[1]Aquatische Biotechnologie, Biofilm Centre, Universität Duisburg – Essen, Universitätsstraße 5, 45141 Essen, Germany.
[2]Department of River Ecology, Helmholtz Centre for Environmental Research-UFZ, Brueckstrasse 3A, 39114 Magdeburg, Germany.

Summary

Biofilm formation and the production of extracellular polymeric substances (EPS) by meso- and thermoacidophilic metal-oxidizing archaea on relevant substrates have been studied to a limited extent. In order to investigate glycoconjugates, a major part of the EPS, during biofilm formation/bioleaching by archaea on pyrite, a screening with 75 commercially available lectins by fluorescence lectin-binding analysis (FLBA) has been performed. Three representative archaeal species, *Ferroplasma acidiphilum* DSM 28986, *Sulfolobus metallicus* DSM 6482[T] and a novel isolate *Acidianus* sp. DSM 29099 were used. In addition, *Acidianus* sp. DSM 29099 biofilms on elemental sulfur were studied. The results of FLBA indicate (i) 22 lectins bound to archaeal biofilms on pyrite and 21 lectins were binding to *Acidianus* sp. DSM 29099 biofilms on elemental sulfur; (ii) major binding patterns, e.g. tightly bound EPS and loosely bound EPS, were detected on both substrates; (iii) the three archaeal species produced various EPS glycoconjugates on pyrite surfaces. Additionally, the substratum induced different EPS glycoconjugates and biofilm structures of cells of *Acidianus* sp. DSM 29099. Our data provide new insights into interactions between acidophilic

*For correspondence. E-mail mario.vera@uni-due.de

Funding Information R. Y. Zhang acknowledges China Scholarship Council (CSC) for financial support (No. 2010637124). Financial support by the "Open Access" DFG-Universität Duisburg-Essen program is acknowledged.

archaea on relevant surfaces and also indicate that FLBA is a valuable tool for in situ investigations on archaeal biofilms.

Introduction

Microbial leaching of metal sulfides (MS) is an expanding biotechnology (Brierley and Brierley, 2013). However, it can also occur as an unwanted natural process called acid rock drainage or acid mine drainage (AMD). This process is accompanied by acidification and heavy metal pollution of water bodies and can cause serious environmental problems (Kalin *et al.*, 2006; Sand *et al.*, 2007). Acidophilic archaea including genera such as *Ferroplasma*, *Acidianus*, *Sulfolobus* and *Metallosphera* play important roles in bioleaching and AMD systems, and have received significant attention for commercial applications (Olson *et al.*, 2003; Golyshina and Timmis, 2005; Rawlings and Johnson, 2007).

The genera *Acidianus* and *Sulfolobus* are thermoacidophiles found in hydrothermal vents or bioleaching systems at temperatures above 60°C. They are capable of oxidizing both iron(II) ions and reduced inorganic sulfur compounds (RISCs). Biological ferric iron regeneration and acidic conditions are crucial for the dissolution of MS (Schippers and Sand, 1999; Sand *et al.*, 2001). Under thermophilic conditions, iron oxidation is accelerated, and the passivation of chalcopyrite ($CuFeS_2$) surfaces by RISCs is nearly eliminated, which has significant importance in the biomining industry.

The mesophilic archaeon *Ferroplasma acidiphilum* was first isolated from a semi-industrial bioleaching reactor processing arsenopyrite in Kazakhstan (Golyshina *et al.*, 2000). It oxidizes iron(II) ions or pyrite in the presence of trace amounts of yeast extract. In addition, all isolated strains of *Ferroplasma* spp. can grow heterotrophically (Dopson *et al.*, 2004). *Ferroplasma* is frequently detected in biomining ecosystems and is considered to be a major player in global iron and sulfur cycles in highly acidic environments (Edwards *et al.*, 2000; Golyshina and Timmis, 2005; Chen *et al.*, 2014).

Biofilms are defined as interface-associated communities of microorganisms embedded in extracellular polymeric substances (EPS). The EPS usually consist of polysaccharides, proteins, lipids and DNA. They are generally subdivided into two types: 'capsular EPS' are tightly bound to cells, while 'colloidal EPS' are loosely bound to cells and can be easily released (e.g. by centrifugation or washing) into the solution (Nielsen and Jahn, 1999). EPS are essential for biofilm structure and function due to their involvement in cellular associations, nutrition exchange and interactions of microorganisms with their bio-physicochemical environment (Wolfaardt et al., 1999; Neu and Lawrence, 2009). EPS are also involved in the attachment and biofilm formation of leaching microorganisms to surfaces of MS, which is an essential step at the start of the leaching process (Vera et al., 2013). Biofilms formed by heterotrophic prokaryotes or phototrophs are usually dynamic structures that can grow to thick three-dimensional macro-communities (Stoodley et al., 2002). In contrast, the majority of metal-oxidizing microorganisms attach directly to the surface of MS, forming monolayer biofilms. By this lifestyle, cells can obtain energy from Iron(II) ions or RISCs, which are released during the dissolution of the MS. Interestingly, two distinct biofilm morphologies were described for Ferroplasma acidarmanus Fer1: A multilayer film was formed on pyrite surfaces after 38 days of incubation, and up to 5 mm-long filaments were found on sintered glass spargers in gas lift bioreactors (Baker-Austin et al., 2010).

Few studies have shown biofilms of archaea, including thermoacidophiles, halophiles and methanogens. The first archaeal biofilm was described for the hyperthermophilic Thermococcus litoralis, which developed in rich media on polycarbonate filters and glass surfaces (Rinker and Kelly, 1996). Pyrococcus furiosus and Methanobacter thermoautotrophicus developed monospecies biofilms on solid surfaces (Näther et al., 2006; Thoma et al., 2008). Bi-species biofilm development of P. furiosus and Methanopyrus kandlerii was shown to be established within less than 24 h on abiotic surfaces (Schopf et al., 2008). Biofilm analysis of three Sulfolobus spp. showed that their structures were different, ranging from simple carpet-like structures in Sulfolobus solfataricus and Sulfolobus tokodaii to high density tower-like structures in Sulfolobus acidocaldarius in static systems. All three species produced EPS containing glucose, galactose, mannose and N-acetylglucosamine (GlcNAc) once biofilm formation was initiated (Koerdt et al., 2010). Biofilm formation by methanogenic archaea under static conditions was studied by confocal laser scanning microscopy (CLSM) and scanning electron microscopy. The three species, Methanosphaera stadtmanae, Methanobrevibacter smithii and Methanosarcina mazei strain Gö1, formed mainly bilayer biofilms on mica surfaces. Nevertheless, the development of multilayer biofilms was also observed (Bang et al., 2014). Biofilm formation of haloarchaea, including species of Halobacterium, Haloferax and Halorubrum, was investigated by a fluorescence-based live cell adhesion assay. Cellular appendages were speculated to be involved in the initial attachment (Fröls et al., 2012). Two types of biofilm structures were detected including carpet-like multilayers and large aggregates adhering to glass surfaces. Similar as occurring in the acidophilic archaea such as Sulfolobus and Ferroplasma (Baker-Austin et al., 2010; Koerdt et al., 2010; Zhang et al., 2014), biofilm development occurs in a surprisingly wide variety in haloarchaea. In addition, EPS like eDNA and various glycoconjugates were found to be present in these biofilms (Fröls et al., 2012).

Lectins are proteins or glycoproteins capable of binding reversibly and specifically to carbohydrates without altering their structures. Fluorescence lectin-binding analysis (FLBA) represents the only option for non-destructive and in situ glycoconjugate analysis and, therefore, is widely used in glycoconjugate/biofilm analysis in combination with other fluorochromes, e.g. specific for nucleic acids (Zippel and Neu, 2011; Bennke et al., 2013; Castro et al., 2014). Furthermore, their multivalency ensures high-affinity binding to the cell surface and biofilm structures containing various glycoconjugates. Only a few lectins combined with nucleic acid dyes have been used in investigations on acidophilic biofilms related to bioleaching and AMD systems. The most frequently used lectin is Concanavalin A (Con A) from the jack-bean, Canavalia ensiformis, binding to mannose and glucose residues (Goldstein et al., 1965). Con A has been used to visualize various acidophilic archaeal and bacterial biofilm cells, e.g. Sulfolobus (Koerdt et al., 2010; Zolghadr et al., 2010; Bellenberg et al., 2012), F. acidiphilum (Zhang et al., 2014) and Metallosphaera hakonensis (Africa et al., 2013). As EPS are complex mixtures consisting of many types of macromolecules, it is impossible to address their complexity with a single staining approach. Even for the similar glycoconjugates, multiple lectin probes have to be used (Neu and Lawrence, 2009). Thus, it is necessary to screen a library of lectins in order to find the ones binding to the glycoconjugates in a particular biofilm (Peltola et al., 2008; Zippel and Neu, 2011; Bennke et al., 2013).

To date, EPS production and biofilms of archaeal species have been investigated only to a limited extent, especially concerning the ones growing in acidic environments (Orell et al., 2013). Nevertheless, it is essential to visualize EPS glycoconjugate identity and distribution on relevant surfaces together with analysis of their chemical composition to understand their function(s) in bioleaching. In the present study, three representative archaeal strains – a euryarchaeote F. acidiphilum DSM 28986 and two crenarchaeota, Sulfolobus metallicus DSM 6482^T and Acidianus sp. DSM 29099 – were selected for FLBA of their EPS glycoconjugates and biofilm structures during bioleaching of pyrite as well as on elemental sulfur in case of Acidianus sp. DSM 29099. In order to image EPS glycoconjugates in these biofilms, 75 commercially available lectins were tested for applicability. This is the first report of EPS glycoconjugate probing by means of FLBA for archaeal biofilms in situ during bioleaching.

Results and discussion

Visualization of attached archaea and biofilms

In previous reports, acridine orange (Fröls *et al.*, 2012), fluorescein (Baker-Austin *et al.*, 2010) and DAPI (Henche *et al.*, 2012; Koerdt *et al.*, 2012) have been used for staining acidophilic archaeal species. In order to display the distribution of archaeal cells as well as to visualize EPS including proteins, nucleic acids and lipophilic compounds in biofilms on pyrite surfaces, six fluorochromes including SybrGreen (Invitrogen, Carlsbad, CA, USA), Syto 9 (Invitrogen), Syto 64 (Invitrogen), SyproRed (Invitrogen), SyproOrange (Invitrogen) and FM4-64 (Invitrogen) were selected to evaluate their potential suitability (Table 1). Sypro stains like SyproRed and SyproOrange were originally developed for measuring protein concentrations in solution or in gels. Later, they were used for flow cytometry studies (Zubkov *et al.*, 1999) and finally for staining the biofilm matrix for CLSM examination (Neu and Lawrence, 1999a; Lawrence *et al.*, 2003). FM-dyes (FM4-64 and FM1-43) are widely used to study endocytosis, vesicle trafficking and organelle organization in living eukaryotic cells (Bolte *et al.*, 2004).

Archaeal cells and biofilms on pyrite. As negative control, the abovementioned dyes including fluoroconjugated lectins were selected randomly to stain sterile pyrite for evaluation of their unspecific binding. Surface structures of sterile, cleaned pyrites showed no unspecific binding of dyes when examined by CLSM (not shown). As shown in Fig. 1A–C, cells of *F. acidiphilum* DSM 28986 attached to pyrite were successfully stained by SybrGreen, Syto 9 and Syto 64. Similarly, cells of *Acidianus* sp. DSM 29099 and *S. metallicus*[T] were also clearly visualized by staining with these dyes (Fig. 2A, B, D and E). In addition, SyproRed stained cells of *F. acidiphilum* DSM 28986

(Fig. 1D), *Acidianus* sp. DSM 29099 (Fig. 2C) and *S. metallicus*[T] (Fig. 2E). FM4-64 stained cells of the *Sulfolobales* (not shown) and *F. acidiphilum* DSM 28986 (Fig. 1E). SyproRed and FM4-64 staining gave clear cell-corresponding signals. Therefore, these fluorochromes were used for counter staining in the following tests for cell localization.

In this study, Sypro was used in order to examine archaeal cell surfaces as well as extracellular features. *Ferroplasma*, *Acidiplasma* and *Thermoplasma*, unlike other Archaea, lack a cell wall (Golyshina *et al.*, 2000; 2009). It has been shown that the cytoplasmic membrane of *F. acidiphilum* is covered with a thin layer of an amorphous, electron-dense surface matrix (Golyshina and Timmis, 2005). The positive Sypro staining indicates that the thin layer of electron-dense material observed could be a proteinaceous layer, although there is no surface layer (S-layer) characterized in *Ferroplasma*. Another explanation could be that Sypro interacts with membrane proteins. In contrast to *F. acidiphilum*, *Acidianus* sp. DSM 29099 and *S. metallicus*[T] possess a cell wall, which is mainly composed of S-layer proteins and anchored by their carboxyl-terminal transmembrane domains to the cytoplasmic membrane (Albers and Meyer, 2011). Obviously, S-layer proteins of these two thermophilic archaeal strains were recognized by SyproRed (Fig. 2C and E).

Besides cell visualization, these protein-, lipid- and nucleic acid-specific dyes should also allow the detection of proteins, lipids and DNA as part of the EPS in biofilms (Neu and Lawrence, 2014). In this study, staining of three archaeal strains by abovementioned fluorochromes was mostly restricted to cells, as no smear or diffuse signals around cells were visible (Figs 1 and 2). This indicates that the EPS components including proteins, lipids and eDNA were not present in colloidal fractions or below their detection limit if assessed by means of CLSM. These findings are in good agreement with the EPS analysis by colorimetric methods. The colloidal EPS of *F. acidiphilum* DSM 28986 as well as *Acidianus* sp. DSM 29099 grown on pyrite mainly contained polysaccharides. In contrast, capsular EPS contained both polysaccharides and pro-

Table 1. List of dyes and their Ex and Em wavelengths and associated binding targets.

Dyes	Specificity	Ex/Em wavelength (nm)	Company
SYTO 9	NA	483/478–488, 500–560	Invitrogen
SYTO 64	NA	483, 599/475–489, 625–700	Invitrogen
SybrGreen	NA	483/475–489, 500–560	Invitrogen
FM4-64	Lipid-rich domain	483, 506/650–790	Invitrogen
SyproRed	Proteins	475, 500/470–480, 580–680	Invitrogen
SyproOrange	Proteins	475/470–480, 520–620	Invitrogen
TRITC or Alexa 488-conjugated lectins	EPS glycoconjugates	490/505–545	EY Laboratories, Inc.
FITC-conjugated lectins	EPS glycoconjugates	490/485–495, 510–600	Sanbio Laboratory/ EY Laboratories, Inc.

Em, emission; Ex, excitation; FITC, fluorescein isothiocyanate; NA, nucleic acids; TRITC, tetramethyl rhodamine isothiocyanate.

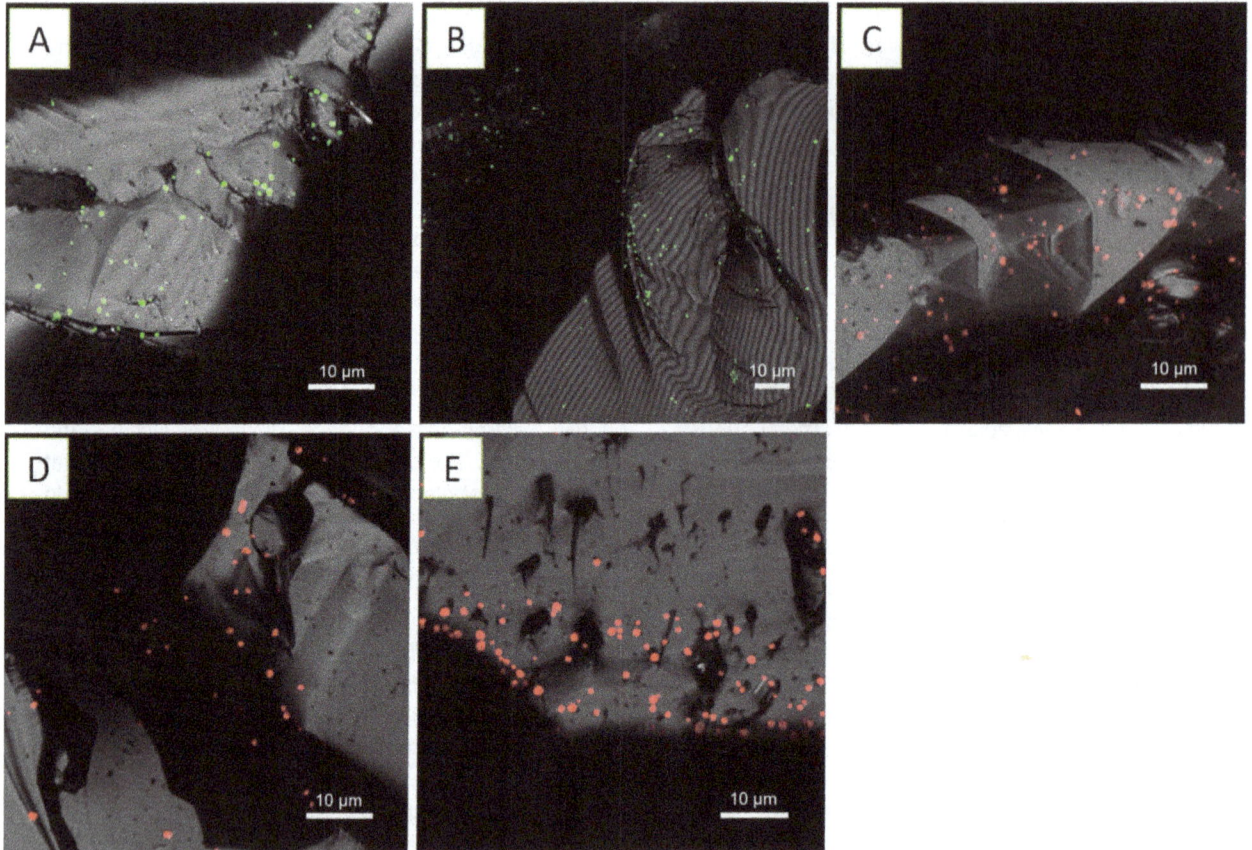

Fig. 1. Maximum intensity projections of *F. acidiphilum* DSM 28986 biofilms on pyrite stained by SybrGreen (A), Syto 9 (B), Syto 64 (C), SyproRed (D) and FM4-64 (E). Color allocation: green = SybrGreen/Syto 9, red = SyproRed/FM4-64. The pyrite surface is shown in reflection mode (= grey).

teins (R. Y. Zhang, unpublished). In this context, extracellular proteins on cell surface were also stained by SyproRed (Figs 1D, 2C and E). We did not detect eDNA in both cases. It is widely accepted that eDNA has a crucial role in biofilm development and dynamics (Whitchurch *et al.*, 2002; Karatan and Watnick, 2009). However, as DNA is a costly molecule for the cell to synthesize, it is reasonable to assume that the chemolithotrophic organisms tested, which obtain little energy by oxidation of pyrite, are not excreting measurable amounts of DNA.

In general, cells of the three species were heterogeneously distributed and developed monolayer biofilms on pyrite surfaces. Large pyrite areas remained uncolonized (~ 90%). Nevertheless, two levels of spatial organization were observed: cells and small clusters of cells (Figs 1 and 2). These results were confirmed by atomic force microscopy (AFM) combined with epifluorescence microscopy (EFM) (Supporting Information Fig. S1). It must be noted that cell attachment by the strains to pyrite did not occur randomly. Cells of *Acidianus* sp. DSM 29099 and *S. metallicus*ᵀ preferentially colonized surface locations with defects (Fig. 2). During the examination of pyrite grains, it became obvious that highly colonized grains

exhibited more scratches, microcracks or grooves as compared with the less colonized ones. More pits were observed when pyrite was leached by cells of *Acidianus* sp. DSM 29099 or *S. metallicus*ᵀ as compared with *F. acidiphilum* DSM 28986 (Figs 1 and 2). *Acidianus* sp. DSM 29099 and *S. metallicus*ᵀ, due to their increased growth temperature and their ability to oxidize RISCs arising from pyrite dissolution, have a much higher pyrite leaching capacity than *F. acidiphilum* DSM 28986 (approximately 25 times, Table 2). In a previous report, cells of *Metallosphaera* and *Sulfolobus* spp. did not exhibit any preferential orientation when they attached to pyrite (Etzel *et al.*, 2008). This maybe ascribed to the use of

Table 2. Comparison of pyrite leaching activities the strains used after 20 days of cultivation.

Strain	Fe total (mg l⁻¹)	Fe III/ Fe II ratio	Temperature
Ferroplasma acidiphilum DSM 28986	255	0.4	37°C
*Sulfolobus metallicus*ᵀ	6353	6	65°C
Acidianus sp. DSM 29099	5913	2.7	65°C

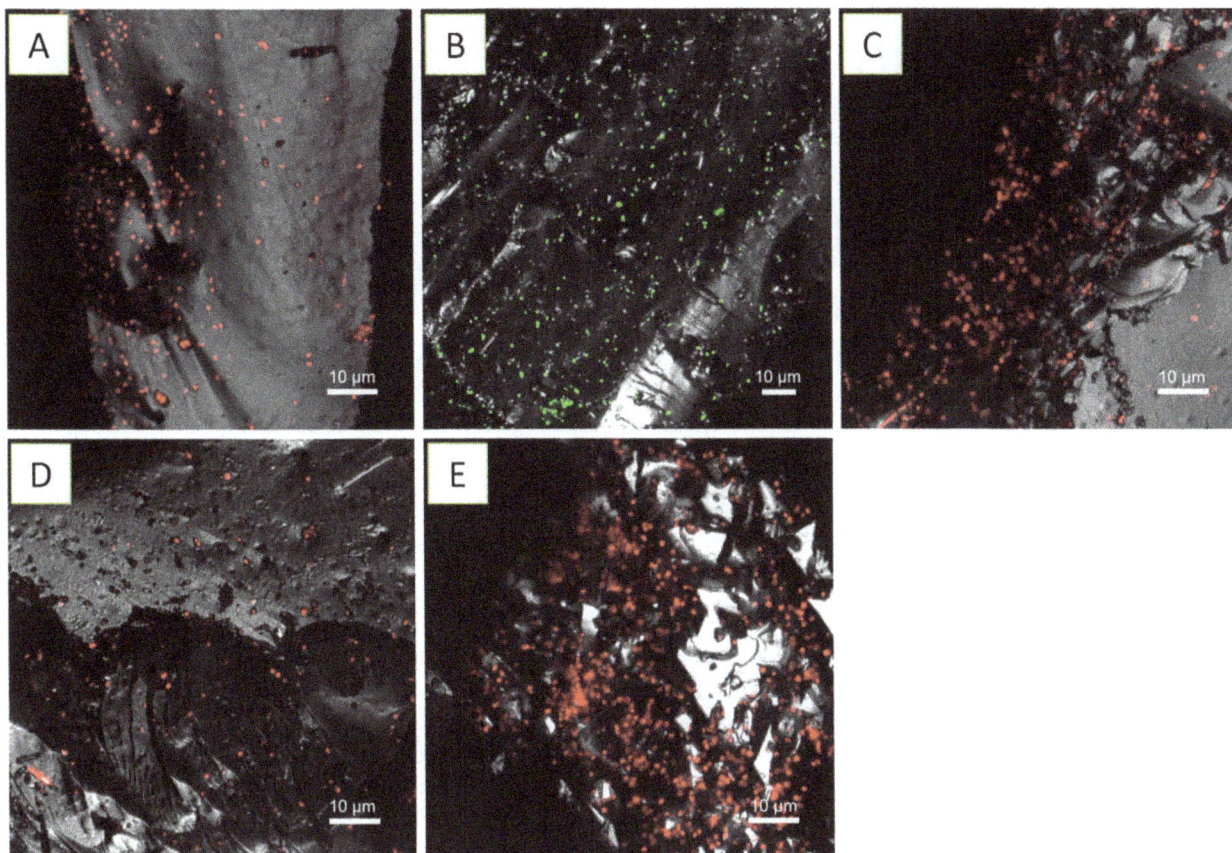

Fig. 2. Maximum intensity projections of *Acidianus* sp. DSM 29099 biofilms on pyrite stained by Syto 64 (A), SybrGreen (B) and SyproRed (C). *Sulfolobus metallicus*[T] biofilms on pyrite stained by Syto 64 (D) and SyproRed (E). Color allocation: green = SybrGreen, red = Syto 64/SyproRed, grey = reflection.

different pyrite qualities having different surface properties (e.g. crystallographic orientation).

Acidianus *sp. DSM 29099 on elemental sulfur.* The first observation of acidophilic microbes attached to elemental sulfur was described for *Acidithiobacillus thiooxidans* by means of electron microscopy (Schaeffer *et al.*, 1963). The attachment of sulfur-oxidizing microbes to sulfur surfaces has been shown to be favoured by the presence of pili, filamentous or glycoglyx materials (Weiss, 1973; Bryant *et al.*, 1984; Blais *et al.*, 1994). These studies focused mainly on bacteria, and usually, samples were pre-fixed by glutaraldehyde and dehydrated before visualization. By directly applying different stains including SybrGreen, Syto 64 and SyproRed, biofilm cells of *Acidianus* sp. DSM 29099 were clearly visualized on elemental sulfur under fully hydrated conditions, as shown in Fig. 3. Biofilms were heterogeneously distributed and characterized as individual groups of cells, thin but large colonies with up to 50 μm in diameter. Cells formed large aggregates or dense biofilms, in particular, on some sites with cracks and grooves. These cell distribution patterns suggest that adhesion does not occur randomly, and

biofilm formation does not proceed uniformly at the sulfur surface. In this case, the presence of cell aggregates suggest that the physical contact of *Acidianus* cells with sulfur is a necessary step for sulfur solubilization, while the upper cells in the aggregates could be oxidizing soluble RISCs.

FLBA of biofilms on pyrite and elemental sulfur

The application of FLBA usually includes a screening of all commercially available lectins for probing their reaction with glycoconjugates of (archaeal) biofilms. With the most suitable lectins (for acidophilic archaea), the production of glycoconjugates during biofilm formation may be monitored. In combination with nucleic acid-specific fluorochromes, samples can be analyzed by multichannel CLSM, which has several advantages in analyzing structuring features of hydrated biofilms (Stewart *et al.*, 1995; Lawrence *et al.*, 1998; Neu and Lawrence, 1999b; 2002). As shown in Table 3, pyrite-grown cells of the three species tested were stainable by 22 (eight for *F. acidiphilum* DSM 28986, eight for *Acidianus* sp. DSM 29099 and 14 for *S. metallicus*[T]) out of 75 lectins. In

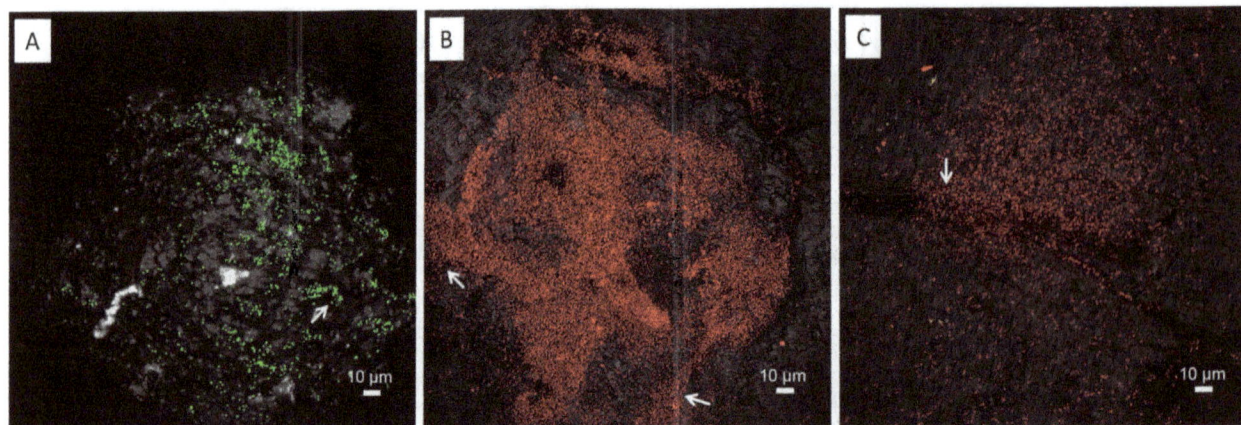

Fig. 3. Maximum intensity projections of biofilms from *Acidianus* sp. DSM 29099 on elemental sulfur. Samples were stained by SybrGreen (A), SyproOrange (B) and Syto 64 (C). Color allocation: green = SybrGreen, red = Syto 64/SyproOrange, grey = reflection. Cells formed a thin biofilm in shallow regions, cracks or holes when sulfur prills were used (A). A clear tendency of cells to attach to physical defects on sulfur coupons is also evident (B and C). Arrows show preferential attachment sites with distortions.

addition, three binding patterns of lectins to EPS glycoconjugates on pyrite surfaces could be differentiated (Table 3). As indicated in Fig. 4A–C (see also Supporting Information Fig. S2), Con A bound to the cell surface of *Acidianus* sp. DSM 29099. Pyrite surfaces free of cells showed no Con A binding, indicating that Con A only reacted with tightly bound EPS of *Acidianus* sp. DSM 29099.

The lectins *Aleuria aurantia* lectin (AAL) and *Limulus polyphemus* agglutinin (LPA) stained biofilm cells of *Acidianus* sp. DSM 29099 (Fig. 4D and E) and *F. acidiphilum* DSM 28986 as well as their surrounding sites (Fig. 5). Unlike the binding of Con A to cells on pyrite surface, these lectins also bound to cell-free EPS on pyrite surfaces. This 'colloidal' EPS binding pattern may allow us to speculate that capsular EPS are probably involved in the initial attachment, and consequently, more EPS were produced in form of colloidal EPS by cells building biofilms on pyrite. Thus, as ocurring in bioleaching bacteria, large areas of the surface may actually be devoid of cells but may be covered by 'colloidal' EPS (Sand *et al.*, 2001). In this context, it has been reported that *T. litoralis* excretes exopolysaccharides into the growth medium and that these may cause a conditioning layer on surfaces (Rinker and Kelly, 1996).

Fig. 6 shows that the cell surfaces of *Acidianus* sp. DSM 29099 and *S. metallicus*^T clearly reacted with the lectin *Griffonia simplicifolia* lectin (GS-I). Within stained biofilms, GS-I signals mostly covered the cells but also filled the space between cells and pyrite surfaces. This cell-associated binding pattern indicates the potential complexity of biofilm structures. It can be concluded that the lectins showing cell-associated binding patterns

reacted with relatively more biofilm components than the ones present just in capsular or colloidal EPS.

Glycoconjugates produced on pyrite. The glycoconjugates present in biofilms of the three archaeal species on pyrite are listed in Table 3. Based on lectin specificity, fucose, glucose, N-acetylgalactosamine (GalNAc), galactose and mannose were found in the three tested species.

Acidianus sp. DSM 29099 and *S. metallicus*^T biofilms on pyrite were shown to possess similar glycoconjugates including fucose, GlcNAc, GalNAc, galactose, mannose and glucose. Both GalNAc and GlcNAc are present in archaeal cell walls. As *Ferroplasma* lacks a cell wall, it is not surprising that we did not detect GlcNAc in these archaea. The binding of the lectin LPA to biofilm cells strongly suggest *F. acidiphilum* DSM 28986 to possess sialic acid residues. Sialic acids are a family of about 50 derivatives of *N*-acetyl or *N*-glycolyl neuraminic acids. They typically occupy the distal end of glycan chains, which makes them suitable for interaction with other cells or with environmental constituents and interfaces. Sialic acids are mainly found in animals and their pathogens, and in certain bacteria (Angata and Varki, 2002; Vimr *et al.*, 2004). They are important components of glycoproteins and glycolipids in animal cell membranes. Studies concerning the presence of sialic acid in archaea are rare (Angata and Varki, 2002; Lewis *et al.*, 2009). By applying reverse-phase high-performance liquid chromatography combined with a fluorescent labelling method using 1,2-diamino-4,5-methylenedioxybenzene dihydrochloride to label the sialic acids (Hara *et al.*, 1989), we were able to confirm the presence of sialic acids in cells of *F. acidiphilum* DSM 28986 (R. Y. Zhang & V. Blanchard, unpublished). Additionally, planktonic cells of this micro-

Table 3. Results of lectin binding assays to extracellular glycoconjugates of three archaeal strains on pyrite.

Lectins[a]	Specificities	Ferroplasma acidiphilum DSM 28986	Acidianus sp. DSM 29099	Sulfolobus metallicus[T]
AAL	Fuc	+ Cell-associated structures	+ Colloidal	+ Cell-associated structures
AIA	Gal; GalNAc	+ Capsular	−	−
Con A	Man; Glc	+ Capsular	+ Capsular	+ Capsular
DGL	Man; Glc	+ Colloidal	−	−
GNA	Man	−	+ Capsular	+ Colloidal
GS-I	Gal; GalNAc	−	+ Cell-associated structures	+ Cell-associated structures
HHA	Man	−	−	+ Capsular
HPA	GalNAc	+ Capsular	+ Colloidal	−
IAA	GalNAc	−	−	+ Cell-associated structures
LEA	GlcNAc	−	+ Cell-associated structures	−
LPA	Sia	+ Colloidal	−	−
MNA-G	Gal	−	−	+ Capsular
MPA	GalNAc	−	+ Colloidal	−
PA-I	Gal	+ Capsular	−	−
PHA-E	Man	+ Colloidal	−	+ Capsular
PHA-L	GalNAc	−	−	+ Capsular
PMA	Man	−	+ Capsular	−
PSA	Man	−	−	+ Capsular
SJA	GalNAc	−	−	+ Culloidal
STA	GluNAc			Capsular
VVA	GalNAc	−	−	+ Cell-associated structures
WFA	GalNAc	−	−	+ Capsular

a. The details of all lectins used in this study are shown in Supporting Information Table S1. Staining and visualization procedures are described in the section Experimental procedures.

+, lectin binding; −, no binding; AIA, *Artocarpus integrifolia* agglutinin; DGL, *Dioclea grandiflora lectin*; GNA, *Galanthus nivalis* agglutinin; HHA, Amaryllis lectin; HPA, *Helix pomatia* agglutinin; IAA, *Iberis amara* agglutinin; LEA, *Lycopersicon esculentum* agglutinin; MNA-G, Morniga G; MPA, *Maclura pomifera* agglutinin; PA-I, *Pseudomonas aeruginosa* lectin I; PHA-E, *Phaseolus vulgaris* agglutinin E; PHA-L, *Phaseolus vulgaris* agglutinin L; PMA, *Polygonatum multiflorum* agglutinin; PSA, *Pisum sativum* agglutinin; SJA, *Sophora japonica* agglutinin; STA, *Solanum tuberosum* agglutinin; VVA, *Vicia villosa* agglutinin; WFA, *Wisteria floribunda* agglutinin.

organism were not stained by LPA. As sialic acid residues are mostly present in biofilms of *F. acidiphilum* DSM 28986 grown on pyrite, it can be assumed that these residues play an important role in attachment and presumably also in the biologically accelerated oxidation of pyrite. Consequently, these results indicate that acidophilic leaching archaea might use different surface compounds (i.e. sialic acids in case of *F. acidiphilum*) for mediating cell–mineral interactions compared with bacterial ones. These are considered to be established by uronic acids complexing iron(III) ions, which are mediating

cell attachment by electrostatic interactions and increase the concentration of the pyrite oxidizing agent iron(III) ions (Sand *et al.*, 2001).

A few lectins including Con A, GS-I, GS-II and WGA have been applied to stain and visualize EPS of *Sulfolobus* spp. WGA was first reported to stain *S. acidocaldarius* and *Sulfolobus shibatae* on polycarbonate membrane filters (Fife *et al.*, 2000). In our assays, this lectin did not bind to any of the strains tested. The lectins Con A, GS-I and GS-II have been used in studies on *S. acidocaldarius*, *S. solfataricus* and

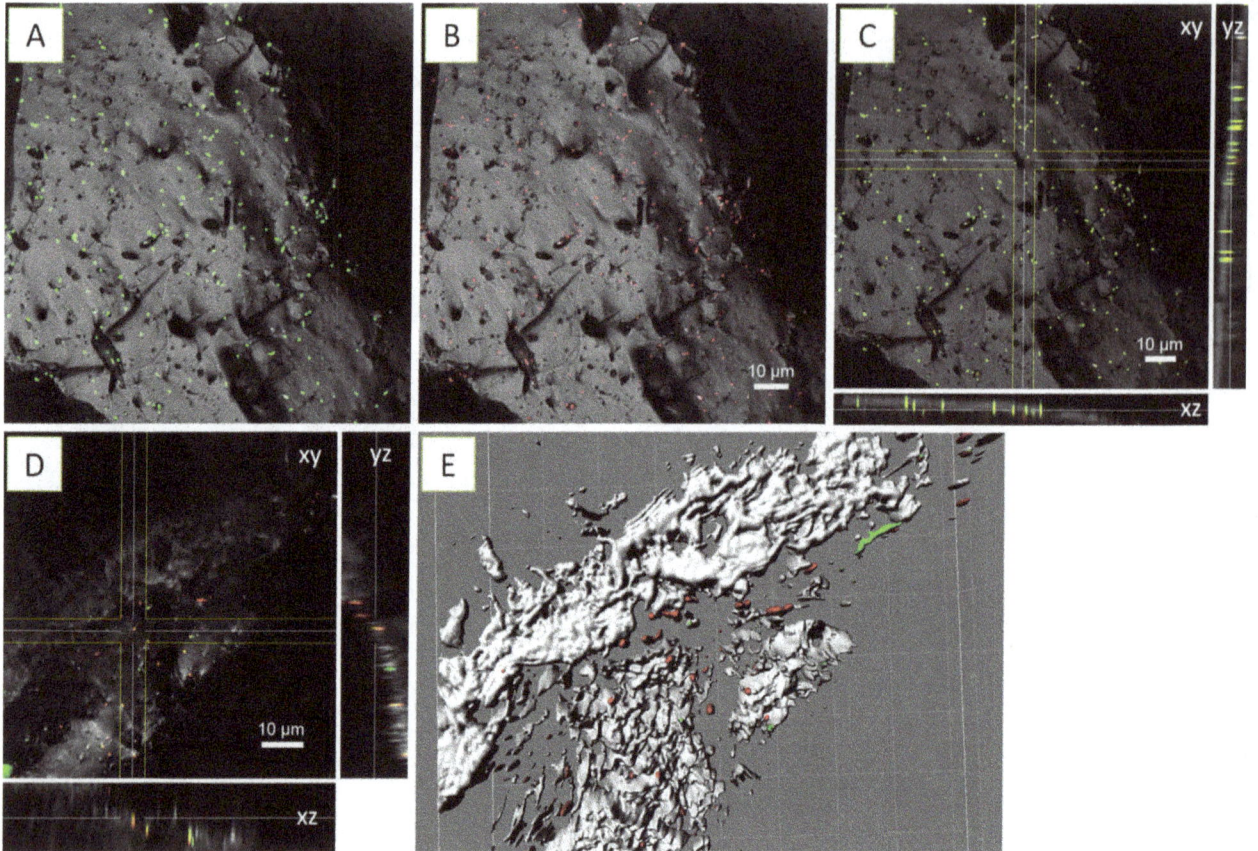

Fig. 4. Maximum intensity projections (A, B), XYZ projection (C, D) and isosurface projection (E) of *Acidianus* sp. DSM 29099 biofilms on pyrite stained by Con A (B) and AAL (D, E), and counter stained by and SybrGreen (A) and Syto 64 (D, E). Color allocation: green = SybrGreen/AAL-fluorescein isothiocyanate (FITC), red = Syto 64/Con A-tetramethyl rhodamine isothiocyanate (TRITC), grey = reflection.

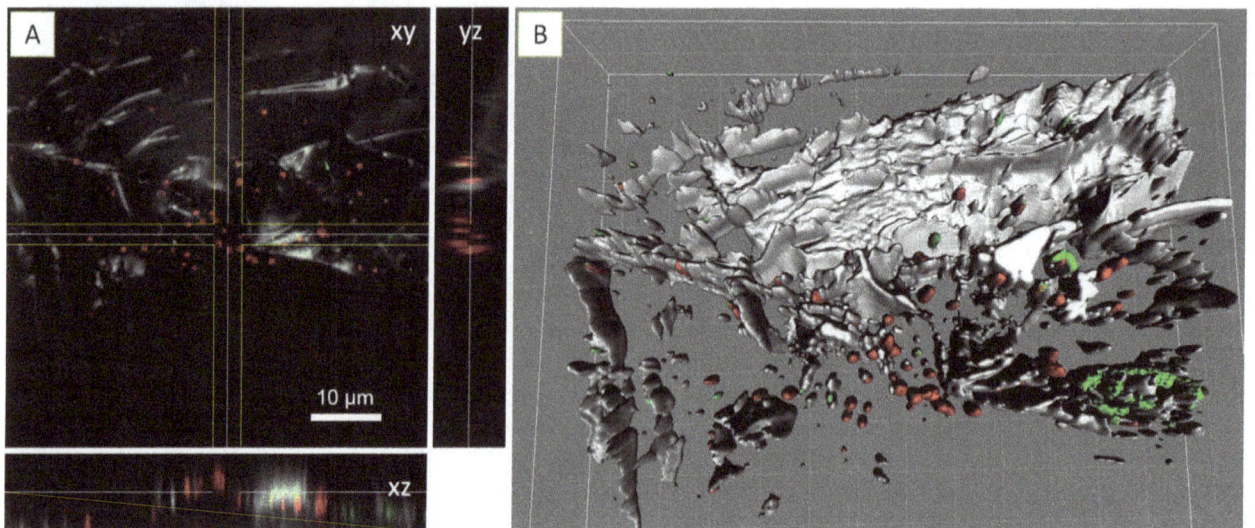

Fig. 5. XYZ projection (A) and isosurface projection (B) of *F. acidiphilum* DSM 28986 biofilms on pyrite stained by LPA-fluorescein isothiocyanate (FITC) and counter stained by FM4-64. Color allocation: green = LPA-FITC, red = FM4-64, grey = reflection. Grid size in B = 10 μm.

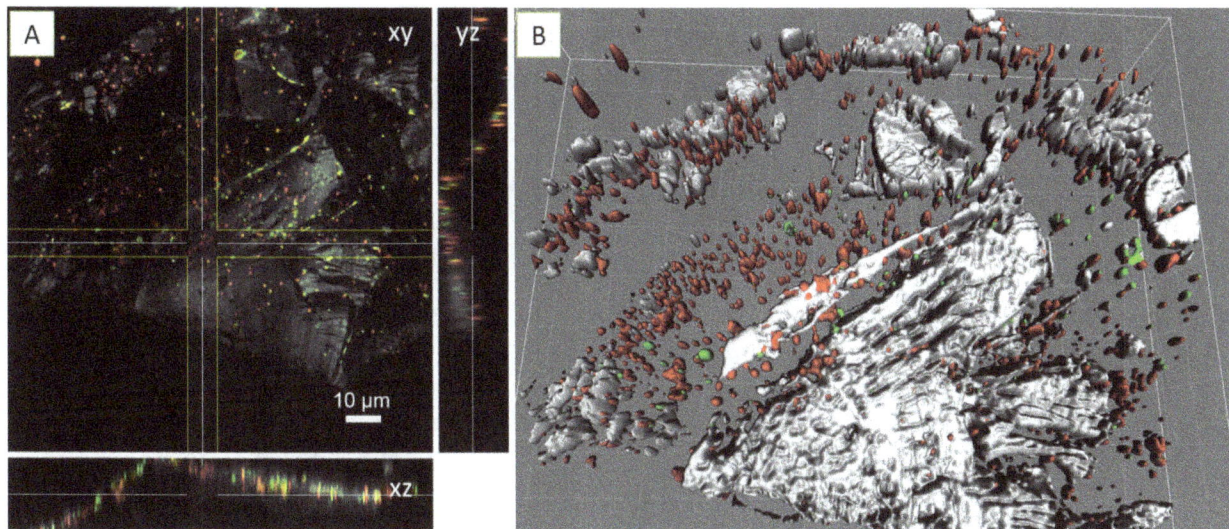

Fig. 6. XYZ projection (A) and isosurface projection (B) of *Acidianus* sp. DSM 29099 biofilms on pyrite stained by GS-I-fluorescein isothiocyanate (FITC) and counter stained by Syto64. Color allocation: green = GS-I-FITC, red = Syto 64, grey = reflection. Grid size in B = 10 μm.

S. tokodaii, including some *S. solfataricus* mutants defective in biofilm formation and production of cell surface appendages (Koerdt *et al.*, 2010; 2012; Zolghadr *et al.*, 2010; Henche *et al.*, 2012). The reaction of these lectins with *Acidianus* sp. DSM 29099 and *S. metallicus*[T] (Tables 3 and 4) strongly indicates that especially Con A and GS-I can be used to monitor biofilm formation of *Sulfolobales*. However, it is important to remark that the abovementioned species are normally grown in complex media (e.g. 0.1% tryptone or 0.2% maltose), while in our experiments, we have used pyrite or elemental

Table 4. Results of lectin binding assays to extracellular glycoconjugates of *Acidianus* sp. DSM 29099 on elemental sulfur.

Lectins	Binding pattern	Lectins	Binding pattern
AAA	Capsular	AAL	Capsular/cell-associated structures
Con A	Capsular/cell-associated structures	DBA	Capsular
ECA	Colloidal	EEA	Capsular
GHA	Capsular	GS-I	Colloidal/cell-associated structures
HHA	Colloidal	HMA	Colloidal
IAA	Colloidal	IRA	Capsular
LAL	Capsular	LBA	Capsular
LcH	Capsular	MOA	Colloidal
PNA	Capsular	PSA	Capsular
SJA	Capsular	TL	Capsular
VGA	Capsular		

DBA, *Dolichos biflorus* agglutinin; ECA, *Erythrina cristagalli* agglutinin; EEA, *Euonymus europaeus* agglutinin; GHA, *Glechoma hederacea* agglutinin; HMA, *Homarus americanus* agglutinin; IRA, *Iris hybrid* agglutinin; LAL, *Laburnum anagyroides lectin*; LBA, *Phaseolus lunatus* agglutinin; LcH, *Lens culinaris* haemagglutinin; MOA, *Marasmium oreades* agglutinin; PNA, Peanut agglutinin; TL, *Tulipa* sp. agglutinin; VGA, *Vicia graminea* agglutinin.

sulfur (in case of *Acidianus* sp. DSM 29099) as energy sources. Under these conditions, most of the carbon for biosynthesis must be fixed from CO_2. Control experiments showed no significant cell growth of *Acidianus* sp. DSM 29099 with 0.02% yeast extract as sole energy source (not shown). As we focused on biofilm formation on pyrite or sulfur surfaces, we cannot rule out that under presence of sufficient amounts of organic carbon, *Acidianus* sp. DSM 29099 may build structurally more complex biofims as described for other *Sulfolobales* (Koerdt *et al.*, 2010).

The lectins AAL and Con A stained cells of the three species used in this study, which is consistent with previous reports that these two lectins have the potential to stain various kinds of biofilms (Neu and Lawrence, 2002; Neu *et al.*, 2002; Strathmann *et al.*, 2002; Staudt *et al.*, 2003; Bellenberg *et al.*, 2012). The monomers glucose, mannose and fucose were also found in EPS fractions of *Acidithiobacillus ferrooxidans* (Gehrke *et al.*, 1998). Thus, leaching bacteria and archaea may have similarities in their EPS composition.

Glycoconjugates produced on elemental sulfur. Twenty-one lectins were shown to bind EPS glycoconjugates of *Acidianus* sp. DSM 29099 (Table 4). In addition, two major binding patterns became evident. Five lectins showed signals covering cells and an extended area around them, indicating their binding to colloidal or loosely bound EPS. In contrast, 16 lectins showed a 'capsular binding pattern', in which signals were only restricted to cells (Fig. 7 and Supporting Information Fig. S2). Among these 21 lectins, only AAL, Con A and GS-I stained EPS glycoconjugates of *Acidianus* sp. DSM 29099 on pyrite (Tables 3 and 4). It indicates that the glycoconjugates, which these three

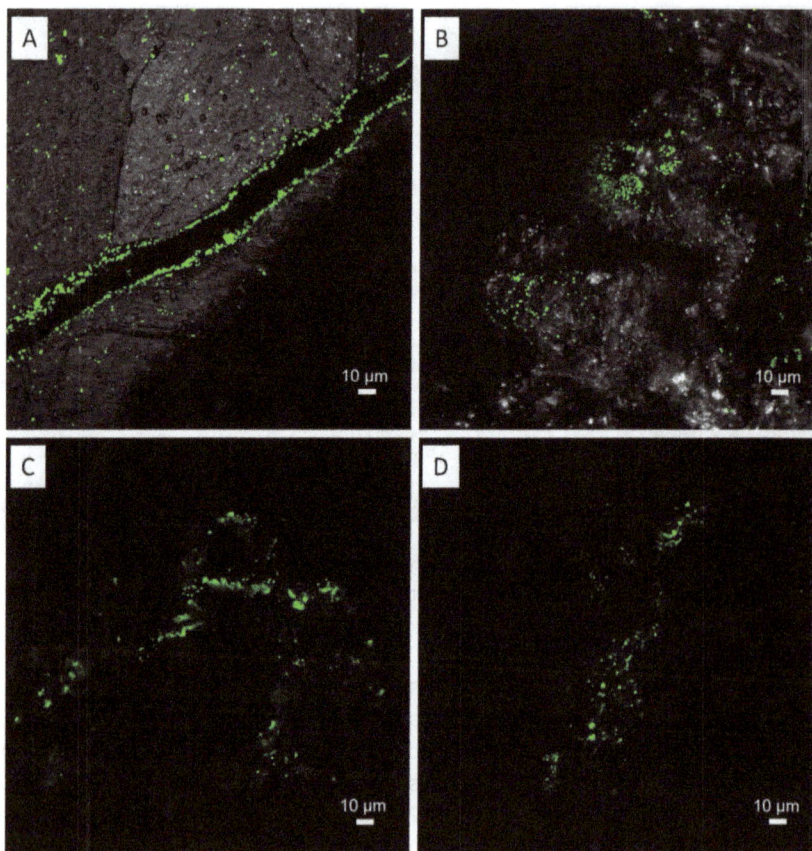

Fig. 7. Maximum intensity projections of biofilms from *Acidianus* sp. DSM 29099 on elemental sulfur. Samples were stained by lectins AAL-Alexa488 (A), Peanut agglutinin (PNA)-fluorescein isothiocyanate (FITC) (B), *Erythrina cristagalli* agglutinin (ECA)-FITC (C) and GS-I (D). Two distinguished lectin binding patterns became visible, tightly bound "capsular" EPS staining (A and B) and loosely bound "colloidal" (C and D). Color allocation: green = lectins, grey = reflection.

lectins recognized (i.e. glucose, mannose, GalNAc and galactose), are common components of *Acidianus* sp. DSM 29099 attached on pyrite and sulfur surfaces. From another point of view, *Acidianus* cells may express different glycoconjugates which correlate with their energy sources. In addition, extracellular proteins of *Sulfolobales* are highly glycosylated. N-glycans usually consist of glucose, mannose, GalNAc and GlcNAc. The high glycosylation density in S-layers could represent an adaptation of these organisms to the high temperature and acidic environment that they naturally encounter (Meyer and Albers, 2013). It has been shown that the loss of one terminal hexose of the N-glycan has effects on cell growth and motility (Jarrell *et al.*, 2014). We suggest that some of the lectins bound also to the N-glycans from the S-layer, which are probably differentially expressed between pyrite and sulfur-grown cells in biofilms of *Acidianus* sp. DSM 29099.

Conclusions

We studied a library of lectins for their potential to visualize and characterize glycoconjugates in acidophilic archaeal

biofilms. The lectin binding tests provided the first hint for the distribution of glycoconjugates that are involved in biofilm formation on pyrite as well as on elemental sulfur. By FLBA, the glycoconjugates in acidophilic archaeal biofilms were characterized as polymers containing sugar moieties like glucose, galactose, mannose, GlcNAc, GalNAc, sialic acid and fucose. Twenty-two lectins were shown to be useful for the study of EPS glycoconjugates of acidophilic archaea during pyrite dissolution. These lectins may be used in future studies for assessment of interactions between various members of microbial bioleaching communities, especially in order to elucidate the role of archaea in detail. In addition, lectins which are species or strain specific (e.g. LPA staining *F. acidiphilum*) may be used as probes to differentiate a target archaeon from others in multi-species biofilm studies.

Experimental procedures

Archaeal strains and cultivation

Ferroplasma acidiphilum DSM 28986 = JCM 30201 (former BRGM4) was isolated from a stirred tank reactor (Bryan *et al.*, 2009). *Sulfolobus metallicus* DSM 6482^T was purchased from DSMZ. Strain *Acidianus* sp. DSM 29099 = JCM 30227, an iron- and sulfur-oxidizer, was isolated from a hot spring at Copahue Volcano, Neuquén, Argentina (R. Y. Zhang & W. Sand, unpublished). All strains were cultivated in MAC medium (Mackintosh, 1978) containing 0.02% yeast extract with an initial pH 1.8 for *F. acidiphilum* DSM 28986 or pH 2.5 for *S. metallicus*^{T} and *Acidianus* sp. DSM 29099 respectively. *Ferroplasma acidiphilum* DSM 28986 was grown at 37°C with 4 g l^{-1} iron(II) ions. *Sulfolobus metallicus*^{T} and *Acidianus* sp. DSM 29099 were grown at 65°C with 10 g l^{-1} elemental sulfur.

Substratum, biofilm formation and pyrite leaching

Pyrite grains with a size of 200–500 μm were selected after grinding and sieving of pyrite cubes from Rioja (Spain). They were cleaned and sterilized as described (Schippers *et al.*, 1996; Schippers and Sand, 1999). For cell attachment assays, 20 g of sterile pyrite grains were incubated with pure cultures of each strain in 300 ml Mac (initial cell concentration 10^8 cells/ml) to allow biofilm development and pyrite dissolution. Iron ions were quantified by the phenanthroline method (Tamura *et al.*, 1974).

Sulfur powder (Roth, Germany) was molten and poured into deionized water with agitation. Sulfur prills with a diameter of 1–3 mm were formed due to rapid cooling. A plate covered with aluminium was used to get a sulfur layer after solidification. Sulfur coupons with a size of approximately 0.5 cm × 0.5 cm × 2 mm were obtained by breaking the sulfur layer. Both sulfur prills and coupons were sterilized at 110°C for 90 min.

Staining

Cell distribution was observed after nucleic acid staining with Syto 9, Syto 64 or SybrGreen. In addition, SyproRed, binding to cellular proteins, or FM4-64, binding to lipid-rich domains (Bolte *et al.*, 2004), were used for counter staining. For glycoconjugate staining, biofilms of each cell population were assayed with lectins conjugated with fluorescein isothiocyanate, Alexa 488 or tetramethyl rhodamine isothiocyanate (Supporting Information Table S1). Briefly, samples were washed with filter-sterilized tap water and incubated with 0.1 mg ml^{-1} lectins for 20 min at room temperature. Afterwards, stained samples were washed three times with filter-sterilized tap water in order to remove unbound lectins. Direct light exposure was avoided. Counter staining was done in a coverwell chamber of 20 mm in diameter with 0.5 mm spacer (Invitrogen). Counter-stained samples were directly observed using CLSM without any further treatment.

CLSM

Examination of stained biofilms was performed by CLSM using a TCS SP5X (Leica, Heidelberg, Germany), controlled by the LASAF 2.4.1 build 6384. The system was equipped with an upright microscope and a super continuum light source (470–670 nm) as well as a 405 nm laser diode. Images were collected with a 63 × water immersion lens with a numerical aperture (NA) of 1.2 and a 63 × water immersible lens with an NA of 0.9. The details of fluorescent dyes along with excitation and emission filters are shown in Table 1. CLSM data sets were recorded in sequential mode in order to avoid cross talk of the fluorochromes between two different channels. Surface topography and texture of the pyrite as well as of the elemental sulfur surface were recorded by using the CLSM in reflection mode.

AFM and EFM

Pyrite slices were rinsed with sterile MAC medium and deionized water. Cells attached to pyrite coupons and their EPS were stained by Syto 9 and by fluorescently labelled Con A. Stained samples were dried at room temperature and visualized by EFM (Zeiss, Germany) combined with AFM (BioMaterial™Workstation, JPK Instruments) for the investigation of cell morphology and distribution of cells on the surfaces of pyrite coupons (Zhang *et al.*, 2014).

Digital image analysis

Fluorescence images were analyzed using an extended version of software IMAGEJ (Abràmoff *et al.*, 2004). Maximum intensity and XYZ projections of three-dimensional data sets were produced with the software IMARIS version 7.3.1 (Bitplane AG, Zurich, Switzerland).

Acknowledgements

We acknowledge Prof. Dr. Barrie Johnson (Bangor University, UK) and Prof. Dr. E. Donati (CONICET-UNLP, La Plata, Argentina) for kindly providing strain *F. acidiphilum* DSM 28986 and samples for the isolation of strain *Acidianus* sp. DSM 29099 respectively. Dr. Véronique Blanchard (Charité Medical University, Berlin, Germany) is gratefully acknowledged for her assistance with monosaccharide analysis. In particular, we would like to thank the reviewer for helpful comments.

Conflict of interest

None declared.

References

Abràmoff, M.D., Magalhães, P.J., and Ram, S.J. (2004) Image processing with ImageJ. *Biophotonics International* **11:** 36–43.

Africa, C.J., van Hille, R.P., Sand, W., and Harrison, S.T.L. (2013) Investigation and *in situ* visualisation of interfacial interactions of thermophilic microorganisms with metal-sulphides in a simulated heap environment. *Miner Eng* **48:** 100–107.

Albers, S.V., and Meyer, B.H. (2011) The archaeal cell envelope. *Nat Rev Microbiol* **9:** 414–426.

Angata, T., and Varki, A. (2002) Chemical diversity in the sialic acids and related α-keto acids: an evolutionary perspective. *Chem Rev* **102:** 439–470.

Baker-Austin, C., Potrykus, J., Wexler, M., Bond, P.L., and Dopson, M. (2010) Biofilm development in the extremely acidophilic archaeon 'Ferroplasma acidarmanus' Fer1. *Extremophiles* **14:** 485–491.

Bang, C., Ehlers, C., Orell, A., Prasse, D., Spinner, M., Gorb, S.N., *et al.* (2014) Biofilm formation of mucosa-associated methanoarchaeal strains. *Front Microbiol* **5:** 353.

Bellenberg, S., Leon-Morales, C.F., Sand, W., and Vera, M. (2012) Visualization of capsular polysaccharide induction in *Acidithiobacillus ferrooxidans*. *Hydrometallurgy* **129:** 82–89.

Bennke, C.M., Neu, T.R., Fuchs, B.M., and Amann, R. (2013) Mapping glycoconjugate-mediated interactions of marine *Bacteroidetes* with diatoms. *Syst Appl Microbiol* **36:** 417–425.

Blais, J.-F., Tyagi, R., Meunier, N., and Auclair, J. (1994) The production of extracellular appendages during bacterial colonization of elemental sulphur. *Process Biochem* **29:** 475–482.

Bolte, S., Talbot, C., Boutte, Y., Catrice, O., Read, N., and Satiat-Jeunemaitre, B. (2004) FM-dyes as experimental probes for dissecting vesicle trafficking in living plant cells. *J Microsc* **214:** 159–173.

Brierley, C.L., and Brierley, J.A. (2013) Progress in bioleaching: part B: applications of microbial processes by the minerals industries. *Appl Microbiol Biotechnol* **97:** 7543–7552.

Bryan, C.G., Joulian, C., Spolaore, P., Challan-Belval, S., El Achbouni, H., Morin, D.H.R., and D'Hugues, P. (2009) Adaptation and evolution of microbial consortia in a stirred tank reactor bioleaching system: indigenous population versus a defined consortium. *Adv Mat Res* **71:** 79–82.

Bryant, R., Costerton, J., and Laishley, E. (1984) The role of *Thiobacillus albertis* glycocalyx in the adhesion of cells to elemental sulfur. *Can J Microbiol* **30:** 81–90.

Castro, L., Zhang, R., Muñoz, J.A., González, F., Blázquez, M.L., Sand, W., and Ballester, A. (2014) Characterization of exopolymeric substances (EPS) produced by *Aeromonas hydrophila* under reducing conditions. *Biofouling* **30:** 501–511.

Chen, Y., Li, J., Chen, L., Hua, Z., Huang, L., Liu, J., *et al.* (2014) Biogeochemical processes governing natural pyrite oxidation and release of acid metalliferous drainage. *Environ Sci Technol* **48:** 5537–5545.

Dopson, M., Baker-Austin, C., Hind, A., Bowman, J.P., and Bond, P.L. (2004) Characterization of *Ferroplasma* isolates and *Ferroplasma acidarmanus* sp. nov., extreme acidophiles from acid mine drainage and industrial bioleaching environments. *Appl Environ Microbiol* **70:** 2079–2088.

Edwards, K.J., Bond, P.L., Gihring, T.M., and Banfield, J.F. (2000) An archaeal iron-oxidizing extreme acidophile important in acid mine drainage. *Science* **287:** 1796–1799.

Etzel, K., Klingl, A., Huber, H., Rachel, R., Schmalz, G., Thomm, M., and Depmeier, W. (2008) Etching of {111} and {210} synthetic pyrite surfaces by two archaeal strains, *Metallosphaera sedula* and *Sulfolobus metallicus*. *Hydrometallurgy* **94:** 116–120.

Fife, D.J., Bruhn, D.F., Miller, K.S., and Stoner, D.L. (2000) Evaluation of a fluorescent lectin-based staining technique for some acidophilic mining bacteria. *Appl Environ Microbiol* **66:** 2208–2210.

Fröls, S., Dyall-Smith, M., and Pfeifer, F. (2012) Biofilm formation by haloarchaea. *Environ Microbiol* **14:** 3159–3174.

Gehrke, T., Telegdi, J., Thierry, D., and Sand, W. (1998) Importance of extracellular polymeric substances from *Thiobacillus ferrooxidans* for bioleaching. *Appl Environ Microbiol* **64:** 2743–2747.

Goldstein, I., Hollerman, C., and Merrick, J.M. (1965) Protein-carbohydrate interaction I. The interaction of polysaccharides with concanavalin A. *BBA-Gen Subjects* **97:** 68–76.

Golyshina, O.V., and Timmis, K.N. (2005) *Ferroplasma* and relatives, recently discovered cell wall-lacking archaea making a living in extremely acid, heavy metal-rich environments. *Environ Microbiol* **7:** 1277–1288.

Golyshina, O.V., Pivovarova, T.A., Karavaiko, G.I., Kondratéva, T.F., Moore, E., Abraham, W.R., *et al.* (2000) *Ferroplasma acidiphilum* gen. nov., sp. nov., an acidophilic, autotrophic, ferrous-iron-oxidizing, cell-wall-lacking, mesophilic member of the *Ferroplasmaceae* fam. nov., comprising a distinct lineage of the Archaea. *Int J Syst Evol Microbiol* **50:** 997–1006.

Golyshina, O.V., Yakimov, M.M., Lünsdorf, H., Ferrer, M., Nimtz, M., Timmis, K.N., *et al.* (2009) *Acidiplasma aeolicum* gen. nov., sp. nov., a euryarchaeon of the family *Ferroplasmaceae* isolated from a hydrothermal pool, and transfer of *Ferroplasma cupricumulans* to *Acidiplasma cupricumulans* comb. nov. *Int J Syst Evol Microbiol* **59:** 2815–2823.

Hara, S., Yamaguchi, M., Takemori, Y., Furuhata, K., Ogura, H., and Nakamura, M. (1989) Determination of mono-*O*-acetylated *N*-acetylneuraminic acids in human and rat sera by fluorometric high-performance liquid chromatography. *Anal Biochem* **179:** 162–166.

Henche, A.L., Koerdt, A., Ghosh, A., and Albers, S.V. (2012) Influence of cell surface structures on crenarchaeal biofilm formation using a thermostable green fluorescent protein. *Environ Microbiol* **14:** 779–793.

Jarrell, K.F., Ding, Y., Meyer, B.H., Albers, S.-V., Kaminski, L., and Eichler, J. (2014) N-Linked glycosylation in archaea: a structural, functional, and genetic analysis. *Microbiol Mol Biol Rev* **78:** 304–341.

Kalin, M., Fyson, A., and Wheeler, W.N. (2006) The chemistry of conventional and alternative treatment systems for the neutralization of acid mine drainage. *Sci Total Environ* **366:** 395–408.

Karatan, E., and Watnick, P. (2009) Signals, regulatory networks, and materials that build and break bacterial biofilms. *Microbiol Mol Biol Rev* **73:** 310–347.

Koerdt, A., Gödeke, J., Berger, J., Thormann, K.M., and Albers, S.V. (2010) Crenarchaeal biofilm formation under extreme conditions. *PLoS ONE* **5:** e14104.

Koerdt, A., Jachlewski, S., Ghosh, A., Wingender, J., Siebers, B., and Albers, S.-V. (2012) Complementation of *Sulfolobus solfataricus* PBL2025 with an α-mannosidase: effects on surface attachment and biofilm formation. *Extremophiles* **16:** 115–125.

Lawrence, J., Neu, T., and Swerhone, G. (1998) Application of multiple parameter imaging for the quantification of algal, bacterial and exopolymer components of microbial biofilms. *J Microbiol Methods* **32:** 253–261.

Lawrence, J., Swerhone, G., Leppard, G., Araki, T., Zhang, X., West, M., and Hitchcock, A. (2003) Scanning transmission X-ray, laser scanning, and transmission electron microscopy mapping of the exopolymeric matrix of microbial biofilms. *Appl Environ Microbiol* **69:** 5543–5554.

Lewis, A.L., Desa, N., Hansen, E.E., Knirel, Y.A., Gordon, J.I., Gagneux, P., *et al.* (2009) Innovations in host and microbial sialic acid biosynthesis revealed by phylogenomic prediction of nonulosonic acid structure. *Proc Natl Acad Sci* **106:** 13552–13557.

Mackintosh, M.E. (1978) *Nitrogen fixation by* Thiobacillus ferrooxidans. *J Gen Microbiol* **105:** 215–218.

Meyer, B.H., and Albers, S.-V. (2013) Hot and sweet: protein glycosylation in Crenarchaeota. *Biochem Soc Trans* **41:** 384–392.

Näther, D.J., Rachel, R., Wanner, G., and Wirth, R. (2006) Flagella of *Pyrococcus furiosus*: multifunctional organelles, made for swimming, adhesion to various surfaces, and cell-cell contacts. *J Bacteriol* **188:** 6915–6923.

Neu, T., and Lawrence, J. (2009) Extracellular polymeric substances in microbial biofilms. In *Microbial Glycobiology: Structures, Relevance and Applications.* Moran, A., Brennan, P., Holst, O., and von Itzstein, M. (eds). San Diego, CA, USA: Elsevier, pp. 735–758.

Neu, T.R., and Lawrence, J.R. (1999a) In situ characterization of extracellular polymeric substances (EPS) in biofilm systems. In *Microbial Extracellular Polymeric Substances: Characterization, Structure and Function.* Wingender, J., Neu, T.R., and Flemming, H.-C. (eds). Berlin, Germany: Springer, pp. 21–47.

Neu, T.R., and Lawrence, J.R. (1999b) Lectin-binding analysis in biofilm systems. *Methods Enzymol* **310:** 145–152.

Neu, T.R., and Lawrence, J.R. (2002) Laser scanning microscopy in combination with fluorescence techniques for biofilm study. In *Encyclopedia of Environmental Microbiology.* Bitton, G. (ed.). New York, USA: Wiley, pp. 1772–1788.

Neu, T.R., and Lawrence, J.R. (2014) Advanced techniques for in situ analysis of the biofilm matrix (structure, composition, dynamics) by means of laser scanning microscopy. In *Microbial Biofilms: Methods and Protocols, Methods.* In *Molecular Biology.* Donelli, G. (ed.). New York, USA: Springer, pp. 43–64.

Neu, T.R., Kuhlicke, U., and Lawrence, J.R. (2002) Assessment of fluorochromes for two-photon laser scanning microscopy of biofilms. *Appl Environ Microbiol* **68:** 901–909.

Nielsen, P.H., and Jahn, A. (1999) Extraction of EPS. In *Microbial Extracellular Polymeric Substances: Characterization, Structure and Function.* Wingender, J., Neu, T.R., and Flemming, H.-C. (eds). Berlin, Germany: Springer, pp. 49–72.

Olson, G., Brierley, J., and Brierley, C. (2003) Bioleaching review part B. *Appl Microbiol Biotechnol* **63:** 249–257.

Orell, A., Fröls, S., and Albers, S.-V. (2013) Archaeal biofilms: the great unexplored. *Annu Rev Microbiol* **67:** 337–354.

Peltola, M., Neu, T.R., Raulio, M., Kolari, M., and Salkinoja-Salonen, M.S. (2008) Architecture of

Deinococcus geothermalis biofilms on glass and steel: a lectin study. *Environ Microbiol* **10:** 1752–1759.

Rawlings, D.E., and Johnson, D.B. (2007) The microbiology of biomining: development and optimization of mineral-oxidizing microbial consortia. *Microbiology* **153:** 315–324.

Rinker, K.D., and Kelly, R.M. (1996) Growth physiology of the hyperthermophilic archaeon *Thermococcus litoralis*: development of a sulfur-free defined medium, characterization of an exopolysaccharide, and evidence of biofilm formation. *Appl Environ Microbiol* **62:** 4478–4485.

Sand, W., Gehrke, T., Jozsa, P.G., and Schippers, A. (2001) Bio) chemistry of bacterial leaching – direct vs. indirect bioleaching. *Hydrometallurgy* **59:** 159–175.

Sand, W., Jozsa, P.-G., Kovacs, Z.-M., Săsăran, N., and Schippers, A. (2007) Long-term evaluation of acid rock drainage mitigation measures in large lysimeters. *J Geochem Explor* **92:** 205–211.

Schaeffer, W., Holbert, P., and Umbreit, W. (1963) Attachment of *Thiobacillus thiooxidans* to sulfur crystals. *J Bacteriol* **85:** 137–140.

Schippers, A., and Sand, W. (1999) Bacterial leaching of metal sulfides proceeds by two indirect mechanisms via thiosulfate or via polysulfides and sulfur. *Appl Environ Microbiol* **65:** 319–321.

Schippers, A., Jozsa, P., and Sand, W. (1996) Sulfur chemistry in bacterial leaching of pyrite. *Appl Environ Microbiol* **62:** 3424–3431.

Schopf, S., Wanner, G., Rachel, R., and Wirth, R. (2008) An archaeal bi-species biofilm formed by *Pyrococcus furiosus* and *Methanopyrus kandleri. Arch Microbiol* **190:** 371–377.

Staudt, C., Horn, H., Hempel, D., and Neu, T. (2003) Screening of lectins for staining lectin-specific glycoconjugates in the EPS of biofilms. In. In *Biofilms in Medicine, Industry and Environmental Technology.* Lens, P., O'Flaherty, V., Moran, A.P., Stoodley, P., and Mahony, T. (eds). London, UK: IWA Publishing, pp. 308–327.

Stewart, P.S., Murga, R., Srinivasan, R., and de Beer, D. (1995) Biofilm structural heterogeneity visualized by three microscopic methods. *Water Res* **29:** 2006–2009.

Stoodley, P., Sauer, K., Davies, D., and Costerton, J.W. (2002) Biofilms as complex differentiated communities. *Annu Rev Microbiol* **56:** 187–209.

Strathmann, M., Wingender, J., and Flemming, H.-C. (2002) Application of fluorescently labelled lectins for the visualization and biochemical characterization of polysaccharides in biofilms of *Pseudomonas aeruginosa. J Microbiol Methods* **50:** 237–248.

Tamura, H., Goto, K., Yotsuyanagi, T., and Nagayama, M. (1974) Spectrophotometric determination of iron (II) with 1, 10-phenanthroline in the presence of large amounts of iron (III). *Talanta* **21:** 314–318.

Thoma, C., Frank, M., Rachel, R., Schmid, S., Näther, D., Wanner, G., and Wirth, R. (2008) The Mth60 fimbriae of *Methanothermobacter thermoautotrophicus* are functional adhesins. *Environ Microbiol* **10:** 2785–2795.

Vera, M., Schippers, A., and Sand, W. (2013) Progress in bioleaching: fundamentals and mechanisms of bacterial metal sulfide oxidation – part A. *Appl Microbiol Biotechnol* **97:** 7529–7541.

Vimr, E.R., Kalivoda, K.A., Deszo, E.L., and Steenbergen, S.M. (2004) Diversity of microbial sialic acid metabolism. *Microbiol Mol Biol Rev* **68:** 132–153.

Weiss, R. (1973) Attachment of bacteria to sulphur in extreme environments. *J Gen Microbiol* **77:** 501–507.

Whitchurch, C.B., Tolker-Nielsen, T., Ragas, P.C., and Mattick, J.S. (2002) Extracellular DNA required for bacterial biofilm formation. *Science* **295:** 1487–1487.

Wolfaardt, G.M., Lawrence, J.R., and Korber, D.R. (1999) Function of EPS. In *Microbial Extracellular Polymeric Substances: Characterization, Structure and Function.* Wingender, J., Neu, T.R., and Flemming, H.-C. (eds). Berlin, Germany: Springer, pp. 171–200.

Zhang, R., Bellenberg, S., Castro, L., Neu, T.R., Sand, W., and Vera, M. (2014) Colonization and biofilm formation of the extremely acidophilic archaeon *Ferroplasma acidiphilum*. *Hydrometallurgy* doi: 10.1016/j.hydromet.2014.07.001.

Zippel, B., and Neu, T. (2011) Characterization of glycoconjugates of extracellular polymeric substances in tufa-associated biofilms by using fluorescence lectin-binding analysis. *Appl Environ Microbiol* **77:** 505–516.

Zolghadr, B., Klingl, A., Koerdt, A., Driessen, A.J., Rachel, R., and Albers, S.-V. (2010) Appendage-mediated surface adherence of *Sulfolobus solfataricus. J Bacteriol* **192:** 104–110.

Zubkov, M.V., Fuchs, B.M., Eilers, H., Burkill, P.H., and Amann, R. (1999) Determination of total protein content of bacterial cells by SYPRO staining and flow cytometry. *Appl Environ Microbiol* **65:** 3251–3257.

Supporting information

Additional Supporting Information may be found in the online version of this article at the publisher's web-site:

Fig. S1. Biofilm cells of *Acidianus* sp. DSM 29099 visualized by atomic force microscope (AFM) combined with epifluorescence microscope (EFM), exhibiting preferential attack on the pyrite lattice along planes. A and B show EFM images of *Acidianus* sp. DSM 29099 biofilms stained by Syto 9 (green) and TRITC-conjugated Con A (red), respectively. C shows AFM scanning corresponding to EFM (A and B). White arrows show a cell cluster. Bars represent 10 μm.

Fig. S2. Biofilm cells of *Acidianus* sp. DSM 29099. A, cells grown on pyrite and stained by TRITC-conjugated Con A (red) and Syto 9 (green), respectively. B, cells grown on elemental sulfur and stained by FITC-conjugated Con A (green). Con A stained cell surfaces and gave a clear 'capsular binding' pattern. Arrows show cell surfaces. Bars represent 5 μm.

Table S1. List of fluorescent labeled lectins used for staining of archaeal biofilms and their binding target.

18

Enhancing photo-catalytic production of organic acids in the cyanobacterium *Synechocystis sp.* PCC 6803 Δ*glgC*, a strain incapable of glycogen storage

Damian Carrieri,[†] Charlie Broadbent, David Carruth, Troy Paddock,[†] Justin Ungerer,[†] Pin-Ching Maness, Maria Ghirardi and Jianping Yu*
Biosciences Center, National Renewable Energy Laboratory, 15013 Denver West Parkway, Golden, CO 80401, USA.

Summary

A key objective in microbial biofuels strain development is to maximize carbon flux to target products while minimizing cell biomass accumulation, such that ideally the algae and bacteria would operate in a photo-catalytic state. A brief period of such a physiological state has recently been demonstrated in the cyanobacterium *Synechocystis sp.* PCC 6803 Δ*glgC* strain incapable of glycogen storage. When deprived of nitrogen, the Δ*glgC* excretes the organic acids alpha-ketoglutarate and pyruvate for a number of days without increasing cell biomass. This study examines the relationship between the growth state and the photo-catalytic state, and characterizes the metabolic adaptability of the photo-catalytic state to increasing light intensity. It is found that the culture can transition naturally from the growth state into the photo-catalytic state when provided with limited nitrogen supply during the growth phase. Photosynthetic capacity and pigments are lost over time in the photo-catalytic state. Reversal to growth state is observed with re-addition of nitrogen nutrient, accompanied by restoration of photosynthetic capacity and pigment levels in the cells. While the overall productivity increased under high light conditions, the ratio of alpha-ketoglutarate/pyruvate is altered, suggesting that carbon partition between the two products is adaptable to environmental conditions.

*For correspondence. E-mail Jianping.Yu@nrel.gov

Funding Information This work is supported by the US Department of Energy, Office of Science Basic Energy Sciences Program (M. G., D. Carrieri, T. P., J. U. and J. Y.), Office of Science Science Undergraduate Laboratory Internship Program (C. B. and D. Carruth) and Energy Efficiency and Renewable Energy, Fuel Cells Technologies Office (P. C. M.).

Introduction

Algae including cyanobacteria are receiving increasing attention for their potential in direct photosynthetic conversion of carbon dioxide into fuels and chemicals. Compared with land plants, algae can achieve higher areal productivity and do not have to use arable land and fresh water, thus do not compete with food production for these valuable natural resources (Dismukes *et al.*, 2008). To realize the potential of photosynthetic microorganisms in renewable chemical production, a strategy to minimize growth while maximizing photosynthetic productivity of target chemicals is ideal (Melis, 2012). In such an approach, a culture would not grow and instead dedicate fixed carbon from photosynthesis to production of target compounds. Towards that vision, we have genetically modified the model cyanobacterium *Synechocystis sp.* PCC 6803 (hereafter referred to as S6803) to make it incapable of glycogen storage by deleting the gene at locus slr1176 (*glgC*), which encodes a glucose-1-phosphate adenylyltransferase. When deprived of nitrogen, the wild-type (WT) strain increases cell biomass with glycogen accumulation, while the mutant instead produces and excretes alpha-ketoglutarate and pyruvate without biomass increase. The mutant continues to photosynthesize at rates similar to WT (although both WT and mutant strains decline in photosynthetic rates after nitrogen removal) and utilize newly fixed atmospheric carbon, instead of stored carbon, to produce the excreted keto acids (Carrieri *et al.*, 2012). Similar findings have also been reported by other researchers and with other strains (Gründel *et al.*, 2012; Hickman *et al.*, 2013). In *Synechococcus elongatus* PCC 7942 and *Synechococcus sp.* PCC 7002, the *glgC* deletion mutant has been used to improve yields of targeted products with expression of heterologous pathways (Davies *et al.*, 2014; Li *et al.*, 2014). Excretion of keto acids by a glycogen synthesis mutant was also reported to occur under mixotrophic conditions with glucose supplementation and is considered a manifestation of overflow metabolism (Gründel *et al.*, 2012). However, the photo-catalytic

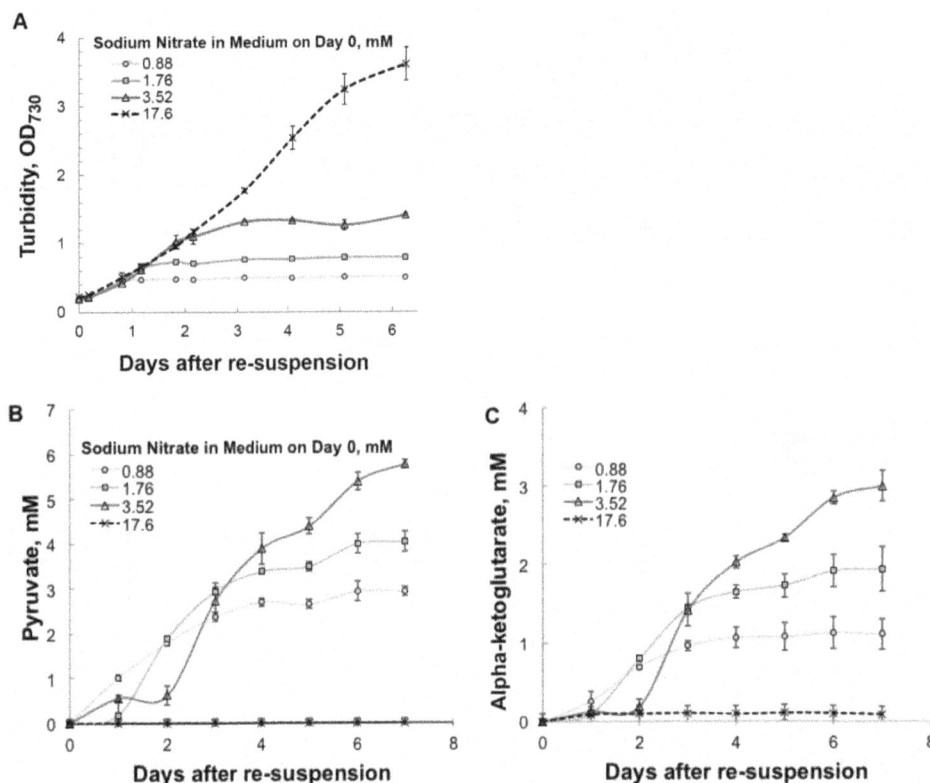

Fig. 1. Accumulated concentration of extracellular pyruvate and alpha-ketoglutarate in batch-grown cultures of S6803 Δ*glgC* with limited nitrate in the inoculum. Cells were grown to an OD_{730} of approximately 1.0 in BG-11 (containing 17.6 mM NaNO₃) under 50 μE* m⁻² s⁻¹ illumination from cool white fluorescent bulbs. These precultures were inoculated into BG-11 media with the indicated nitrate content and cultivated under the same illumination. Cultures were sampled regularly to measure (A) cell density (as OD_{730}), (B) extracellular pyruvate and (C) extracellular alpha-ketoglutarate.

conversion process declines over several days and remains to be optimized for enhanced productivity.

This study aims to understand the relationship between the growth state and the photo-catalytic state observed in nitrogen starvation conditions. The approach is to identify physiological parameters involved in the transition into and out of photo-catalysis as well as the adaptability of cells in photo-catalysis to changes in environmental conditions such as increasing light intensity.

Results and discussion

Our initial communication reporting the ability of S6803 Δ*glgC* to produce extracellular organic acids noted a severe decline in whole-chain photosynthetic capacity, a decline that was identical in rate and magnitude to that of WT. In S6803 WT, the decreased photosynthetic rates correlated with a diminishing rate of glycogen storage, while in the Δ*glgC*, concentrations of exogenous organic acids pyruvate and alpha-ketoglutarate did not increase after 3 days of continuous photo-catalysis (Carrieri *et al.*, 2012). A strong decline in photosynthetic capacity under nitrogen starvation is well documented in cyanobacteria

and results in dramatic transcriptional and physiological responses, which affect, among other pathways, photosynthesis and carbon fixation (Sauer *et al.*, 2001; Krasikov *et al.*, 2012). However, we had found that during photoautotrophic growth in the presence of nitrate, S6803 Δ*glgC* did not produce detectable amounts of extracellular organic acids. Because NO_3^- is the sole nitrogen source in standard BG11 medium, we expected that nitrogen availability was critical to production of organic acids.

We reasoned that by providing a limiting concentration of nitrate in growth medium, cells of S6803 Δ*glgC* might assimilate all available nitrogen and transition without further manipulation into the photo-catalytic production of pyruvate and alpha-ketoglutarate. We found that while cultures in standard BG-11, which contains 17.6 mM NaNO₃, grow to cell densities beyond 3.0 OD_{730} without excreting organic acids, growth of cells cultivated in modified medium containing 5–20% of this original nitrate concentration was arrested in a nitrate concentration dependent manner, coincident with the start of pyruvate and alpha-ketoglutarate production (Fig. 1).

This finding shows that S6803 Δ*glgC* can naturally transition from growth state to photo-catalytic state and offers

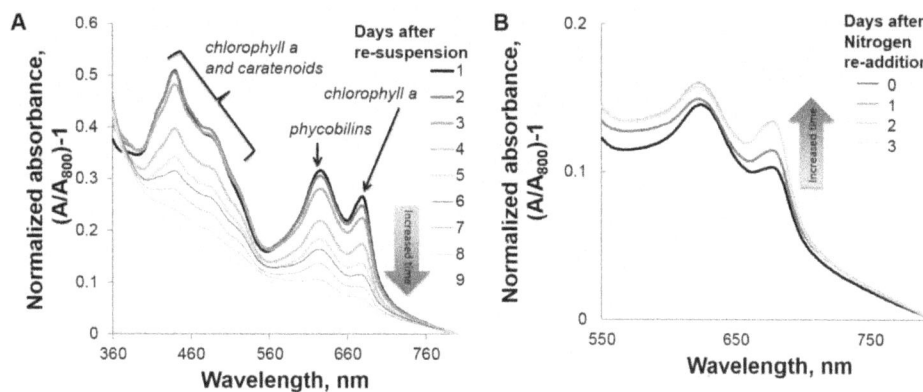

Fig. 2. Whole-cell spectra of nitrate-free cultures of S6803 $\Delta glgC$ (A) over 9 days of photo-catalysis and (B) upon re-addition of 2 mM NaNO₃ following 6 days of photo-catalysis. Cultures grown for 6 days in nitrate-limited medium in A are the same as those in B, labelled 0 days after nitrogen re-addition. Illumination was supplied by cool white fluorescent bulbs at an intensity of 50 μE* m⁻² s⁻¹.

a solution to an otherwise insurmountable problem for large-scale production of organic acids from this strain by nitrogen removal. Harvesting and washing, followed by re-suspension of cells in nitrate-free medium at a production-scale, would likely be prohibitively costly and energy intensive.

Nevertheless, the approach taken in Fig. 1 does not address the problem of cellular photo-catalyst degradation over time. We observed degradation of pigments in whole cells of S6803 $\Delta glgC$ over time by taking whole-cell absorption spectra daily during the photo-catalysis phase. Figure 2A shows that after prolonged and continuous photo-catalysis (approximately 3 days after nitrogen removal in this experiment), cultures begin to lose all pigments involved in photosynthesis. This general decline in carotenoids, chlorophyll a, and bilins within phycobilisome is different from the rapid degradation of pigments associated with nitrogen limited cyanobacterial wild-types. In wild-type cyanobacteria, nitrogen deprivation results in degradation of phycobilisome within hours and a much slower decline in chlorophyll a (Sauer et al., 2001; Carrieri et al., 2012; Krasikov et al., 2012). It is not yet understood why glgC deletion mutants do not rapidly degrade their phycobilisomes under nitrogen-depleted conditions.

Nevertheless, similar to what is reported for wild-type cells (e.g. Krasikov et al., 2012), we found that the level of pigments in nitrogen-free cultures of S6803 $\Delta glgC$ can increase again upon nitrate re-addition. Even after 6 days of illumination of nitrate-free cultures of S6803 $\Delta glgC$, pigments increased in cultures upon addition of 2 mM NaNO₃ (Fig. 2B). This suggested that perhaps the photo-catalyst could be regenerated in S6803 $\Delta glgC$.

We therefore reverted to our original conditions for production of organic acids by S6803 $\Delta glgC$ and compared them with cultures in which 2 mM NaNO₃ was added after three or four days of photo-catalysis, as shown in Fig. 3.

As expected, re-addition of nitrate allowed for cell growth to resume (as indicated by turbidity measurements in Fig. 3A), followed by production of more organic acids pyruvate (3B) and alpha-ketoglutarate (3C) than in control cultures.

These data suggest that the photo-catalytic state is reversible at the culture level and that photobioreactors producing pyruvate and alpha-ketoglutarate with S6803 $\Delta glgC$ could likely be regenerated for at least two cycles of photo-catalysis. We should note, however, that, based on our data, we cannot distinguish whether the newly generated cells after nitrate re-addition were responsible for recovery of pigments and photo-catalytic activity of cultures or individual cells recovered from inactivity to production. The latter explanation is plausible, considering that in the cyanobacterium S. elongatus PCC 7942, long-term nitrogen deprivation of WT cells leads to a survivable but low-level photosynthetic state (Sauer et al., 2001). Regeneration of the whole-cell photo-catalyst could inspire further research on this physiological state and further optimization towards development of a high carbon efficiency process for photosynthetic production of fuels and chemicals.

Finally, we supposed that under certain conditions, light should be limiting for photo-catalytic production of organic acids and that higher light flux should result in faster generation of products. This result has been observed in many other cyanobacterial systems (e.g. Halfmann et al., 2014). To test this hypothesis, we cultured S6803 $\Delta glgC$ cells under three different light intensities and re-suspended them in nitrate-free medium at an optical density of approximately 3.0 (OD₇₃₀). Figure 4 illustrates that indeed higher light fluxes can be used to drive faster production of organic acids. We note that the cultures used in this experiment were at a higher optical density than shown for Figs 1 and 3, explaining the higher yields of organic acids. There is also a longer period of linear

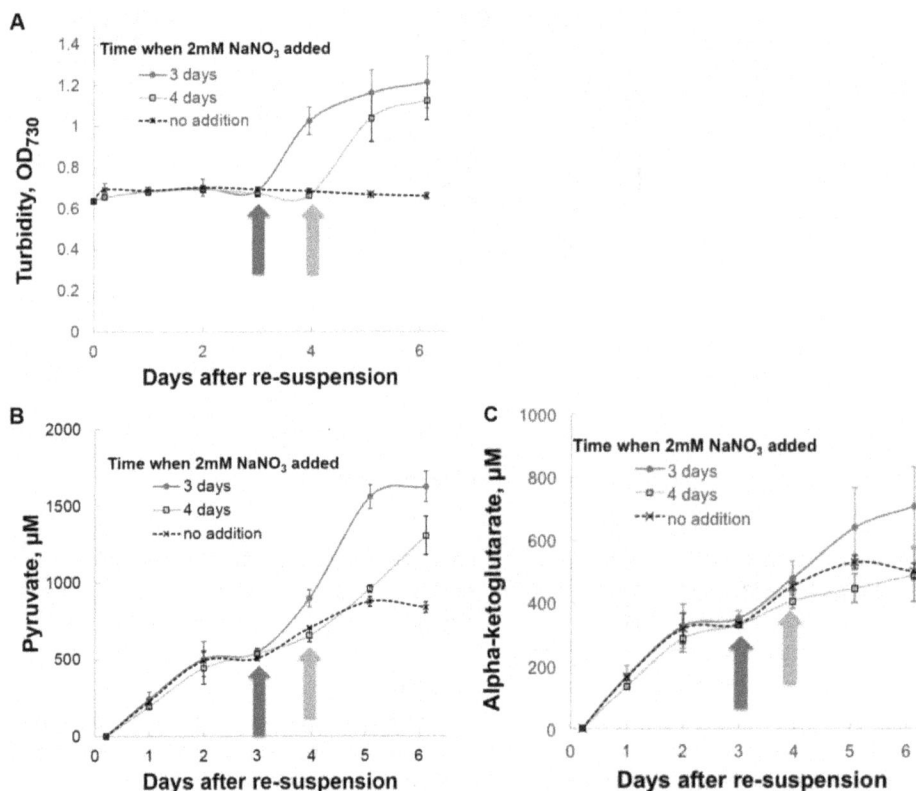

Fig. 3. Regeneration of photo-catalytic production of pyruvate and alpha-ketoglutarate from nitrate-free cell suspensions of S6803 Δ*glgC* by addition of 2 mM NaNO₃ after 3 or 4 days of photo-catalysis. Illumination was supplied by cool white fluorescent bulbs at an intensity of 50 μE* m⁻² s⁻¹. Cells were grown to an OD₇₃₀ of approximately 1.0 in BG-11 (containing 17.6 mM NaNO₃) under 50 μE* m⁻² s⁻¹ from cool white fluorescent bulbs and re-suspended after washing in nitrate-free BG-11 to an OD₇₃₀ of approximately 0.6. Cultures were sampled regularly to measure (A) cell density (as OD₇₃₀), (B) extracellular pyruvate and (C) extracellular alpha-ketoglutarate.

production of acids (longer living photo-catalyst) in Fig. 4 that may be because of diminished photo-degradation of the cells due to shading effects from high cell densities. The improved productivity was realized only with cultures adapted to the higher light conditions prior to nitrogen starvation (Fig. S1). This observation suggests that once in the photo-catalytic state, the cells are not able to increase productivity under higher light intensities.

Notably, the improvements in production of pyruvate are higher than for alpha-ketoglutarate upon increasing light intensity, as shown in the products ratios (Fig. 4). This suggests that carbon partition between the two products is adaptable to changes in environmental conditions. One possible mechanism is that activity of enzymes beyond pyruvate in the citric acid cycle become limiting at higher light fluxes. Alternatively, other factors such as the ATP/NADPH ratio could vary between different light intensities, re-directing carbon fluxes towards pathways that require more ATP or NADPH (Pattanayak *et al.*, 2014). Further research is needed to understand the regulation of carbon partition in this system, which could lead to strategies to produce more of one or the other organic acid.

Conclusions

We have extended the initial discovery and demonstrated some key features for enhancing photo-catalytic production of organic acids in nitrogen starved cultures of S6803 Δ*glgC*. We have shown that cultures can transition from growth state to photo-catalytic state when inoculated in medium with limited concentrations of nitrate, that the photo-catalytic state can be reversed by re-addition of nitrogen nutrient and that light levels can be increased to drive faster production of organic acids when presumably limiting, accompanied by changes in the ratio of alpha-ketoglutarate/pyruvate. These features suggest that improved understanding of photo-catalytic state and regulation of carbon partition may be useful to commercial applications.

Experimental procedures

Cell culture

Cultures of S6803 Δ*glgC* were grown in modified BG-11 medium comprised of 1X standard BG11 freshwater nutrients (except for nitrogen source) and two additions: 20 mM

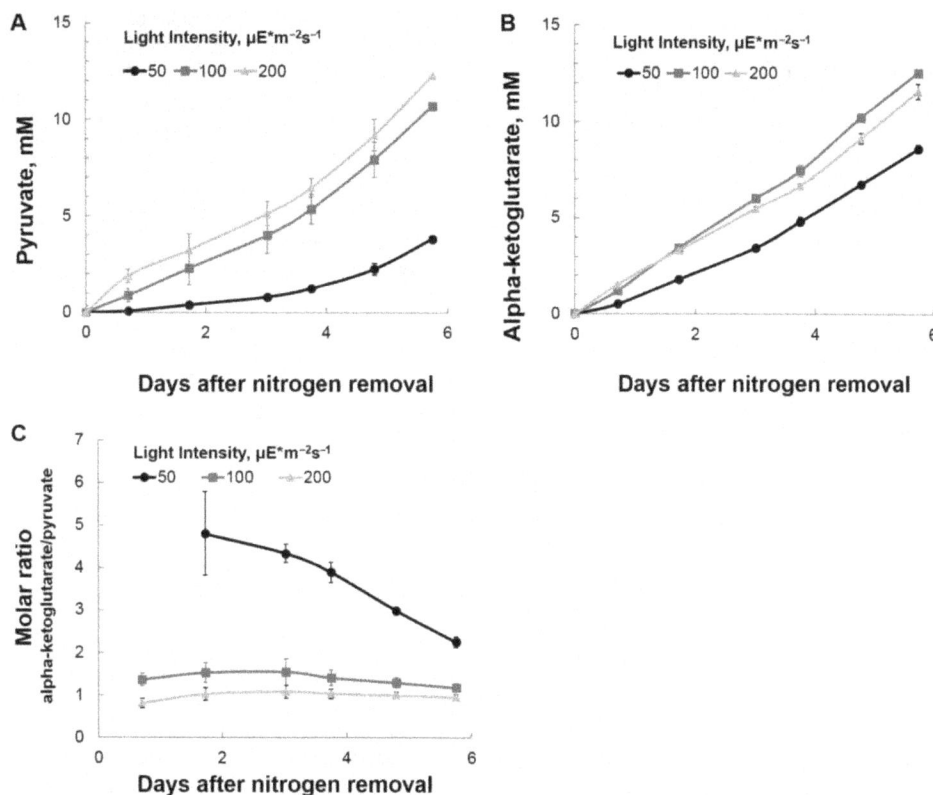

Fig. 4. Effect of continuous illumination intensity on photo-catalytic production of (A) pyruvate and (B) alpha-ketoglutarate, and (C) the calculated molar ratio of alpha-ketoglutarate/pyruvate yield. Cultures of S6803 $\Delta glgC$ were grown under 24 h of illumination from cool white LEDs at the indicated light intensities, harvested and washed in nitrate-free BG-11, and re-suspended at a cell density of 3.1 OD_{730}. The new suspensions were returned to the same conditions in which they were grown and sampled daily to measure extracellular metabolites. The molar ratio of products from cultures illuminated at 50 $\mu E^* m^{-2} s^{-1}$ had high uncertainty (because of detection limits of high-performance liquid chromatography for pyruvate) prior to 1 day after nitrogen removal and is therefore not shown in panel C.

sodium phosphate (pH 7.1) and 5 $\mu g\ ml^{-1}$ gentamicin. The media were typically prepared without $NaNO_3$, which was added separately to the indicated final concentrations as needed following preculture biomass accumulation in complete medium (17.6 mM $NaNO_3$), harvesting by centrifugation (4000 × g) and washing with nitrogen free BG-11. For all experiments, cultures were maintained in 250 ml Erlenmeyer flasks in growth chambers in the presence of 5% CO_2 enriched air atmosphere. Continuous (24 h) illumination was provided by cool white fluorescent bulbs with a light flux of ~50 $\mu E^* m^{-2} s^{-1}$ at the surface of the flasks or by cool white LEDs at light intensities as high as 300 $\mu E^* m^{-2} s^{-1}$ at the surface of the flasks.

Determination of organic acids

Aliquots (1 ml) of culture were harvested and centrifuged at 13 000 × g. Supernatants were filtered through 0.20 micron filters prior to injection (100 µl) in an Agilent Technology 1200 Series HPLC fitted with Bio-Rad Aminex HPX-87H column, and eluting with 5 mM sulfuric acid at a flow rate of 0.6 ml min⁻¹ and detecting with UV detector at 210 nm. Peak retention times and signal integrals were compared against known standard solutions prepared in modified BG11 media with addition of purchased standards of sodium alpha-

ketoglutarate and sodium pyruvate (Sigma-Aldrich, USA). No peaks were observed by high-performance liquid chromatography other than from alpha-ketoglutarate and pyruvate that were attributable to biologically produced compounds. All organic acid measurements were determined as the average of at least three independently grown cultures from identical conditions with error bars in figures representing ± 1 standard deviation.

Optical density and whole-cell spectra

Culture aliquots (1 ml) were measured in a Beckman Coulter DU 800 Spectrophotometer for whole-cell absorbance wavelength scans or scattering at 730 nm. All optical density measurements were determined as the average of at least three independently grown cultures from identical conditions with error bars in figures representing ± 1 standard deviation.

Conflict of interest

D. Carrieri and T. Paddock are now employed at Matrix Genetics, LLC, a company that engineers cyanobacteria strains of commercial interest.

References

Carrieri, D., Paddock, T., Maness, P.-C., Seibert, M., and Yu, J. (2012) Photo-catalytic conversion of carbon dioxide to organic acids by a recombinant cyanobacterium incapable of glycogen storage. *Energ Environ Sci* **5:** 9457–9461.

Davies, F.K., Work, V.H., Beliaev, A.S., and Posewitz, M.C. (2014) Engineering limonene and bisabolene production in type and a glycogen-deficient mutant of *Synechococcus* sp. PCC 7002. *Front Bioeng Biotechnol* **2:** 1–11.

Dismukes, G.C., Carrieri, D., Bennette, N., Ananyev, G.M., and Posewitz, M.C. (2008) Aquatic phototrophs: efficient alternatives to land-based crops for biofuels. *Curr Opin Biotechnol* **19:** 235–240.

Gründel, M., Scheunemann, R., Lockau, W., and Zilliges, Y. (2012) Impaired glycogen synthesis causes metabolic overflow reactions and affects stress responses in the cyanobacterium *Synechocystis* sp. PCC 6803. *Microbiology* **158:** 3032–3043.

Halfmann, C., Gu, L., and Zhou, R. (2014) Engineering cyanobacteria for the production of a cyclic hydrocarbon fuel from CO_2 and H_2O. *Green Chem* **16:** 3175–3185.

Hickman, J.W., Kotovic, K.M., Miller, C., Warrener, P., Kaiser, B., Jurista, T., *et al.* (2013) Glycogen synthesis is a required component of the nitrogen stress response in *Synechococcus elongatus* PCC 7942. *Algal Res* **2:** 98–106.

Krasikov, V., von Wobeser, E.A., Dekker, H.L., Huisman, J., and Matthijs, H.C.P. (2012) Time-series resolution of gradual nitrogen starvation and its impact on photosynthesis in the cyanobacterium *Synechocystis* PCC 6803. *Physiol Plant* **145:** 426–439.

Li, X., Shen, C.R., and Liao, J.C. (2014) Isobutanol production as an alternative metabolic sink to rescue the growth deficiency of the glycogen mutant of *Synechococcus elongatus* PCC 7942. *Photosynth Res* **120:** 301–310.

Melis, A. (2012) Photosynthesis-to-fuels: from sunlight to hydrogen, isoprene, and botryococcene production. *Energ Environ Sci* **5:** 5531–5539.

Pattanayak, G.K., Phong, C.P., and Rust, M.J. (2014) Rhythms in energy storage control the ability of the cyanobacterial circadian clock to reset. *Curr Biol* **24:** 1934–1938.

Sauer, J., Schreiber, U., Schmid, R., Völker, U., and Forchhammer, K. (2001) Nitrogen starvation-induced chlorosis in *Synechococcus* PCC 7942. Low-level photosynthesis as a mechanism of long-term survival. *Plant Physiol* **126:** 233–243.

Supporting information

Additional Supporting Information may be found in the online version of this article at the publisher's web-site:

Fig. S1. Pyruvate and alpha-ketoglutarate production of ΔglgC cells under varying illumination following growth in lowest indicated light flux (60 μE m^{-2} s^{-1}). Cultures were first grown to mid-log phase under 60 μE m^{-2} s^{-1} light and re-suspended in nitrate-free medium at an optical density (OD730) equal to 0.65 and placed under illumination at the light fluxes indicated in units of μE m^{-2} s^{-1} (60, 130 or 300 μE m^{-2} s^{-1}).

Permissions

The contributors of this book come from diverse backgrounds, making this book a truly international effort. This book will bring forth new frontiers with its revolutionizing research information and detailed analysis of the nascent developments around the world.

We would like to thank all the contributing authors for lending their expertise to make the book truly unique. They have played a crucial role in the development of this book. Without their invaluable contributions this book wouldn't have been possible. They have made vital efforts to compile up to date information on the varied aspects of this subject to make this book a valuable addition to the collection of many professionals and students.

This book was conceptualized with the vision of imparting up-to-date information and advanced data in this field. To ensure the same, a matchless editorial board was set up. Every individual on the board went through rigorous rounds of assessment to prove their worth. After which they invested a large part of their time researching and compiling the most relevant data for our readers.

The editorial board has been involved in producing this book since its inception. They have spent rigorous hours researching and exploring the diverse topics which have resulted in the successful publishing of this book. They have passed on their knowledge of decades through this book. To expedite this challenging task, the publisher supported the team at every step. A small team of assistant editors was also appointed to further simplify the editing procedure and attain best results for the readers.

Apart from the editorial board, the designing team has also invested a significant amount of their time in understanding the subject and creating the most relevant covers. They scrutinized every image to scout for the most suitable representation of the subject and create an appropriate cover for the book.

The publishing team has been an ardent support to the editorial, designing and production team. Their endless efforts to recruit the best for this project, has resulted in the accomplishment of this book. They are a veteran in the field of academics and their pool of knowledge is as vast as their experience in printing. Their expertise and guidance has proved useful at every step. Their uncompromising quality standards have made this book an exceptional effort. Their encouragement from time to time has been an inspiration for everyone.

The publisher and the editorial board hope that this book will prove to be a valuable piece of knowledge for researchers, students, practitioners and scholars across the globe.

List of Contributors

María Isabel González-Siso
Grupo de Investigación EXPRELA, Departamento de Bioloxía Celular e Molecular, Facultade de Ciencias, Universidade da Coruña, Campus de A Coruña, 15071-A Coruña, Spain

Alba Touriño
Grupo de Investigación EXPRELA, Departamento de Bioloxía Celular e Molecular, Facultade de Ciencias, Universidade da Coruña, Campus de A Coruña, 15071-A Coruña, Spain

ÁngelVizoso
Grupo de Investigación EXPRELA, Departamento de Bioloxía Celular e Molecular, Facultade de Ciencias, Universidade da Coruña, Campus de A Coruña, 15071-A Coruña, Spain

Ángel Pereira-Rodríguez
Grupo de Investigación EXPRELA, Departamento de Bioloxía Celular e Molecular, Facultade de Ciencias, Universidade da Coruña, Campus de A Coruña, 15071-A Coruña, Spain

EstherRodríguez-Belmonte
Grupo de Investigación EXPRELA, Departamento de Bioloxía Celular e Molecular, Facultade de Ciencias, Universidade da Coruña, Campus de A Coruña, 15071-A Coruña, Spain

Manuel Becerra
Grupo de Investigación EXPRELA, Departamento de Bioloxía Celular e Molecular, Facultade de Ciencias, Universidade da Coruña, Campus de A Coruña, 15071-A Coruña, Spain

María Esperanza Cerdán
Grupo de Investigación EXPRELA, Departamento de Bioloxía Celular e Molecular, Facultade de Ciencias, Universidade da Coruña, Campus de A Coruña, 15071-A Coruña, Spain

Yun Bai
Institute of Resource Biology and Biotechnology, Department of Biotechnology, College of Life Science and Technology
Key Laboratory of Molecular Biophysics Ministry of Education, Huazhong University of Science and Technology, Wuhan 430074, China

Peng-Peng Zhou
Institute of Resource Biology and Biotechnology, Department of Biotechnology, College of Life Science and Technology
Key Laboratory of Molecular Biophysics Ministry of Education, Huazhong University of Science and Technology, Wuhan 430074, China

Pei Fan
Institute of Resource Biology and Biotechnology, Department of Biotechnology, College of Life Science and Technology
Key Laboratory of Molecular Biophysics Ministry of Education, Huazhong University of Science and Technology, Wuhan 430074, China

Yuan-Min Zhu
Institute of Resource Biology and Biotechnology, Department of Biotechnology, College of Life Science and Technology
Key Laboratory of Molecular Biophysics Ministry of Education, Huazhong University of Science and Technology, Wuhan 430074, China

Yao Tong
Institute of Resource Biology and Biotechnology, Department of Biotechnology, College of Life Science and Technology
Key Laboratory of Molecular Biophysics Ministry of Education, Huazhong University of Science and Technology, Wuhan 430074, China

Hong-bo Wang
Institute of Resource Biology and Biotechnology, Department of Biotechnology, College of Life Science and Technology
Key Laboratory of Molecular Biophysics Ministry of Education, Huazhong University of Science and Technology, Wuhan 430074, China

Long-Jiang Yu
Institute of Resource Biology and Biotechnology, Department of Biotechnology, College of Life Science and Technology
Key Laboratory of Molecular Biophysics Ministry of Education, Huazhong University of Science and Technology, Wuhan 430074, China

Micol Bellucci
School of Civil Engineering and Geosciences, Newcastle University, Newcastle upon Tyne, NE1 7RU, UK
Dipartimento di Scienze Agrarie, Alimentari ed Ambientali, Università di Foggia, via Napoli 25, Foggia, 71122 Italy

Irina D. Ofit¸eru
School of Chemical Engineering and Advanced Materials, Merz Court, Newcastle University, Newcastle upon Tyne, NE1 7RU, UK
Chemical Engineering Department, University Politehnica of Bucharest, Polizu 1-7, Bucharest 011061, Romania

LucianoBeneduce
Dipartimento di Scienze Agrarie, Alimentari ed Ambientali, Università di Foggia, via Napoli 25, Foggia, 71122 Italy

David W. Graham
School of Civil Engineering and Geosciences, Newcastle University, Newcastle upon Tyne, NE1 7RU, UK

Ian M. Head
School of Civil Engineering and Geosciences, Newcastle University, Newcastle upon Tyne, NE1 7RU, UK

Thomas P. Curtis
School of Civil Engineering and Geosciences, Newcastle University, Newcastle upon Tyne, NE1 7RU, UK

Michael Bott
Institute of Bio- and Geosciences, IBG-1: Biotechnology, Forschungszentrum Jülich, Jülich D-52428, Germany

Akio Chiba
Departments of Bacteriology and Infectious Disease and Control, The Jikei University
School of Medicine, 3–25-8, Nishi-Shimbashi, Minato-ku, Tokyo 105-8461, Japan

Shinya Sugimoto
Departments of Bacteriology and Infectious Disease and Control, The Jikei University
School of Medicine, 3–25-8, Nishi-Shimbashi, Minato-ku, Tokyo 105-8461, Japan

Fumiya Sato
Infectious Disease and Control, The Jikei University School of Medicine, 3–25-8, Nishi-Shimbashi, Minato-ku, Tokyo 105-8461, Japan

Seiji Hori
Infectious Disease and Control, The Jikei University School of Medicine, 3–25-8, Nishi-Shimbashi, Minato-ku, Tokyo 105-8461, Japan

Yoshimitsu Mizunoe
Departments of Bacteriology and Infectious Disease and Control, The Jikei University
School of Medicine, 3–25-8, Nishi-Shimbashi, Minato-ku, Tokyo 105-8461, Japan

Sara P. Cuellar-Bermudez
Cátedra de Bioprocesos Ambientales, Centro del Agua Para América Latina y el Caribe, nstituto Tecnológico y de Estudios Superiores de Monterrey, Monterrey, Nuevo Leon 64849, Mexico

Iris Aguilar-Hernandez
Cátedra de Bioprocesos Ambientales, Centro del Agua Para América Latina y el Caribe, nstituto Tecnológico y de Estudios Superiores de Monterrey, Monterrey, Nuevo Leon 64849, Mexico

Diana L. Cardenas-Chavez
Cátedra de Bioprocesos Ambientales, Centro del Agua Para América Latina y el Caribe, nstituto Tecnológico y de Estudios Superiores de Monterrey, Monterrey, Nuevo Leon 64849, Mexico

Nancy Ornelas-Soto
Cátedra de Bioprocesos Ambientales, Centro del Agua Para América Latina y el Caribe, nstituto Tecnológico y de Estudios Superiores de Monterrey, Monterrey, Nuevo Leon 64849, Mexico

Miguel A. Romero-Ogawa
Cátedra de Bioprocesos Ambientales, Centro del Agua Para América Latina y el Caribe, nstituto Tecnológico y de Estudios Superiores de Monterrey, Monterrey, Nuevo Leon 64849, Mexico

Roberto Parra-Saldivar
Cátedra de Bioprocesos Ambientales, Centro del Agua Para América Latina y el Caribe, nstituto Tecnológico y de Estudios Superiores de Monterrey, Monterrey, Nuevo Leon 64849, Mexico

Daniel Dobslaw
Department of Biological Waste Air Purification, Institute of Sanitary Engineering, Water Quality and Solid Waste Management, University of Stuttgart, Bandtäle 2, Stuttgart, D-70569, Germany

Karl-Heinrich Engesser
Department of Biological Waste Air Purification, Institute of Sanitary Engineering, Water Quality and Solid Waste Management, University of Stuttgart, Bandtäle 2, Stuttgart, D-70569, Germany

María-Eugenia Guazzaroni
Departamento de Química, FFCLRP

Rafael Silva-Rocha
Departamento de Bioquímica e Imunologia, FMRP, University of São Paulo, Ribeirão Preto, SP, Brazil

Richard John Ward
Departamento de Química, FFCLRP

Jackson Z. Lee
Department of Civil and Environmental Engineering, Colorado School of Mines, Golden, CO, USA

Andrew Logan
Nutrinsic, Corp., Aurora, CO, USA

Seth Terry
Nutrinsic, Corp., Aurora, CO, USA

John R. Spear
Nutrinsic, Corp., Aurora, CO, USA

Hongxing Li
State Key Laboratory of Microbial Technology, Shandong University, Jinan 250100, China

Meiling Wu
State Key Laboratory of Microbial Technology, Shandong University, Jinan 250100, China

Lili Xu
State Key Laboratory of Microbial Technology, Shandong University, Jinan 250100, China

Jin Hou
State Key Laboratory of Microbial Technology, Shandong University, Jinan 250100, China

Ting Guo
State Key Laboratory of Microbial Technology, Shandong University, Jinan 250100, China
Guangzhou Sugarcane Industry Research Institute, Guangzhou 510316, China

Xiaoming Bao
State Key Laboratory of Microbial Technology, Shandong University, Jinan 250100, China

Yu Shen
State Key Laboratory of Microbial Technology, Shandong University, Jinan 250100, China
Guangzhou Sugarcane Industry Research Institute, Guangzhou 510316, China

Jinxue Luo
Research Center for Eco-Environmental Sciences, Chinese Academy of Sciences, Beijing, China
School of Biological Sciences, 60 Nanyang Drive, SBS-01N-27 Singapore, 637551, Advanced Environmental Biotechnology Centre

Jinsong Zhang
Singapore Membrane Technology Centre, Nanyang Environment and Water Research Institute

Robert J. Barnes
Advanced Environmental Biotechnology Centre

Xiaohui Tan
Advanced Environmental Biotechnology Centre

Diane McDougald
School of Biological Sciences, 60 Nanyang Drive, SBS-01N-27 Singapore, 637551, Centre for Marine Bio-Innovation, School of Biotechnology and Biomolecular Sciences, The University of New South Wales, Sydney, NSW, Australia
Singapore Centre on Environmental Life Sciences Engineering, Nanyang Technological University, Singapore

Anthony G.Fane
Singapore Membrane Technology Centre, Nanyang Environment and Water Research Institute

Guoqiang Zhuang
Research Center for Eco-Environmental Sciences, Chinese Academy of Sciences, Beijing, China

Staffan Kjelleberg
School of Biological Sciences, 60 Nanyang Drive, SBS-01N-27 Singapore, 637551, Centre for Marine Bio-Innovation, School of Biotechnology and Biomolecular Sciences, The University of New South Wales, Sydney, NSW, Australia
Singapore Centre on Environmental Life Sciences Engineering, Nanyang Technological University, Singapore

Yehuda Cohen
School of Biological Sciences, 60 Nanyang Drive, SBS-01N-27 Singapore, 637551, Singapore Centre on Environmental Life Sciences Engineering, Nanyang Technological University, Singapore

Scott A. Rice
School of Biological Sciences, 60 Nanyang Drive, SBS-01N-27 Singapore, 637551, Centre for Marine Bio-Innovation, School of Biotechnology and Biomolecular Sciences, The University of New South Wales, Sydney, NSW, Australia
Singapore Centre on Environmental Life Sciences Engineering, Nanyang Technological University, Singapore

Thuy Thu Nguyen
Research group for Physiology and Applications of Microorganisms (PHAM group) at Center for Life Science Research, Departments of Microbiology and Biochemistry, Faculty of Biology, Vietnam National University – University of Science, Nguyen Trai 334, Thanh Xuan, Hanoi, Vietnam

Tha Thanh Thi Luong
Research group for Physiology and Applications of Microorganisms (PHAM group) at Center for Life Science Research, Departments of Microbiology and Biochemistry, Faculty of Biology, Vietnam National University – University of Science, Nguyen Trai 334, Thanh Xuan, Hanoi, Vietnam

Phuong Hoang Nguyen Tran
Research group for Physiology and Applications of Microorganisms (PHAM group) at Center for Life Science Research, Departments of Microbiology and Biochemistry, Faculty of Biology, Vietnam National University – University of Science, Nguyen Trai 334, Thanh Xuan, Hanoi, Vietnam

Ha Thi Viet Bui
Research group for Physiology and Applications of Microorganisms (PHAM group) at Center for Life Science Research, Departments of 2Microbiology and 3Biochemistry, Faculty of Biology, Vietnam National University – University of Science, Nguyen Trai 334, Thanh Xuan, Hanoi, Vietnam
Microbiology and Biochemistry, Faculty of Biology, Vietnam National University – University of Science, Nguyen Trai 334, Thanh Xuan, Hanoi, Vietnam

Huy Quang Nguyen
Research group for Physiology and Applications of Microorganisms (PHAM group) at Center for Life Science Research, Departments of Microbiology and Biochemistry, Faculty of Biology, Vietnam National University – University of Science, Nguyen Trai 334, Thanh Xuan, Hanoi, Vietnam
Biochemistry, Faculty of Biology, Vietnam National University – University of Science, Nguyen Trai 334, Thanh Xuan, Hanoi, Vietnam

Hang Thuy Dinh
Laboratory of Microbial Ecology, Institute of Microbiology and Biology, Vietnam National University, Xuan Thuy 144, Cau Giay, Hanoi, Vietnam

Byung Hong Kim
Korea Institute of Science and Technology, Hwarangno 14-gil, 5 Seongbuk-gu, Seoul 136-791, Korea
Fuel Cell Institute, National University of Malaysia, Bangi 43600 UKM, Selangor, Malaysia
School of Municipal and Environmental Engineering, Harbin Institute of Technology, 73 Huanghe Road, Nangang District, Harbin 150090, China

Hai The Pham
Research group for Physiology and Applications of Microorganisms (PHAM group) at Center for Life Science Research, Departments of Microbiology and Biochemistry, Faculty of Biology, Vietnam National University – University of Science, Nguyen Trai 334, Thanh Xuan, Hanoi, Vietnam

Microbiology and Biochemistry, Faculty of Biology, Vietnam National University – University of Science, Nguyen Trai 334, Thanh Xuan, Hanoi, Vietnam

Claudia Schmidt-Dannert,
Department of Biochemistry,Molecular Biology and Biophysics, University of Minnesota, 1479 Gortner Avenue, St. Paul, MN 55108, USA

Xiao-Mei Su
Department of Environmental Engineering, College of Environmental and Resource Sciences, Zhejiang University, Hangzhou 310058, China

Yin-Dong Liu
Department of Environmental Engineering, College of Environmental and Resource Sciences, Zhejiang University, Hangzhou 310058, China

Muhammad Zaffar Hashmi
Department of Environmental Engineering, College of Environmental and Resource Sciences, Zhejiang University, Hangzhou 310058, China

Lin-Xian Ding
College of Geography and Environmental Science, Zhejiang Normal University, Jinhua 321004, China

Chao-Feng Shen
Department of Environmental Engineering, College of Environmental and Resource Sciences, Zhejiang University, Hangzhou 310058, China

Mohan Raj Subramanian
Center for Bioprocessing Research and Development, Department of Civil and Environmental Engineering, South Dakota School of Mines and Technology, Rapid City, SD 57701, USA

Suvarna Talluri
Center for Bioprocessing Research and Development, Department of Civil and Environmental Engineering, South Dakota School of Mines and Technology, Rapid City, SD 57701, USA

Lew P. Christopher
Center for Bioprocessing Research and Development, Department of Civil and Environmental Engineering, South Dakota School of Mines and Technology, Rapid City, SD 57701, USA

Department of Civil and Environmental Engineering, South Dakota School of Mines and Technology, Rapid City, SD 57701, USA

Karina Verdel-Aranda
Evolution of Metabolic Diversity Laboratory, Unidad de Genómica Avanzada (Langebio), Cinvestav-IPN, Km 9.6 Libramiento Norte, Irapuato, Guanajuato CP36822, México

Susana T. López-Cortina
Facultad de Ciencias Químicas, Universidad Autónoma de Nuevo León, San Nicolás de los Garza, Nuevo León, México

David A. Hodgson
School of Life Sciences, University Warwick, Coventry, UK

Francisco Barona-Gómez
Evolution of Metabolic Diversity Laboratory, Unidad de Genómica Avanzada (Langebio), Cinvestav-IPN, Km 9.6 Libramiento Norte, Irapuato, Guanajuato CP36822, México

R. Y. Zhang
Aquatische Biotechnologie, Biofilm Centre, Universität Duisburg – Essen, Universitätsstraße 5, 45141 Essen, Germany

T. R. Neu
Department of River Ecology, Helmholtz Centre for Environmental Research-UFZ, Brueckstrasse 3A, 39114 Magdeburg, Germany

S. Bellenberg
Aquatische Biotechnologie, Biofilm Centre, Universität Duisburg – Essen, Universitätsstraße 5, 45141 Essen, Germany

U. Kuhlicke
Department of River Ecology, Helmholtz Centre for Environmental Research-UFZ, Brueckstrasse 3A, 39114 Magdeburg, Germany

W. Sand
Aquatische Biotechnologie, Biofilm Centre, Universität Duisburg – Essen, Universitätsstraße 5, 45141 Essen, Germany

M. Vera
Aquatische Biotechnologie, Biofilm Centre, Universität Duisburg – Essen, Universitätsstraße 5, 45141 Essen, Germany

Damian Carrieri
Biosciences Center, National Renewable Energy Laboratory, 15013 Denver West Parkway, Golden, CO 80401, USA

Charlie Broadbent
Biosciences Center, National Renewable Energy Laboratory, 15013 Denver West Parkway, Golden, CO 80401, USA

David Carruth
Biosciences Center, National Renewable Energy Laboratory, 15013 Denver West Parkway, Golden, CO 80401, USA

Troy Paddock
Biosciences Center, National Renewable Energy Laboratory, 15013 Denver West Parkway, Golden, CO 80401, USA

Justin Ungerer
Biosciences Center, National Renewable Energy Laboratory, 15013 Denver West Parkway, Golden, CO 80401, USA

Pin-Ching Maness
Biosciences Center, National Renewable Energy Laboratory, 15013 Denver West Parkway, Golden, CO 80401, USA

Maria Ghirardi
Biosciences Center, National Renewable Energy Laboratory, 15013 Denver West Parkway, Golden, CO 80401, USA

Jianping Yu
Biosciences Center, National Renewable Energy Laboratory, 15013 Denver West Parkway, Golden, CO 80401, USA